21 Springer Series in Solid-State Sciences

Edited by Hans-Joachim Queisser

Springer Series in Solid-State Sciences

Editors: M. Cardona P. Fulde H.-J. Queisser

Volume 40 **Semiconductor Physics** – An Introduction   By K. Seeger

Volume 41 **The LMTO Method**   By H.L. Skriver

Volume 42 **Crystal Optics with Spatial Dispersion and the Theory of Excitations**
By V.M. Agranovich and V.L. Ginzburg

Volume 43 **Resonant Nonlinear-Interactions of Light with Matter**
By A.E. Kaplan, V.S. Butylkin, Yu.G. Khronopulo, and E.I. Yakubovich

Volumes 1 – 39 are listed on the back inside cover

B. K. Vainshtein
V. M. Fridkin   V. L. Indenbom

Modern Crystallography II

Structure of Crystals

With 345 Figures,

Springer-Verlag Berlin Heidelberg New York 1982

Professor Dr. *Boris K. Vainshtein*
Professor Dr. *Vladimir M. Fridkin*
Professor Dr. *Vladimir L. Indenbom*

Institute of Crystallography, Academy of Sciences of the USSR, 59 Leninsky prospect,
SU-117333 Moscow, USSR

Series Editors:

Professor Dr. Manuel Cardona
Professor Dr. Peter Fulde
Professor Dr. Hans-Joachim Queisser

Max-Planck-Institut für Festkörperforschung, Heisenbergstrasse 1
D-7000 Stuttgart 80, Fed. Rep. of Germany

Title of the original Russian edition:
Sovremennaia kristallografiia; Structura Kristallov
© by "Nauka" Publishing House, Moscow 1979

ISBN 3-540-10517-4 Springer-Verlag Berlin Heidelberg New York
ISBN 0-387-10517-4 Springer-Verlag New York Heidelberg Berlin

Library of Congress Cataloging in Publication Data. Main entry under title: Modern crystallography. (Springer series in solid-state sciences ; 15, 21) Translation of Scvremennaia kristallografiia. Includes bibliographies and indexes. Contents: 1. Vaĭnshteĭn, B. K. Symmetry of crystals. Methods of structural crystallography. – 2. Vaĭnshteĭn, B. K., Fridkin, V. M. Indenbom, V. L. Structure of crystals. 1. Crystallography. I. Vaĭnshteĭn, Boris Konstantinovich. II. Series: Springer series in solid-state sciences ; 15, 21, etc. QD905.2.S6813 548 80-17797 AACRl

Typesetting: K + V Fotosatz, 6124 Beerfelden

Offset printing: Beltz Offsetdruck, 6944 Hemsbach/Bergstr. Bookbinding: J. Schäffer OHG, 6718 Grünstadt.
2153/3130-5 4 3 2 1 0

Modern Crystallography

in Four Volumes*

I   Symmetry of Crystals. Methods of Structural Crystallography
II   Structure of Crystals
III   Formation of Crystals
IV   Physical Properties of Crystals

Editorial Board:
B. K. Vainshtein (Editor-in-Chief)   **A. A. Chernov**   **L. A. Shuvalov**

Foreword

Crystallography—the science of crystals—has undergone many changes in the course of its development. Although crystals have intrigued mankind since ancient times, crystallography as an independent branch of science began to take shape only in the 17th—18th centuries, when the principal laws governing crystal habits were found, and the birefringence of light in crystals was discovered. From its very origin crystallography was intimately connected with mineralogy, whose most perfect objects of investigation were crystals. Later, crystallography became associated more closely with chemistry, because it was apparent that the habit depends directly on the composition of crystals and can only be explained on the basis of atomic-molecular concepts. In the 20th century crystallography also became more oriented towards physics, which found an ever-increasing number of new optical, electrical, and mechanical phenomena inherent in crystals. Mathematical methods began to be used in crystallography, particularly the theory of symmetry (which achieved its classical completion in space-group theory at the end of the 19th century) and the calculus of tensors (for crystal physics).

---

* Published in *Springer Series in Solid-State Sciences*, I: Vol. 15; II: Vol. 21; III: Vol. 36; IV: Vol. 37

Early in this century, the newly discovered x-ray diffraction by crystals made a complete change in crystallography and in the whole science of the atomic structure of matter, thus giving a new impetus to the development of solid-state physics. Crystallographic methods, primarily x-ray diffraction analysis, penetrated into materials sciences, molecular physics, and chemistry, and also into many other branches of science. Later, electron and neutron diffraction structure analyses became important since they not only complement x-ray data, but also supply new information on the atomic and the real structure of crystals. Electron microscopy and other modern methods of investigating matter—optical, electronic paramagnetic, nuclear magnetic, and other resonance techniques—yield a large amount of information on the atomic, electronic, and real crystal structures.

Crystal physics has also undergone vigorous development. Many remarkable phenomena have been discovered in crystals and then found various practical applications.

Other important factors promoting the development of crystallography were the elaboration of the theory of crystal growth (which brought crystallography closer to thermodynamics and physical chemistry) and the development of the various methods of growing synthetic crystals dictated by practical needs. Man-made crystals became increasingly important for physical investigations, and they rapidly invaded technology. The production of synthetic crystals made a tremendous impact on the traditional branches: the mechanical treatment of materials, precision instrument making, and the jewelry industry. Later it considerably influenced the development of such vital branches of science and industry as radiotechnics and electronics, semiconductor and quantum electronics, optics, including nonlinear optics, acoustics, etc. The search for crystals with valuable physical properties, study of their structure, and development of new techniques for their synthesis constitute one of the basic lines of contemporary science and are important factors of progress in technology.

The investigation of the structure, growth, and properties of crystals should be regarded as a single problem. These three intimately connected aspects of modern crystallography complement each other. The study, not only of the ideal atomic structure, but also of the real defect structure of crystals makes it possible to conduct a purposeful search for new crystals with valuable properties and to improve the technology of their synthesis by using various techniques for controlling their composition and real structure. The theory of real crystals and the physics of crystals are based on their atomic structure as well as on the theoretical and experimental investigations of elementary and macroscopic processes of crystal growth. This approach to the problem of the structure, growth, and properties of crystals has an enormous number of aspects, and determines the features of modern crystallography.

The branches of crystallography and their relation to adjacent fields can be represented as a diagram showing a system of interpenetrating branches which have no strict boundaries. The arrows show the relationship between the branches, indicating which branch influences the activity of the other, although, in fact, they are usually interdependent.

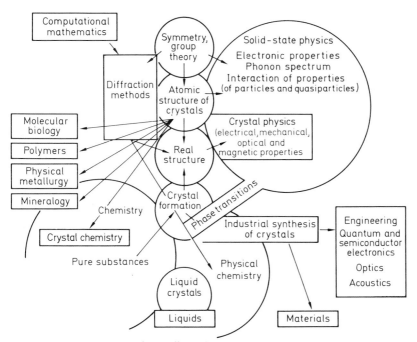

Branches of crystallography and its relation to other sciences

Crystallography proper occupies the central part of the diagram. It includes the theory of symmetry, the investigation of the structure of crystals (together with diffraction methods and crystal chemistry), and the study of the real structure of crystals, their growth and synthesis, and crystal physics.

The theoretical basis of crystallography is the theory of symmetry, which has been intensively developed in recent years.

The study of the atomic structure has been extended to extremely complicated crystals containing hundreds and thousands of atoms in the unit cell. The investigation of the real structure of crystals with various disturbances of the ideal crystal lattices has been gaining in importance. At the same time, the general approach to the atomic structure of matter and the similarity of the various diffraction techniques make crystallography a science not only of the structure of crystals themselves, but also of the condensed state in general.

The specific applications of crystallographic theories and methods allow the utilization of structural crystallography in physical metallurgy, materials science, mineralogy, organic chemistry, polymer chemistry, molecular biology, and the investigation of amorphous solids, liquids, and gases. Experimental and theoretical investigations of crystal growth and nucleation processes and their development draw on advances in chemistry and physical chemistry and, in turn, contribute to these areas of science.

Crystal physics deals mainly with the electrical, optical, and mechanical properties of crystals closely related to their structure and symmetry, and adjoins solid-state physics, which concentrates its attention on the analysis of laws defining the general physical properties of crystals and the energy spectra of crystal lattice.

The first two volumes are devoted to the structure of crystals, and the last two, to the growth of crystals and their physical properties. The authors present the material in such a way that the reader can find the basic information on all important problems of crystallography. Due to the limitation of space the exposition of some sections is concise, otherwise many chapters would have become separate monographs. Fortunately, such books on a number of crystallographic subjects are already available.

The purpose of such an approach is to describe all the branches of crystallography in their interrelation, thus presenting crystallography as a unified science to elucidate the physical meaning of the unity and variety of crystal structures. The physico-chemical processes and the phenomena taking place in the course of crystal growth and in the crystals themselves are described, from a crystallographic point of view, and the relationship of properties of crystals with their structure and conditions of growth is elucidated.

This four-volume edition is intended for researchers working in the fields of crystallography, physics, chemistry, and mineralogy, for scientists studying the structure, properties, and formation of various materials, for engineers and those engaged in materials science technology, particularly in the synthesis of crystals and their use in various technical devices. We hope that this work will also be useful for undergraduate and graduate students at universities and higher technical colleges studying crystallography, solid-state physics, and related subjects.

Modern Crystallography is written by a large group of authors from the Institute of Crystallography of the USSR Academy of Sciences, who benefited from the assistance and advice of many other colleagues. The English edition of all four volumes of *Modern Crystallography* is being published almost simultaneously with the Russian edition. The authors have included in the English edition some of the most recent data. In several instances some additions and improvements have been made.

B.K. Vainshtein

Preface

This volume *Modern Crystallography II* is devoted to the structure of crystals. It is a sequel to *Modern Crystallography I* which covers the symmetry of crystals, their geometry, and the methods for investigating the atomic and real structures of crystals. This volume *Modern Crystallography II*, discusses the concepts of the atomic and electron structures of the crystal lattice, their dynamics, and the real microscopic structure of crystals. The structure of crystals develops in the course of their formation and growth under definite thermodynamic conditions. These problems will be treated in *Modern Crystallography III*. On the other hand, the formed crystal structure determines the physical properties of crystals, which will be elucidated in *Modern Crystallography IV*.

Chapter 1 of this volume expounds the fundamentals of the theory of the atomic structure of crystals and of the chemical bonding between atoms, and the principal concepts of crystal chemistry, of the geometric and symmetry principles of the formation of crystal structures and of lattice energy.

Chapter 2 reviews the main classes of crystalline structures — elements, inorganic and organic compounds, metals and alloys. The description of the structure of polymers, liquid crystals, as well as crystals and macromolecules of biological origin, which have been practically neglected in most of the books on crystallography, are given more attention than usual. These first two chapters were written by B.K. Vainshtein.

Chapter 3 presents the fundamentals of the electron theory and band structure of the crystal lattice.

Chapter 4 is devoted to lattice dynamics and phase transitions. It considers the vibrations of atoms in crystals, their heat capacity and thermal conductivity, and also the relationship of the thermodynamic characteristics and symmetry of crystals with phase transitions. Chapters 3 and 4 were written by V.M. Fridkin, except for Section 4.8, which was prepared by E.B. Loginov. These chapters may serve as a bridge linking crystallography with solid-state physics; they supply microscopic substantiation of the treatment of many physical properties of crystals.

Finally, Chapter 5 considers the structure of crystals not as idealized systems consisting of atoms with the strict three-dimensional periodicity, but as the real structure with various defects. The principal types of lattice imperfections are classified and analyzed, particular attention being given to the dislocation theory. This chapter will aid in understanding *Modern Crystallography III* which is concerned with the mechanisms of real crystal structure formation. At the same time it prepares the reader for *Modern Crystallography IV* dealing with the physical properties of crystals, many of which — especially the mechanical properties — essentially depend on the real structure. Chapter 5 was written by V.L. Indenbom.

The authors are grateful to L.A. Feigin, A.M. Mikhailov, V.V. Udalova, G.N. Tishchenko, L.I. Man, E.M. Voronkova, and many other colleagues for their effective aid in the preparation of this volume.

Moscow, April 1982

B.K. Vainshtein
V.M. Fridkin
V.L. Indenbom

Contents

1. Principles of Formation of the Atomic Structure of Crystals

1.1 The Structure of Atoms 1
    1.1.1 A Crystal as an Assembly of Atoms 1
    1.1.2 Electrons in an Atom 3
    1.1.3 Multielectron Atoms and the Periodic System 8
1.2 Chemical Bonding Between Atoms 15
    1.2.1 Types of Chemical Bonding 15
    1.2.2 Ionic Bond 19
    1.2.3 Covalent Bond. Valence-Bond Method 26
    1.2.4 Hybridization. Conjugation 29
    1.2.5 Molecular-Orbital (MO) Method 33
    1.2.6 Covalent Bond in Crystals 39
    1.2.7 Electron Density in a Covalent Bond 44
    1.2.8 Metallic Bond 49
    1.2.9 Weak (van der Waals) Bonds 53
    1.2.10 Hydrogen Bonds 55
    1.2.11 Magnetic Ordering 58
1.3 Energy of the Crystal Lattice 61
    1.3.1 Experimental Determination of the
            Crystal Energy 61
    1.3.2 Calculation of the Potential Energy 62
    1.3.3 Organic Structures 67
1.4 Crystallochemical Radii Systems 69
    1.4.1 Interatomic Distances 69
    1.4.2 Atomic Radii 70
    1.4.3 Ionic Radii 74
    1.4.4 The System of Atomic-Ionic Radii of a
            Strong Bond 83
    1.4.5 System of Intermolecular Radii 86
    1.4.6 Weak- and Strong-Bond Radii 88
1.5 Geometric Regularities in the Atomic Structure of Crystals . 90

1.5.1   The Physical and the Geometric Model of
        a Crystal 90
1.5.2   Structural Units of a Crystal 90
1.5.3   Maximum-Filling Principle 92
1.5.4   Relationship Between the Symmetry of Structural
        Units and Crystal Symmetry 93
1.5.5   Statistics of the Occurrence of Space Groups 97
1.5.6   Coordination 99
1.5.7   Classification of Structures According to the
        Dimensionality of Structural Groupings 101
1.5.8   Coordination Structures 101
1.5.9   Relationship Between Coordination and
        Atomic Sizes 102
1.5.10 Closest Packings 103
1.5.11 Structures of Compounds Based on Close
        Packing of Spheres 107
1.5.12 Insular, Chain and Layer Structures 111
1.6   Solid Solutions and Isomorphism 114
1.6.1   Isostructural Crystals...................... 114
1.6.2   Isomorphism 114
1.6.3   Substitutional Solid Solutions 115
1.6.4   Interstitial Solid Solutions 120
1.6.5   Modulated and Incommensurate Structures 123
1.6.6   Composite Ultrastructures 124

2. Principal Types of Crystal Structures
2.1   Crystal Structures of Elements 127
2.1.1   Principal Types of Structures of Elements 127
2.1.2   Crystallochemical Properties of Elements 136
2.2   Intermetallic Structures 137
2.2.1   Solid Solutions and Their Ordering 138
2.2.2   Electron Compounds 140
2.2.3   Intermetallic Compounds 142
2.3   Structures with Bonds of Ionic Nature 143
2.3.1   Structures of Halides, Oxides and Salts 143
2.3.2   Silicates 147
2.3.3   Superionic Conductors 156
2.4   Covalent Structures 158
2.5   Structure of Complex and Related Compounds 165
2.5.1   Complex Compounds 165
2.5.2   Compounds with Metal Atom Clusters 169

    2.5.3   Metal-Molecular Bonds (π Complexes of
            Transition Metals) 170
    2.5.4   Compounds of Inert Elements 172
2.6 Principles of Organic Crystal Chemistry 172
    2.6.1   The Structure of Organic Molecules 173
    2.6.2   Symmetry of Molecules 177
    2.6.3   Packing of Molecules in a Crystal 179
    2.6.4   Crystals with Hydrogen Bonds 187
    2.6.5   Clathrate and Molecular Compounds 189
2.7 Structure of High-Polymer Substances 190
    2.7.1   Noncrystallographic Ordering 190
    2.7.2   Structure of Chain Molecules of High Polymers .. 191
    2.7.3   Structure of a Polymer Substance 196
    2.7.4   Polymer Crystals 196
    2.7.5   Disordering in Polymer Structures 200
2.8 Structure of Liquid Crystals 204
    2.8.1   Molecule Packing in Liquid Crystals 204
    2.8.2   Types of Liquid-Crystal Ordering 205
2.9 Structures of Substances of Biological Origin 214
    2.9.1   Types of Biological Molecules 214
    2.9.2   Principles of Protein Structure 216
    2.9.3   Fibrous Proteins 224
    2.9.4   Globular Proteins 226
    2.9.5   Structure of Nucleic Acids 253
    2.9.6   Structure of Viruses 261

3. Band Energy Structure of Crystals 274
3.1 Electron Motion in the Ideal Crystal 275
    3.1.1   Schrödinger Equation and Born-Karman
            Boundary Conditions 275
    3.1.2   Energy Spectrum of an Electron 280
3.2 Brillouin Zones 282
    3.2.1   Energy Spectrum of an Electron in the Weak-Bond
            Approximation 282
    3.2.2   Faces of Brillouin Zones and the Laue Condition . 285
    3.2.3   Band Boundaries and the Structure Factor 287
3.3 Isoenergetic Surfaces. Fermi Surface and Band Structure . 288
    3.3.1   Energy Spectrum of an Electron in the
            Strong-Bond Approximation 289
    3.3.2   Fermi Surfaces 291

4. Lattice Dynamics and Phase Transitions

4.1 Atomic Vibrations in a Crystal 294
    4.1.1 Vibrations of a Linear Atomic Chain 294
    4.1.2 Vibration Branches 296
    4.1.3 Phonons 297
4.2 Heat Capacity, Thermal Expansion, and Thermal
    Conductivity of Crystals 299
    4.2.1 Heat Capacity 299
    4.2.2 Linear Thermal Expansion 300
    4.2.3 Thermal Conductivity 301
4.3 Polymorphism. Phase Transitions 302
    4.3.1 Phase Transitions of the First and Second Order .. 304
    4.3.2 Phase Transitions and the Structure 304
4.4 Atomic Vibrations and Polymorphous Transitions 308
4.5 Ordering-Type Phase Transitions 312
4.6 Phase Transitions and Electron–Phonon Interaction 315
    4.6.1 Contribution of Electrons to the Free
          Energy of the Crystal 316
    4.6.2 Interband Electron–Phonon Interaction 317
    4.6.3 Photostimulated Phase Transitions 321
    4.6.4 Curie Temperature and the Energy Gap Width ... 322
4.7 Debye's Equation of State and Grüneisen's Formula 323
4.8 Phase Transitions and Crystal Symmetry 325
    4.8.1 Second-Order Phase Transitions 325
    4.8.2 Description of Second-Order Transitions with
          an Allowance for the Symmetry 328
    4.8.3 Phase Transitions Without Changing the Number of
          Atoms in the Unit Cell of a Crystal 331
    4.8.4 Changes in Crystal Properties on Phase Transitions 334
    4.8.5 Properties of Twins (Domains) Forming on Phase
          Transformations 335
    4.8.6 Stability of the Homogeneous State of the
          Low-Symmetry Phase 336

5. The Structure of Real Crystals 338

5.1 Classification of Crystal Lattice Defects 339
5.2 Point Defects of the Crystal Lattice 340
    5.2.1 Vacancies and Interstitial Atoms 340
    5.2.2 Role of Impurities, Electrons, and Holes 346
    5.2.3 Effect of External Influences 347
5.3 Dislocations. 349

5.3.1 Burgers Circuit and Vector 350
5.3.2 Elastic Field of Straight Dislocation 353
5.3.3 Dislocation Reactions 358
5.3.4 Polygonal Dislocations 360
5.3.5 Curved Dislocations 365
5.4 Stacking Faults and Partial Dislocations 367
5.5 Continuum Description of Dislocations 375
5.5.1 Dislocation-Density Tensor 375
5.5.2 Example: A Dislocation Row 376
5.5.3 Scalar Dislocation Density 378
5.6 Subgrain Boundaries (Mosaic Structures) in Crystals 378
5.6.1 Examples of Subgrain Boundaries:
        A Tilt Boundary and a Twist Boundary. 378
5.6.2 The Dislocation Structure of the Subgrain
        Boundary in General 380
5.6 3 Subgrain Boundary Energy 384
5.6.4 Incoherent Boundaries 384
5.7 Twins 387
5.7.1 Twinning Operations 387
5.7.2 Twinning With a Change in Crystal Shape 390
5.7.3 Twinning Without a Change in Shape 394
5.8 Direct Observation of Lattice Defects 396
5.8.1 Ionic Microscopy 396
5.8 2 Electron Microscopy 397
5.8.3 X-Ray Topography 402
5.8.4 Photoelasticity Method 408
5.8.5 Selective Etching Method 410
5.8 6 Investigation of the Crystal Surface 411

Bibliography 413

References 419

Subject Index 429

1. Principles of Formation of the Atomic Structure of Crystals

The atoms in crystals are in direct contact, and the interactions of their outer electron shells produce chemical bonding. The condition for the formation of a crystal structure is a sufficiently low temperature, so that the potential energy of attraction between atoms considerably exceeds the kinetic energy of their thermal motion. The diversity of structures of crystals is determined by their composition and by the individual chemical characteristics of their constituent atoms which define the nature of the chemical bond, the electron density distribution, and the geometric arrangement of the atoms within the unit cell. Many important characteristics of crystal structures − interatomic distances, coordination, etc. − can be described on the basis of the geometrical model in which the atoms are represented as rigid spheres of definite radii.

1.1 The Structure of Atoms

1.1.1 A Crystal as an Assembly of Atoms

Each atom in a crystal structure is bound to its nearest neighbors at definite distances. It also interacts with the next nearest atoms, both directly and through the first nearest neighbors, then with still further atoms, and so on, i.e., in end effect with the entire structure. The formation of crystals always results from the collective interaction of atoms, although it may sometimes be regarded approximately as the result of pair interactions. The assembly of atoms in the crystal lattice is also united by its intrinsic system of thermal vibrations.

The concepts of the chemical bond, valence, etc., play an important part in crystal chemistry and the theory of crystal structures. They have mainly been developed to explain the structure of molecules, whereas in a crystal the interaction of the whole set of atoms must be taken into consideration. Let us take, for instance, the simplest compound NaCl. If we write down its formula as Na^+Cl^- (in accordance with the valences of these atoms and the concepts of the positive charge of Na and negative of Cl), we obtain "molecules" Na^+Cl^-, of which the crystal seems to be made up. In fact, the structure of NaCl, as well as most of the ionic, covalent, and metallic structures, has no molecules at all; the formula NaCl is realized in the high-symmetry packing of these atoms, each of which is surrounded by 6 neighbors of a different sort, then, further on, by 12 of the same sort, etc. On the other hand, in a number of crystals the molecules, or some other stable atomic groupings, retain their individuality. In this case the weaker forces binding these groupings into a crystal structure have to be explained. There are also many intermediate cases.

The task of "calculating" the concrete crystal structure is extremely complicated. However, generalization of a huge amount of chemical data on interatomic bonds and valence, quantum-mechanical information on the nature of atomic interactions and on cooperative interactions of particles, and, above all, diffraction and other investigations have made it possible to formulate a number of laws and regularities describing and governing the general principles and peculiarities of various crystal structures. Some of them are rigorously substantiated by theory and have a quantitative expression, while the others are semiempirical or qualitative. Nevertheless, on the whole these theoretical and empirical data, which comprise the subject of the theory of the atomic structure of crystals, or crystal chemistry, ensure reliable orientation in the tremendous diversity of crystal structures, explain many of their peculiarities, and enable one to calculate and predict some of their properties.

This chapter begins by explaining the general principles of the formation of chemical bonds in molecules and crystals. Then it considers the main geometric laws of the construction of crystal structures and some general questions of crystal chemistry. After that we discuss the basic types of crystal structures and classify them according to the nature of their chemical bonds. We also consider the structures of polymers, liquid crystals, and biological substances.

The chemical bond between atoms in molecules and crystals is realized by the electrons of their outer shells. Therefore, although atoms in a crystal always differ, to some extent, from free atoms, this difference actually refers exclusively to their outer shells. The interaction of these shells, the changes in their structure, and the redistribution of electrons from atom to atom or into the common "electron gas" of the crystal − all these actually reflect different aspects or possibilities of approximate description of the state of atoms and electrons assembled in the crystal structure. Even the simplest − ionic − type of bond, when oppositely charged ions attract each other electrostatically, requires, in the first place, an explanation of the reasons why it is "profitable" for one atom to give up an electron and become a cation, and for another to accept it and also become an ion, but with the opposite sign, i.e., an anion. Then problems arise concerning, for example, the equilibrium of the repulsive and attractive forces.

The stability of the inner shells of atoms, the relative stability of the outer shells, and the trend towards the formation of some stable states at the chemical bond in molecules and crystals are the starting points from which chemistry and crystal chemistry begin their analysis of the possible arrangement of atoms. Therefore we must first of all recall some basic information on the structure of atoms.

1.1.2 Electrons in an Atom

Electrons in an atom are in the spherically symmetric Coulomb field of the nucleus Ze/r; they interact, repelling each other owing to electrostatic forces, and obey the Pauli principle. The stationary states of the electrons in an atom are described by the Schrödinger equation $H\Psi = E\Psi$. Wave function $\Psi(x, y, z)$, which is the solution to the Schrödinger equation is generally complex, and the square of modulus $|\Psi|^2 = \Psi\Psi^*$ gives the probability density of finding electrons at point x, y, z, i.e., the electron density of the atom $\rho(x, y, z)$.

Let us consider the hydrogen atom. Here the Schrödinger equation in the spherical coordinates has the form

$$\left(\frac{h^2}{8\pi^2 m} \nabla^2 + \frac{e^2}{r}\right) \psi(r, \theta, \varphi) + E\psi(r, \theta, \varphi) = 0 \tag{1.1}$$

and is solved analytically.

Each of the set of solutions

$$\psi_{nlm}(r, \theta, \varphi) = R_{nl}(r) Y_{lm}(\theta, \varphi) = R_{nl}(r) \Theta_{lm}(\theta) \Phi_m(\varphi) \tag{1.2}$$

is characterized by a definite combination of quantum numbers n (principal quantum number), l (orbital quantum number), and m (magnetic quantum number), and by the eigenvalue of the energy

$$E_n = -\frac{2\pi^2 m e^4}{n^2 h^2}. \tag{1.3}$$

The wave functions ψ of electrons in an atom — of one electron in a hydrogen atom or of each electron in a multielectron atom — which give the distribution of the electrons in an atom, are called atomic orbitals by analogy to Bohr's atomic orbitals AO. With a given principal quantum number n, n^2 combinations according to the following scheme are possible:

$$n = 1, 2, \ldots; \quad l = 0, 1, \ldots, (n-1); \quad m = 0, \pm 1, \pm 2, \ldots, \pm l. \tag{1.4}$$

With a given n the energy of each one of a set of states is the same, i.e., the states are degenerate.

The solution ψ (1.2) consists of radial, R, and angular, Y, components and may have a different symmetry according to the quantum numbers. The sym-

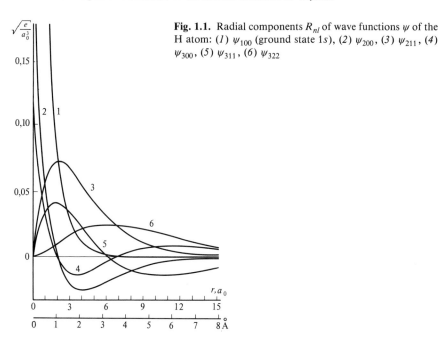

Fig. 1.1. Radial components R_{nl} of wave functions ψ of the H atom: (1) ψ_{100} (ground state 1s), (2) ψ_{200}, (3) ψ_{211}, (4) ψ_{300}, (5) ψ_{311}, (6) ψ_{322}

metry depends on the type of Y_{lm} and is described by one of the point groups of symmetry or antisymmetry. Figure 1.1 shows the radial components R_{nl} for the first states of the H atom. Figure 1.2 represents some orbitals of the H atom. In states with $l = 0$ (m is then also zero), $Y_{00} = 1$; these states are spherically symmetric and called s states (Fig. 1.2a). For instance, the orbital with the lowest energy (the ground state) has the form

$$\psi_{1s} = \frac{1}{\sqrt{\pi a_0^3}} \exp\left(-\frac{r}{a_0}\right), \tag{1.5}$$

where $a_0 = 0.529$ Å is the Bohr radius, the atomic unit of length. States with $l = 1$ are called p states, with $l = 2$, d states, and with $l = 3$, f states.

At $l \geqslant 1$ the orbitals are no longer spherically symmetric. Thus, at $l = 1$ three solutions ($m = 0, \pm 1$) of (1.1) are possible. Generally, angular components are complex, but real combinations may be constructed from them. So three real p orbitals are obtained,

$$p_x = \frac{\sqrt{3}}{2\sqrt{\pi}} R_{n1}(r) \sin \theta \cos \varphi, \qquad p_y = \frac{\sqrt{3}}{2\sqrt{\pi}} R_{n1}(r) \sin \theta \sin \varphi,$$

$$p_z = \frac{\sqrt{3}}{2\sqrt{\pi}} R_{n1}(r) \cos \theta.$$

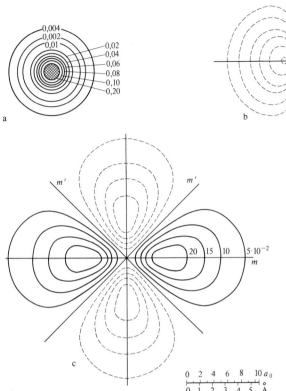

Fig. 1.2a–c. Cross sections of some functions of the H atom in contour lines of equal values ($\sqrt{e/a_0^3}$). The lines are dashed at the negative values. (a) ψ_{100} (1s), (b) ψ_{210} (2p); (c) ψ_{322} ($d_{x^2-y^2}$). The functions have different symmetries; ψ_{100} is spherically symmetric, ψ_{210} is cylindrical and anti-symmetric, (∞/m'), ψ_{322} is tetragonal and antisymmetric (4/mm'm). The m' planes are nodal; ψ is equal to zero on them [1.1]

These orbitals have a cylindrical symmetry and are elongated either along the x, or the y, or the z axis (Fig. 1.2b). These functions reduce to zero on the plane of antisymmetry m' (nodal surface) perpendicular to the axis of cylindrical symmetry; their absolute values are mirror equal, but opposite in sign on the both sides of m', i.e., they are described by the antisymmetry point group ∞/m'. The orbitals with $l = 2$ (d states) have a more complex configuration (Fig. 1.2c). The orbitals with $l = 3$ (f states) are still more complex.

All the orbitals are normalized and orthogonal, i.e., they obey the conditions

$$\int \psi_i \psi_j^* dv = \begin{cases} 1 \text{ for } i = j, \\ 0 \text{ for } i \neq j. \end{cases} \tag{1.6}$$

The first of these relations shows that for each orbital the integral of the probability density ψ^2 of the presence of an electron, taken over the entire volume, is equal to unity, and the second, that for any pair of different orbitals integral (1.6) is equal to zero, which reflects the properties of orbital symmetry. The square of the modulus of the wave function

$$|\psi|^2 = \psi\psi^* = \rho(r) \tag{1.7}$$

is the electron density of an atom in the corresponding state, $\rho(r)$ being expressed by the number of electrons per unit volume. The models of electron density distribution in a hydrogen atom for the ground ($1s$) and other states are shown in Fig. 1.3.

Let us consider the function of radial distribution of electrons in the atom $D(r)$, which is obtained by integrating $|\psi|^2$ over the angles. As a result, the dependence on $Y_{lm}(\theta,\varphi)$ disappears, and the dependence on r is given by $R_{nl}(r)$.

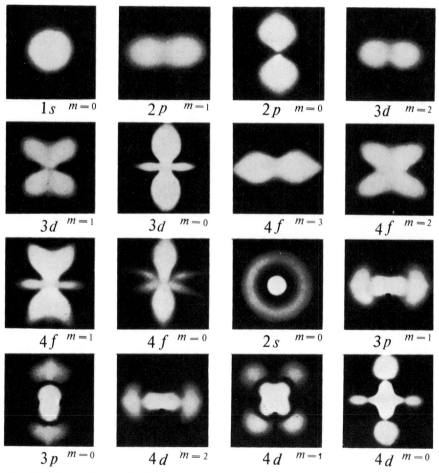

Fig. 1.3. Model of space distribution of the electron cloud in the hydrogen atom in different states [1.2]

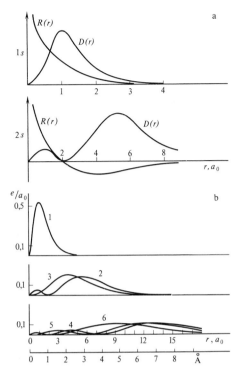

Fig. 1.4a, b. Radial distribution functions of the electron density in the H atom: (a) relationship between the radial component $R(r)$ of the wave function and the radial distribution $D(r)$ of functions ψ_{1s} and ψ_{2s}, (b) functions $D(r)$ for different states n, l: (1) 1,0; (2) 2,0; (3) 2,1; (4) 3,0; (5) 3,1; (6) 3,2. (cf. Fig. 1.1)

Thus, for the hydrogen atom and any multielectron atom the radial distribution of electrons in the n, l state is described by the function

$$D_{nl}(r) = 4\pi r^2 R_{nl}^2(r) \tag{1.8}$$

so that $D(r)dr$ gives the total number of these electrons in a spherical layer of radius r and thickness dr (Fig. 1.4a)[1]. The hydrogen atom contains one electron while, in general, an atom contains Z electrons; therefore,

$$\int |\psi|^2 dv = \int \rho(r) dv_r = \int D(r) dr = Z. \tag{1.9}$$

The graphs of radial functions for the hydrogen atom (Fig. 1.4b) clearly show the peculiarities of the radial distribution of the electrons. The maximum of $D(r)$ for $n = 1$ – the ground state of hydrogen – lies at $r = a_0$, the Bohr radius of the first orbital.

The above laws about the structure of H orbitals hold true for the motion of electrons in any central field and, therefore, can be used for describing the structure of the outer shells of any atom if we assume that the inner electrons, together with the nucleus, form some fixed, spherically symmetric system.

---

[1] By multiplying $\rho(r)$ (1.7) and $D(r)$ (1.8) by e we can express these functions in terms of the charge.

1.1.3 Multielectron Atoms and the Periodic System

The task of calculating multielectron wave functions, i.e., of finding the
energy levels and the distribution of the electron density of atoms or ions is
complicated and cannot be solved analytically, in distinction to the hydrogen
atom. The most rigorous method of finding its solution was proposed by
D. R. Hartree in 1928 and improved by V. A. Fock in 1930; it is known as the
self-consistent field method taking account of the exchange. Each electron of
an atom is described by its one-electron wave function, the orbital, and is
assumed to be in the potential field of the nucleus and of all the other electrons. It can be assumed with sufficient accuracy that this potential is spherically symmetric, and therefore the types of solutions with respect to the s, p,
d, and f states are preserved. The orbitals have the same form as for the H
atom (Fig. 1.5). One should also take into consideration the electron spin and
the Pauli principle; each n, l, and m orbit may have not one, but two electrons
with opposite spins. Therefore, in addition, factor $m_s = \pm 1/2$, describing the
spin coordinate, is introduced into the formula of the orbital $\psi(2)$; then
ψ_{nlmm_s} is called a spin orbital.

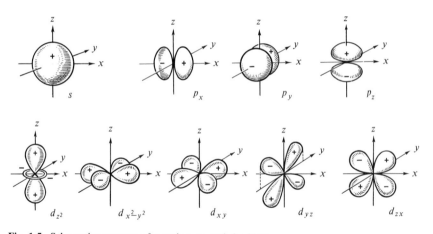

Fig. 1.5. Schematic structure of atomic s, p, and d orbitals

For a system of N electrons, the full wave function Ψ can be constructed
from N spin orbitals $\psi_i(\xi)$, where i is the number of the state, and j is a set of
quantum numbers [see below (1.33)]. Functions $\psi_i(\xi_j)$ are orthonormalized
according to (1.6). As a result, we obtain sets of equations whose solution by
successive approximations determines both the wave functions and the field
correlated with these functions, i.e., the self-consistent field. Beginning with
the helium atom, which has two electrons, exchange terms of Coulomb interaction appear in such equations.

By now, computer calculations of wave functions have been carried out for all the atoms and a number of ions (Figs. 1.6, 7). As the number of electrons increases, orbitals with larger quantum numbers are filled. It is seen from Fig. 1.6 that despite the increase in the number of electrons in the shell with increasing Z, the shells contract within a given period because of the increased Coulomb attraction to the nucleus. For each of the atoms, transitions of electrons from the ground to the excited states, with other quantum numbers and energies[2], are possible. Therefore calculations for excited states have also been carried out for some atoms.

Using the formal combinatorics of quantum numbers and knowing the arrangement of the corresponding energy levels of atoms (which predetermines their filling with electrons), it is possible to explain all the basic regularities of Mendeleyev's periodic system. Figure 1.8 gives the sequence of the filling of the levels, and Table 1.1 their systematics. It should be emphasized that while for small n the calculation of the levels, at least with respect to their energy values, is comparatively simple, at large quantum numbers it becomes highly complicated and is carried out with due regard for experimental spectroscopic and other data.

In the helium atom He ($Z = 2$) which follows hydrogen, the ground state can accommodate a second electron, but with an opposite spin, on the same $1s$ orbit, which yields a stable two-electron, the so-called K shell.

Subsequent filling of energy levels by electrons according to the scheme of Fig. 1.8 is possible at $n = 2$, which gives, at $l = 0$, the completion of the $2s$ level of Li and Be ($Z = 3, 4$), and six more possibilities at $l = 1$ and $m = 0$, $+1$, -1, which correspond to $2p$ states of the atoms of B, C, N, O, F, and Ne ($Z = 5 - 10$). At first, the three p states with parallel spins are filled, those of B, C, and N; then the same states are filled by electrons with antiparallel spins. Thus we obtain the second period of Mendeleyev's system, which is completed by a stable eight-electron shell (L shell) (see Fig. 1.6). The next level, $3s$, is spaced far from $2p$. Further filling according to the scheme of Fig. 1.8 will lead to the third period ($n = 3$, Na – Ar) of the eight elements with $Z = 11 - 18$; the corresponding outer shell is called the M shell (Fig. 1.7b, c). This is followed by two large periods, each of which consists of 18 elements, $n = 4$, K – Kr, $Z = 19 - 36$; $n = 5$, Rb – Xe, $Z = 37 - 54$ (outer N and O shells, respectively). The next period contains 32 elements, $n = 6$, Cs – Rn, $Z = 55 - 86$. The last period, $n = 7$, begins with $Z = 87$. Table 1.1 shows the distribution of the electrons in atoms according to their s, p, d, and f levels. The valencies of the transition elements, Sc – Ni and Y – Pd, rare-earth elements (lanthanides), elements of the platinum group, and transuranic elements, including actinides, are due to the fact that it is energetically advantageous to fill the incomplete inner shells, while the number of electrons on the outermost orbitals may remain constant.

---

[2] The chemical-bond theory uses different energy units, electron volt (eV), kilocalories per mole (kcal/mol), kilojoules per mole (kJ/mol); the atomic energy unit, 1 a.e.u. = 27.7 eV; 1 eV = 23.06 kcal/mol = 96.48 kJ/mol, 1 kcal = 4.184 kJ.

Electron density

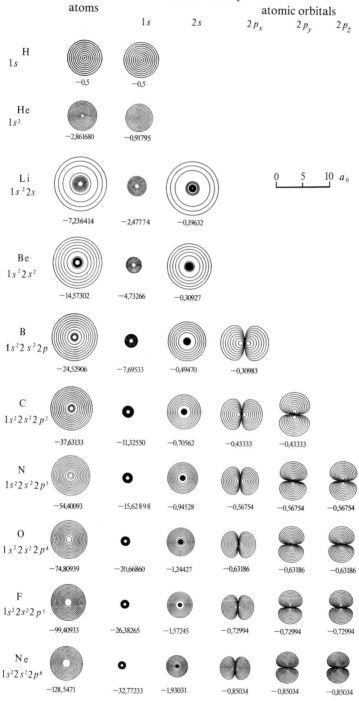

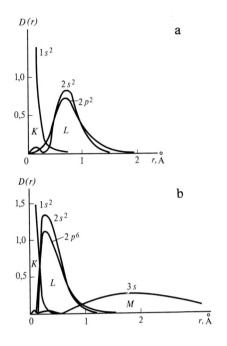

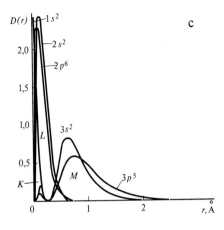

Fig. 1.7a–c. Radial distribution functions $D(r)$ for the electron shells of the C (a), Na (b), and Cl (c) atoms

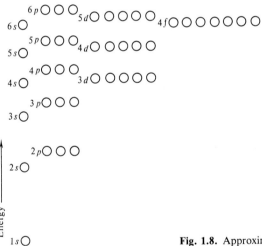

Fig. 1.8. Approximate energy levels of atomic orbitals

Fig. 1.6. Structures of electron shells (squares of wave functions) for the first ten atoms of the periodic system in the ground state, and the electron density of all the constituent orbitals [1.3].

The highest density contour line corresponds to the value $1 e/a_0^3$, the value of the succeeding contours is decreased by a factor of two. The last line corresponds to $4.9 \times 10^{-4} e/a_0^3$. The total and orbital energies are indicated in atomic energy units (1 a.e.u. = 27.7 eV)

Table 1.1. Electron configurations and orbital radii r_0 of elements [1.4]

Z	Element	Electron configuration	Outer shell	r_0	Z	Element	Electron configuration	Outer shell	r_0
1	H	$1s^1$	$1s$	0.529	51	Sb	$5s^25p^3$	$5p$	1.140
2	He	$1s^2$ *	$1s$	0.291	52	Te	$5s^25p^4$	$5p$	1.111
					53	I	$5s^25p^5$	$5p$	1.044
3	Li	$2s^1$	$2s$	1.586	54	Xe	$5s^25p^6$ *	$5p$	0.986
4	Be	$2s^2$	$2s$	1.040					
5	B	$2s^22p^1$	$2p$	0.776	55	Cs	$6s^1$	$6s$	2.518
6	C	$2s^22p^2$	$2p$	0.620	56	Ba	$6s^2$	$6s$	2.060
7	N	$2s^22p^3$	$2p$	0.521	57	La	$5d^16s^2$	$6s$	1.915
8	O	$2s^22p^4$	$2p$	0.450	58	Ce	$4f^26s^2$	$6s$	1.978
9	F	$2s^22p^5$	$2p$	0.396	59	Pr	$4f^36s^2$	$6s$	1.942
10	Ne	$2s^22p^6$ *	$2p$	0.354	60	Nd	$4f^46s^2$	$6s$	1.912
					61	Pm	$4f^56s^2$	$6s$	1.882
11	Na	$3s^1$	$3s$	1.713	62	Sm	$4f^66s^2$	$6s$	1.854
12	Mg	$3s^2$	$3s$	1.279	63	Eu	$4f^76s^2$	$6s$	1.826
13	Al	$3s^23p^1$	$3p$	1.312	64	Gd	$4f^75d^16s^2$	$6s$	1.713
14	Si	$3s^23p^2$	$3p$	1.068	65	Tb	$4f^96s^2$	$6s$	1.775
15	P	$3s^23p^3$	$3p$	0.919	66	Dy	$4f^{10}6s^2$	$6s$	1.750
16	S	$3s^23p^4$	$3p$	0.810	67	Ho	$4f^{11}6s^2$	$6s$	1.727
17	Cl	$3s^23p^5$	$3p$	0.725	68	Er	$4f^{12}6s^2$	$6s$	1.703
18	Ar	$3s^23p^6$ *	$3p$	0.659	69	Tu	$4f^{13}6s^2$	$6s$	1.681
					70	Yb	$4f^{14}6s^2$	$6s$	1.658
19	K	$4s^1$	$4s$	2.162	71	Lu	$5d^16s^2$	$6s$	1.553
20	Ca	$4s^2$	$4s$	1.690	72	Hf	$5d^26s^2$	$6s$	1.476
21	Sc	$3d^14s^2$	$4s$	1.570	73	Ta	$5d^36s^2$	$6s$	1.413
22	Ti	$3d^24s^2$	$4s$	1.477	74	W	$5d^46s^2$	$6s$	1.360
23	V	$3d^34s^2$	$4s$	1.401	75	Re	$5d^56s^2$	$6s$	1.310
24	Cr	$3d^54s^1$	$4s$	1.453	76	Os	$5d^66s^2$	$6s$	1.266
25	Mn	$3d^54s^2$	$4s$	1.278	77	Ir	$5d^76s^2$	$6s$	1.227
26	Fe	$3d^64s^2$	$4s$	1.227	78	Pt	$5d^96s^1$	$6s$	1.221
27	Co	$3d^74s^2$	$4s$	1.181	79	Au	$6s^1$	$6s$	1.187
28	Ni	$3d^84s^2$	$4s$	1.139	80	Hg	$6s^2$	$6s$	1.126
29	Cu	$3d^{10}4s^1$	$4s$	1.191	81	Tl	$6s^26p^1$	$6p$	1.319
30	Zn	$3d^{10}4s^2$	$4s$	1.065	82	Pb	$6s^26p^2$	$6p$	1.215
31	Ga	$4s^24p^1$	$4p$	1.254	83	Bi	$6s^26p^3$	$6p$	1.130
32	Ge	$4s^24p^2$	$4p$	1.090	84	Po	$6s^26p^4$	$6p$	1.212
33	As	$4s^24p^3$	$4p$	0.982	85	At	$6s^26p^5$	$6p$	1.146
34	Se	$4s^24p^4$	$4p$	0.918	86	Rn	$6s^26p^6$*	$6p$	1.090
35	Br	$4s^24p^5$	$4p$	0.851					
36	Kr	$4s^24p^6$ *	$4p$	0.795	87	Fr	$7s^1$	$7s$	2.447
					88	Ra	$7s^2$	$7s$	2.042
37	Rb	$5s^1$	$5s$	2.287	89	Ac	$6d^17s^2$	$7s$	1.895
38	Sr	$5s^2$	$5s$	1.836	90	Th	$6d^27s^2$	$7s$	1.788
39	Y	$4d^15s^2$	$5s$	1.693	91	Pa	$5f^26d^17s^2$	$7s$	1.804
40	Zr	$4d^25s^2$	$5s$	1.593	92	U	$5f^36d^17s^2$	$7s$	1.775
41	Nb	$4d^45s^1$	$5s$	1.589	93	Np	$5f^46d^17s^2$	$7s$	1.741
42	Mo	$4d^55s^1$	$5s$	1.520	94	Pu	$5f^47s^2$	$7s$	1,784
43	Tc	$4d^55s^2$	$5s$	1.391	95	Am	$5f^77s^2$	$7s$	1.757
44	Ru	$4d^75s^1$	$5s$	1.410	96	Cm	$5f^76d^17s^2$	$7s$	1.657
45	Rh	$4d^85s^1$	$5s$	1.364	97	Bk	$5f^86d^17s^2$	$7s$	1.626
46	Pd	$4d^{10}$ *	$4d$	0.567	98	Cf	$5f^96d^17s^2$	$7s$	1.598
47	Ag	$5s^1$	$5s$	1.286	99	Es	$5f^{10}6d^17s^2$	$7s$	1.576
48	Cd	$5s^2$	$5s$	1.184	100	Fm	$5f^{11}6d^17s^2$	$7s$	1.557
49	In	$5s^25p^1$	$5p$	1.382	101	Md	$5f^{12}6d^17s^2$	$7s$	1.527
50	Sn	$5s^25p^2$	$5p$	1.240	102	No	$5f^{13}6d^17s^2$	$7s$	1.581

* The atoms below contain this shell

The electron density of the atom as a whole, ρ_{at}, is a rapidly decreasing function (Fig. 1.9a). It may be characterized by radial functions (1.8) of the electron density of each shell (Fig. 1.7), while the total radial electron density of the atom is their sum (Fig. 1.9b)

$$D(r) = 4\pi r^2 \sum R_{nl}^2(r) . \tag{1.10}$$

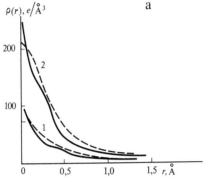

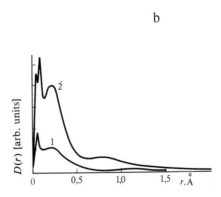

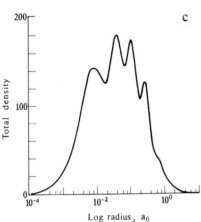

Fig. 1.9a – c. Shapes of the peaks of the electron density $\rho(r)$ for Mg and Rb atoms (a), and the radial electron density distribution $D(r)$ of these atoms (b). (a) the dashed line indicates the electron density of the atom smeared out by its thermal vibrations in the lattice, (b) the peaks are the superposition of a series of orbitals on each other (2s, 3s, 3p; cf. Fig. 1.7), (c) radial electron density distribution for uranium, radius on logarithmic scale [1.4]

The "shell" structure of the atom is clearly seen only in the radial function $D(r)$ (Fig. 1.9b), which differs from the real electron density $\rho(r)$ (Fig. 1.9a), by integration over the angles, which yields the resulting factor $4\pi r^2$. The maxima of the radial function represent the superposition of the squares of atomic orbitals with similar radii.

The function of the electron density of the atom as a whole, as well as the functions describing the structure of separate shells, extends to infinity. But in practice these functions damp out rather rapidly. The electron distribution in an atom can be characterized by the rms radius of the atom

$$\overline{r^2} = \int D(r)r^2 dr/Z, \qquad \langle r \rangle_{at} = \sqrt{\overline{r^2}}, \tag{1.11}$$

where Z is determined by (1.9). When the number Z of electrons in atom increases, so does the charge $+Ze$ of the nucleus to which they converge. As a result, $\langle r \rangle_{at}$ falls off. According to the statistical theory of the atom

$$\langle r \rangle_{at} \sim Z^{-1/3}. \tag{1.12}$$

This statistical dependence undergoes a number of fluctuations depending on the filling of the shells and the forming of new ones (Fig. 1.10).

Each shell can be characterized by the radius corresponding to the position of the maximum of its radial function, the so-called orbital radius

$$r_{oi} = r\{\max [R_i^2(r)r^2]\}. \tag{1.13}$$

It would be natural to take the orbital radius r_o of the outer shell as the characteristic of the size of the atom. Table 1.1 gives the electron configurations of atoms and the orbital radii r_o of the outer shells after *Waber, Cromer* [1.4]. Similar results were obtained by *Bratsev* [1.5].

As mentioned, in spite of the increase in the number of electrons in a given shell its orbital radius diminishes because of the growing Coulomb attraction to the nucleus (Fig. 1.10). When a new shell appears, its r_o is much larger than that of the preceding shell; then, with growing Z the new shell contracts to the nucleus again. Thus, with an increase in Z the value of r_o changes

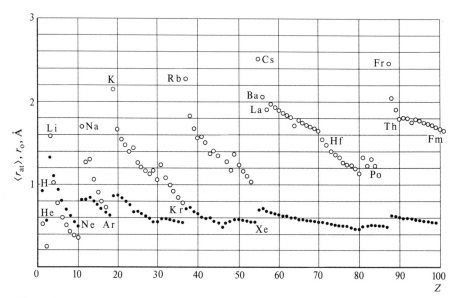

Fig. 1.10. Mean-square atomic radii $\langle r_{at} \rangle$ (*dark circles*) and outer shell orbital radii r_o (*light circles*) of neutral atoms vs atomic number Z

"toothwise", decreasing inside the periods and rising abruptly as new shells appear, which also affects the variation in $\langle r \rangle_{at}$ (1.11). On the average, r_0 increases with atomic number Z.

A knowledge of the electron structure of atoms is important for many branches of crystallography. Apart from the fact that it supplies information on the geometric characteristics of atoms, it is a starting point for the chemical-bond theory and permits calculating the functions of atomic scattering of x rays and electrons, etc. In particular, a decrease in $\langle r \rangle_{at}$ with increasing Z means that on the Fourier maps of the electron density of crystals the peaks of ρ_{at} of the heavier atoms will always be sharp. The peaks are additionally smeared out by the thermal motion of the atoms (Fig. 1.9a, dashed line; see also [Ref. 1.6, Sect. 4.1.5]), but the thermal motion, too, is less for the heavy atoms than for light ones.

The experimental data obtained by x-ray structure analysis confirm these theoretical predictions. In these investigations the electron density of the crystal obtained by Fourier series summation manifests itself as a set of electron density peaks of separate atoms $\rho(r) = \sum \rho_{at j}(r - r_j)$, and not as a set of their radial functions. The experimental results of investigations into the distribution of $\rho(r)$ in crystals are in good agreement with the theory. Since, however, the electron density of each atom falls off rapidly and is, additionally, smeared out in the crystal by the thermal motion (Fig. 1.9a), individual atomic shells are not resolved on ordinary Fourier syntheses of the distribution of $\rho(r)$. Nevertheless, it is possible to reveal the outer shells and to investigate the electron redistribution, due to the chemical bond, with the aid of Fourier difference syntheses [Ref. 1.6, Sect. 4.7.10]; this will be discussed in more detail further on.

1.2 Chemical Bonding Between Atoms

1.2.1 Types of Chemical Bonding

A chemical bond arises when atoms draw closer together and is due to interaction of their outer electron shells. The formation of such a bond (in a molecule or a crystal) means a reduction in the total and, hence, the potential energy of the system. Theory must explain not only the binding of atoms, but all the experimental data on valence; it must explain the directional character of bonds, which occurs in some cases, and their saturation; it must gives values of binding energy which agree with experiment; and, finally, it must serve as a basis for calculating the properties of molecules and crystals. The binding of atoms is the result of the electrostatic interaction of the nuclei and all their electrons and can only be explained on the basis of quantum mechanics. The main principles of the theory are verified by consideration and investigation of simple molecules. At the same time, the analysis and calculation of complex multiatomic systems involve considerable mathematical difficulties and require some simplifying assumptions. This entire complex of questions is the subject of quantum chemistry and the quantum theory of the solid state.

According to the traditional terminology, distinction is made betwen ionic (heteropolar), covalent (homopolar), metallic, and van der Waals types of bonds between atoms, as well as one more special type, the so-called hydrogen bond. The first three types are stronger than the last two. As we shall see further on, this classification into bond types, especially between the first three, is somewhat conventional. All the types of strong chemical bonding are due to the interaction between the outer orbitals of mutually approaching atoms and the formation of common electron states of the new system, i.e., a molecule or a crystal. Although the functions describing the electron distribution in the object are continuous throughout its volume, the distribution has certain features for each type of bond. An increase in the electron concentration in some atoms and a decrease in others result in Coulomb attraction, which is an *ionic* bond. If the shared outer electrons are concentrated predominantly on orbitals spatially fixed relative to the bonded atoms, we speak of a *covalent* bond. If the outer electrons are collective, i.e., distributed throughout the crystal lattice, we have a *metallic* bond.

An isolated atom has a discrete system of energy levels. In a system of N widely spaced atoms each level is essentially N-multiply degenerate. The change in the system of these levels as the atoms draw closer together amounts to eliminating the degeneracy due to interaction which leads to the splitting of the levels. Since there are very many levels, they merge into continuous bands when the atoms form a crystal. The nature of the electron energy spectrum of a crystal depends both on its constituent atoms and on the distances between them (see Chap. 3). Thus, for metals (Fig. 1.11a) the levels merge into a continuous band with vacant levels over the electrons filling the lower part in the bands. With a covalent or ionic bond (Fig. 1.11b) there is a gap in the crystal energy spectrum between the filled lower band and the following ones.

Since the spectrum depends on the interatomic distances, the nature of the bond in certain crystals may change at phase transitions, especially for those due to the pressure. At very high pressures all crystals are metallized.

The simplest qualitative explanation can be given for the ionic bond, which is ascribed to the electrostatic attraction of oppositely charged ions. But all the other bond types, including the ionic, can only be fully understood from the standpoints of quantum-mechanical theory.

It should also be borne in mind that along with those crystals in which the bond can largely be described according to one of the three indicated types, the bond in many other crystals is of an intermediate nature; in this case one can conventionally single out, for example, the ionic and covalent components. There are also a number of special cases, such as complex compounds.

The molecular crystals are somewhat different. The bond between atoms inside the molecules is covalent, and the association of the molecules into a crystal is due to weak van der Waals forces or hydrogen bonds.

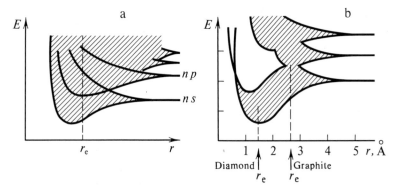

Fig. 1.11a, b. Change in the energy spectrum of electrons as the atoms come closer together. (a) in a metallic crystal levels merge into a continuous band as the atoms come together up to the equilibrium distance r_e, (b) in a carbon crystal the spectrum depends on the arising polymorphous modification, diamond has a slit between the bands, while graphite does not

Any type of interaction between atoms in a crystal or a molecule can be approximately described (and in many cases accurately enough) by the model of central forces with the aid of a potential interaction energy function $u(r)$ (Fig. 1.12) found theoretically or empirically. This curve is characterized by the following principal parameters. The condition $du/dr = 0$ defines the distance r_e of the minimum from the origin, i.e., the equilibrium interatomic distance, and the energy $u_e(r_e)$ of a given interatomic bond at absolute zero. This refers to the interaction of a pair of atoms. The equilibrium values of the interatomic distances in a crystal lattice are defined by the condition $dU/dr = 0$, where U is the energy of the lattice as a whole, which can be calculated from the energy of pair interactions $u(r)$ (see Sect. 1.3.2). The second derivative $d^2u(r)/dr^2$ of the curve $u(r)$ at the minimum characterizes the bond "rigidity". This manifests itself macroscopically in the phonon spectrum of the crystal, its elasticity, so that high frequencies and large elastic constants correspond to more rigid bonds. The steep rise of the curve towards the lesser r characterizes the mutual "impenetrability" of the atoms. The corresponding

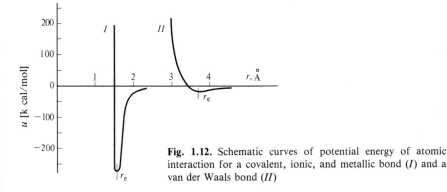

Fig. 1.12. Schematic curves of potential energy of atomic interaction for a covalent, ionic, and metallic bond (*I*) and a van der Waals bond (*II*)

macroscopic characteristic of crystals – their compressibility – makes it possible to select empirically an analytical expression for describing the repulsion forces. This steep rise enables one to operate in crystal chemistry with atom models in the form of rigid spheres touching each other in a crystal structure. The asymmetry of the potential well – a flatter shape of the curve $u(r)$ towards larger r – increases the amplitude and anharmonism of thermal vibrations with temperature, which augments the interatomic distances, i.e., leads to thermal expansion of the crystal. The vibrational anharmonism also explains various nonlinear effects in the acoustics and optics of crystals. Note that for a covalent bond the central-force approximation, although describing the interaction in the main, is insufficient because the covalent bond forces are directional (see Sect. 2.4).

The nature of the chemical bond is expressed in the macroscopic physical properties of crystals. The stronger the bond (the interatomic distances are usually short), the better are the mechanical properties, such as hardness and elasticity, the lower is the coefficient of thermal expansion, and the higher the melting point. The covalent crystals (for instance, diamond) or ion-covalent crystals (for instance, oxides of the type MgO, Al_2O_3) are the strongest and hardest. Yet even among them a transition to relatively lower strength crystals with a lower melting point (the melting temperature range is $\sim 2500 - 500°C$) takes place as the bonding weakens and the interatomic distances increase. A still wider discrepancy of the characteristics of mechanical properties occurs in metals, which include very hard and high-melting as well as low-melting metals, and even those liquid at normal temperatures (mercury). But a specific feature of metals is their high plasticity. The molecular crystals of organic compounds are the weakest and softest, and have the lowest melting point; they also have high coefficients of thermal expansion due to the weakness of the van der Waals bond.

The electrical properties of crystals are determined by the energy spectrum of their electrons (see [Ref. 1.6, Chap. 3]). Ionic crystals are, as a rule, dielectrics; covalent crystals, dielectrics or semiconductors; metals, conductors; and molecular crystals, dielectrics.

The type of the chemical bond is also manifested in the optical properties. A large refractive index is characteristic of ionic and covalent crystals, which are usually transparent in the visible or infrared region, but can also be colored, i.e., absorb light in certain regions of the visible spectrum. For ionic crystals this is due to the cations of the transition or rare-earth elements. Molecular crystals are usually transparent. On the other hand, metals possess a metalic luster, i.e., they are opaque and reflect light well.

We give here the most general characteristics of the manifestation of the chemical bond types in the physical properties of crystals. Further on, in this and subsequent volumes, the question will be treated in more detail. As we have already said, many properties of crystals certainly depend not only on the ideal atomic structure and the type of chemical bond in it, but also on the defects in its real structure. Let us now consider the basic types of chemical bonding. As mentioned above, the quantum-mechanical approach makes it possible to explain and accurately calculate all the types of bonds. At the same time an ionic bond is basically characterized by simple Coulomb interaction; we, therefore, begin our consideration of it from the classical standpoint, followed by the fundamentals of quantum-mechanical treatment.

1.2.2 Ionic Bond

When, during interaction of the outer electrons of neighboring atoms, the electron distribution shifts from one to the other, charged ions appear which are attracted electrostatically. This bond is called ionic and is usually found in relatively simple structures consisting of atoms of typical metals and nonmetals.

If a diatomic molecule with a bond of an ionic nature is formed, it naturally has a dipole moment

$$M = rZ', \tag{1.14}$$

where Z' is the ion charge, and r is the distance between the ion centers. In a crystal, an ion residing in a field of oppositely charged ions causes polarization of the electron shell, i. e., its slight deformation corresponding to the symmetry of the surrounding field. Then a net dipole moment of the structure may arise (for instance, in ferroelectrics) or may not (for instance, in structures of the NaCl type).

The possibility of the formation of a chemical bond by an atom and its nature depend to a large extent on the stability of the outer electron shell of the atom. This factor can be characterized by the ionization potential of the atom I^+, i. e., by the energy required to detach the first valence electron. The lowest ionization potentials I^+ are characteristic of alkali and alkaline-earth metals (4 – 5 eV), and the highest, of noble gases and halides (12 – 24 eV). Thus, metal atoms readily give up their outer electrons, and their ionization energy is low; after this a stable inner shell remains. On the other hand, atoms of nonmetals tend to add electrons, especially if this addition leads to the formation of stable electron shells of the noble-gases type. In that case, addition of one lacking (for the filling of the outer shell) electron causes a release of energy, the affinity energy I^-. This is most prominently manifested for halogens: I^- is 3.5 eV for F, 3.7 eV for Cl, and 3.7 eV for Br. Oxygen readily captures one electron, $I^- = 3.4$ eV, but the addition of a second electron causes a predominance of electrostatic energy of repulsion and is energetically unprofitable.

Consider the formation of an ionic bond on the example of Na^+ and Cl^-. The ionization energy of Na is 5.1 eV, and the electron affinity of Cl is 3.7 eV. Hence, the formation of a pair of such ions requires an energy of 1.4 eV. The distance between these ions in the molecule is about 2.5 Å, and the electrostatic energy of attraction ~ 10 eV. This value greatly exceeds 1.4 eV, which explains the bond between Na^+ and Cl^-. The same is true for a crystal, although the calculation of the electrostatic energy is then more complicated. Thus, in general, the energy of a system built up of ions attracted to each other is on the whole lower than the sum of the initial energies of the neutral atoms, although the ionization energy of metallic atoms has the opposite sign in the overall balance.

The interaction of the two ions is Coulombic, but when they touch one another the forces of repulsion between the electron shells come into play. The repulsive potential is described by a dependence of the type br^{-n}, where $n = 6-9$ and parameters b and n can be found from the compressibility of the crystals. Thus,

$$u_{ion}(r) = -Z_1'Z_2'e^2r^{-1} + br^{-n}, \tag{1.15}$$

where Z_1' and Z_2' are the effective charges of the ions. In a more general form, this relation can be rewritten

$$u(r) = -ar^{-m} + br^{-n} \quad (n > m); \tag{1.16}$$

then it can also be used for describing other types of bonds, naturally with different values of the constants. The condition of the minimum of $du/dr = 0$ gives equilibrium values of r_e and $u_e = u(r_e)$

$$r_e = \sqrt[n-m]{\frac{nb}{ma}}, \quad u_e = \frac{-a}{r_e^m}\left(1 - \frac{m}{n}\right). \tag{1.17}$$

Now we can express a and b and, hence, $u(r)$ in terms of r_e and u_e,

$$u(r) = u_e \frac{nm}{n-m}\left[-\frac{1}{m}\left(\frac{r_e}{r}\right)^m + \frac{1}{n}\left(\frac{r_e}{r}\right)^n\right]. \tag{1.18}$$

Here, for the ionic bond $m = 1$. The exponential dependence $\exp(-\alpha r)$ proves to be more accurate than the power dependence for describing the repulsive potential in ionic crystals [see (1.35)], the value of α^{-1} being about 0.35 Å. The interaction in ionic crystals according to (1.15) must be supplemented by the van der Waals forces operating between the ions and the polarization of the ions in each other's field.

The curve of the potential energy of interaction for the ionic bond is shown in Fig. 1.12. The ionic bond energy usually equals about 100 kcal/mol, being, for instance, 137 kcal/mol for gaseous LiF, and 88 kcal/mol for NaBr.

As discussed above, when atoms join together into a molecule or a crystal mainly owing to the ionic bond, covalent interaction, caused by the joining of the outer electrons, also invariably takes place. Even in crystals of the most prominent representatives of ionic bonding, such as halides of alkali metals, a certain (although very small) fraction of the binding energy is due to covalent interaction.

Chemists and crystal chemists have always been interested in how to describe the tendency of atoms to form an ionic bond and, if the bond is not purely ionic, but ionic covalent, how to estimate the ionic fraction of the bond. The concept of electronegativity (EN) of elements is used for this purpose. L. Pauling has given a semiempirical scale of EN on the basis of thermochemical data (see Table 1.2).

When two atoms combine, the outer electrons shift towards the one whose EN is higher. Naturally, EN is higher for anions than for cations; its values decrease when descending along the groups of the periodic system. The difference between the electronegativities of the combining atoms, Δ_{EN}, roughly characterizes the fraction and energy of the ionic component. According to Pauling, at $\Delta_{EN} \approx 3.0$ the bond is almost completely (by ~90%) ionic, and at $\Delta_{EN} < 1$ the ionic component is less than 20%. The energy of the ionic component is approximately $30 \cdot \Delta_{EN}^2$ kcal/mol.

The concept of electronegativity for atoms can be determined more rigorously in terms of their ionization potentials I^+ and the electron attractive energy I^- for the corresponding valence state (examples of these values for some atoms in one of their states together with the corrected Pauling's EN values are given in Table 1.2) (see, for instance [1.8, 9]). In this approach the electronegativity χ is expressed as one-half of the sum of the ionization

Table 1.2. Electronegativities EN, ionization potentials I^+, and the affinity I^- of certain atoms to an electron*

Element	EN	Orbital	I^+ [eV]	I^- [eV]	Element	EN	Orbital	I^+ [eV]	I^- [eV]
H	2.2	s	13.60	0.75	Cl	3.2	p	15.03	3.82
Li	1.0	s	5.39	0.82	K	0.8	s	4.34	1.46
Be	1.6	σ	9.92	3.18	Ca	1.1	s	7.09	2.26
B	2.0	s	14.91	5.70	Sc	1.3	σ	7.21	4.03
C	2.6	tetr	14.61	1.34	Cr	1.6	–	–	–
N	3.0	p	13.94	0.84	Fe	1.8	–	–	–
O	3.1	p	17.28	1.46	Zn	1.6	–	–	–
F	4.0	p	20.86	3.50	Br	3.0	p	13.10	3.54
Na	0.9	s	5.14	0.47	Rb	0.8	s	4.18	0
Mg	1.2	σ	7.10	1.08	Sn	1.8	p	6.94	0.87
Al	1.6	p	6.47	1.37	Te	2.3	p	11.04	2.58
Si	1.9	tetr	11.82	2.78	I	2.6	p	12.67	3.23
P	2.2	σ	10.73	1.42	Cs	0.7	–	–	–
S	2.6	p	12.39	2.38	Ba	0.9	–	–	–

* Only some values of I^+ and I^-, corresponding to the most frequently encountered states of a given atom, are listed. For other types of atomic orbitals (and also depending on hybridization) these values are different, for instance, $I^+ = 21.01$ and $I^- = 8.01$ for the s state of carbon

potential and the attractive energy

$$\chi = (I^+ + I^-)/2 . \tag{1.19}$$

Pauling's EN and χ (1.19) can be correlated by a certain normalization. The ionicity of bond ε is defined by the equation

$$\varepsilon = \frac{(I^+ + I^-)_{cat} - (I^+ + I^-)_{an}}{(I^+ - I^-)_{cat} + (I^+ - I^-)_{an}} = \frac{2(\chi_{cat} - \chi_{an})}{(I^+ - I^-)_{cat} + (I^+ - I^-)_{an}} , \tag{1.20}$$

which contains the corresponding values for cations and anions. According to (1.20) ε is 0.82 for NaCl, 0.83 for LiF, 0.92 for KCl, and 0.94 for RbCl, whereas for HCl it is 0.18.

The physical characteristic describing the ionic contribution to the bond is the effective charge of the ion Z'.

Let us consider experimental x-ray data on the electron density distribution in a NaCl crystal (Fig. 1.13). These data indicate that the electrons are redistributed from Na to Cl; according to different authors, the number of electrons in Na^+ is $10.3 - 10.15$ (11 in a neutral atom) and $17.7 - 17.85$ in Cl (as against 17 in neutral atom), i.e., the effective charge is about $0.8\,e$ [1.11]. In a rather large region between the atoms the electron density falls off practically to zero. For LiF these data are $2.1\,e$ for Li and 0.9 for F, i.e., the effective charge is equal to $0.9\,e$ [1.12]. The effective charges in ionic crystals can also be estimated on the basis of x-ray and IR spectra, the dielectric constant, and other methods. Thus, for MgF_2, CaCl, and $MgCl_2$ estimation of

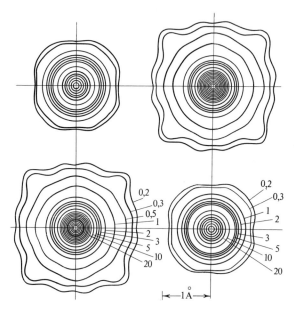

Fig. 1.13. Electron density of a NaCl crystal. Cross section of three-dimensional function $\rho(x,y,z)$ at $z = 0$ through the centers of the Na and Cl atoms [1.10]

the effective charges of anions leads to values of about 0.7 e, and of cations, $1.2 - 1.4$ e. In silicates, the effective charge of Mg is $1.5 - 1.0$ e; Al, 2.0 e, and Si (according to different estimates), $1.0 - 2.0$ e. The oxygen ion in oxides and silicates has an effective negative charge of $0.9 - 1.1$ e.[3]

The effective charge of the ion Z' may be related to the formal valence of the atom n by factor ε, which thus determines the degree of ionization of the atom

$$Z' = \varepsilon n. \qquad (1.21)$$

The degree of ionization ε in (1.21), which is found experimentally, practically coincides with the value of ε in (1.20), defined as the ionicity of the bond.

So, for univalent ions the degree of ionization ε and effective charges Z' are close to unity. At the same time, for divalent and, all the more so, trivalent ions, "integral" ionization $\varepsilon = 1$, which would lead to values of charge $Z' = n$, equal to the valence, takes place very rarely.

Symbols of the type O^{2-}, Cr^{3+}, Nb^{4+}, etc., which are often found in the literature, should only be understood as indicating the formal value of the valence, not as the actual value of the charge, which is always smaller and is not integral[3a]. Occasionally encountered symbols of the type C^{4-}, Te^{6+}, etc., have no physical meaning.

We shall now treat the electron distribution in ions. The filling of the orbitals in the anions of the crystal does not differ essentially from that in free atoms. For example, quantum-mechanical calculation of electron distribution in MgO (such calculations are described in Sect. 1.2.6) showed that in this dielectric crystal, which has four valence zones, one of them is associated with the s state, and the other three, with three p states of the oxygen atom (see Fig. 1.28a).

On ionization of a metallic atom to a cation, the filled inner shell becomes the outer shell with a much smaller orbital radius than the initial outer shell, experiencing almost no change upon ionization. Thus, $r_0(Na) = 1.713$ Å and $r_0(Na^+) = 0.278$ Å. On the other hand, the orbital radii of the outer shells of the anions, which have received the electrons needed for their complete filling, coincide almost exactly with the orbital radii of the outer shells of the same neutral atoms. For instance, $r_0(F) = 0.396$ Å, $r_0(F^-) = 0.400$ Å, $r_0(Cl) = 0.725$ Å, $r_0(Cl^-) = 0.742$ Å, $r_0(Br) = 0.851$ Å, and $r_0(Br^-) = 0.869$ Å. In other words, the electron structure of the anions practically coincides with that of the neutral atom, but their outer electron shell becomes "denser".

---

[3] Note that the concept of the effective charge in a partly ionic bond is somewhat conventional, since it depends on the choice of the range of integration of the charge corresponding to each atom and also to the electrons of the bond.

[3a] Exceptions are some rare cases when a metallic ion is surrounded by strong electronegative anions F^-, such is the case of Cu^{4+} in Cs_2CuF_6 [1.12a].

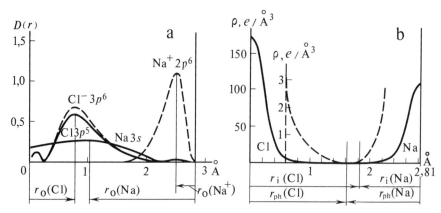

Fig. 1.14a, b. Comparison of (**a**) the theoretical radial density $D(r)$ distribution of the outer valence orbitals of Na^+ and Cl^- ions (*dashed lines*) and neutral Na and Cl atoms (*solid lines*) with (**b**) the experimentally obtained distribution of electron density $\rho(r)$ between these atoms [one-dimensional cross section of $\rho(x00)$ from Fig. 1.13]. The dashed curve in (**b**) referring to the atom periphery is scaled up. r_o: orbital radii, r_i: effective ionic radii, r_{ph}: physical ionic radii

Now, using NaCl as an example again, we consider the radial density distribution $D(r)$ (1.8, 10) for these atoms, taking into account the distance between them in the crystal. To do this, we plot the theoretically calculated radial functions of the outer shells both for neutral atoms and for the ions Na^+ and Cl^-, laying them off towards one another from the ends of a segment equal to the interatomic distance in the crystal (Fig. 1.14a, cf. Fig. 1.7). It should be borne in mind that such a plot is only conventional; it pinpoints fairly well the site where the wave functions overlap, but does not reveal the actual nature of the overlapping because, in fact, it is the wave functions ψ (which rapidly fall off with increasing r) that overlap, and not the radial functions $D(r)$ (1.8, 10) containing a factor r^2 which causes the maxima to appear. As to the position of the maxima, the situation is as follows. For neutral atoms, the maximum of the radial function for the outer $3s$ shell of Na coincides in its position with that for the outer $3p^5$ shell of Cl. For the cation Na^+, the outer shell is now $2p^6$, whose orbital radius is equal to 0.278 Å, and the maximum of the electron density of its former $3s$ shell was transformed and "incorporated" into the $3p$ shell of the anion Cl^- after ionization, actually remaining at the same distance from the center of the Na atom. Since the resulting maxima of the indicated functions coincide, then, approximately,

$$d(Na^+Cl^-) \approx r_o(Na) + r_o(Cl), \qquad (1.22)$$

i.e., the distance between ions Na^+ and Cl^- is approximately equal to the sum of the orbital radii of the neutral atoms.

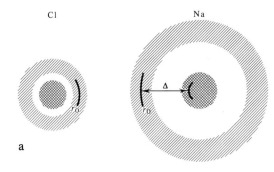

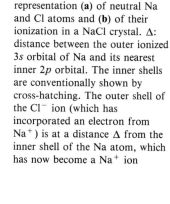

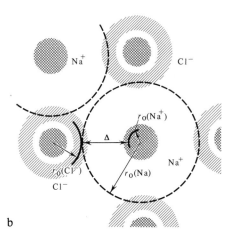

Cl   Na

a

b

Fig. 1.15a, b. Schematic
representation (**a**) of neutral Na
and Cl atoms and (**b**) of their
ionization in a NaCl crystal. Δ:
distance between the outer ionized
$3s$ orbital of Na and its nearest
inner $2p$ orbital. The inner shells
are conventionally shown by
cross-hatching. The outer shell of
the Cl^- ion (which has
incorporated an electron from
Na^+) is at a distance Δ from the
inner shell of the Na atom, which
has now become a Na^+ ion

If we take the sum $r_0(Na^+) + r_0(Cl^-)$, it turns out to be much less than
the observed distance $d(A^+/B^-)$ and differs from it by approximately Δ,
i.e., is the distance between the outer orbital r_0 of the metallic atom and the
nearst inner orbital r_0, which becomes outer on ionization

$$d(A^+B^-) = r_0(A^+) + r_0(B^-) + \Delta . \qquad (1.23)$$

The values of Δ [Å] for certain metals are:

Li	Be	Na	Mg	Al	K	Ca	Rb	Sr
1.40	0.90	1.43	1.03	1.09	1.57	1.15	1.55	1.14

The physical meaning of Δ values is as follows. In accordance with Pauli's
principle and the energy levels determined by solving Schrödinger's equation,
Δ values indicate the shortest distances from the inner cation orbital at which
(and not less than at Δ) the next electrons can be positioned, irrespective of
whether they form the outer orbital of the neutral metallic atom, or whether
they are electrons of the outer orbital of a neighboring anion. Ionization of

the metal atom (Na in our example) in the crystal actually takes place, since its outer electron has been incorporated in the anion shell. But the distance from this electron to the Na nucleus has not actually changed; it is regulated by the value of Δ. We emphasize once more that the scheme of superposition of the radial functions (Fig. 1.14) indicates only the *site* of the overlapping of the orbitals. The picture of ionization and cation – anion contacts is schematized in Fig. 1.15.

On $\rho(r)$ maps obtained by x-ray analysis (Figs. 1.13, 14), the "border" of the Cl atom corresponds to the site of the overlapping of the maxima of the radial functions, while the region of low $\rho(r)$ lies between the outer shells of the ions Na^+ and Cl^- and is determined by the values of Δ_{Na} (Fig. 1.15).

The foregoing refers to the outer shell of univalent cations, when it is practically fully ionized. In di- and tri-valent cations, not necessarily all the electrons leave the outer orbit; only some of them are incorporated in the anion shells as in the above-discussed case. The schemes of the superposition of radial density functions (Fig. 1.14) and the effect of the quantity Δ (Fig. 1.15), which regulate the interatomic distances, remain in force. But the outer orbital of the cation, at a distance corresponding to r_0, may retain some of the electrons (as shown for Na in Fig. 1.15, see also Fig. 1.53). Here, a bond of no longer purely ionic nature corresponds to the real overlapping and rearrangement of the outer orbitals of the cation and anion; a covalent interaction also arises between the ions (we shall discuss it below).

1.2.3 Covalent Bond. Valence-Bond Method

The chemical bond between neutral atoms in molecules and crystals cannot be explained in such a simple, classical way as the ionic bond. At the same time, a covalent bond is typical of most molecules and many crystals.

With the accumulation of a huge amount of experimental data in chemistry and crystal chemistry and the development of the theory of atomic structure it became more and more apparent that the covalent bond is due to the interaction of the outer valence electrons of atoms drawing closer together. The covalent bond is usually defined as a directed chemical bond realized by pairs of electrons. For each chemical valence (denoted by a dash in the scheme) there are two corresponding electrons (two dots on the scheme)

$$H-Cl \qquad O=O \qquad N\equiv N \qquad H-\overset{\displaystyle H}{\underset{\displaystyle H}{\overset{|}{\underset{|}{C}}}}-H,$$

$$H\ :\!\ddot{\underset{..}{Cl}}\!:\qquad \ddot{\underset{..}{O}}\!:\ :\!\ddot{\underset{..}{O}}\qquad :N\!:\ :\!:N\!:\qquad H:\overset{\displaystyle H}{\underset{\displaystyle H}{\overset{}{\underset{}{C}}}}\!:H\ .$$

Such a two-electron bond is stable; from the area surrounding the atoms, assuming that their own valence electrons and those of the added atoms are common, one can single out octets of electrons. From the quantum-mechanical standpoint these formal rules are qualitatively interpreted as a trend to form stable orbitals with pairs of electrons having antiparallel spins.

The first quantum-mechanical explanation of the chemical bond was given in 1927 by W. Heitler and F. London, who carried out a quantitative calculation of the molecular ion H_2^+ and the molecule H_2 of hydrogen. The vigorous development of these concepts and their application to the explanation of the structure of molecules and crystals is due to the work of Pauling and many other researchers.

Two (or more) atoms spaced at very large distances can be regarded as isolated stable systems with their intrinsic orbitals and energy levels. If they start to approach each other, then, beginning with a certain distance, interaction between them takes place, and a new common system with characteristics of its own arises. To calculate such a system, certain functions ψ are chosen as initial. By varying them the solution is improved, the criterion being the attainment and decreasing of the energy minimum. Any functions can, generally speaking, be chosen as the initial ones, but it is natural to use atomic orbitals for this purpose and solve the problem by the methods of perturbation theory.

Heitler and London considered the hydrogen molecule, which consists of protons a and b and electrons 1 and 2. The wave functions of the separate atoms are ψ_{a1} and ψ_{b2}. When they draw closer together, electron 1 can interact with proton b, and electron 2, with proton a. The wave function of the atoms which have drawn together has the form

$$\Psi = c_1 \psi_{a1} \psi_{b2} + c_2 \psi_{a2} \psi_{b1} . \tag{1.24}$$

The electrons interact with both nuclei. The Schrödinger equation for this system has two solutions with energies (supplementary to the initial energy of the hydrogen atoms not drawn together)

$$E = \frac{H_{11} \pm H_{12}}{1 \pm S_{12}}, \quad \text{where} \tag{1.25}$$

$$H_{11} = H_{22} = \iint e^2 \left(\frac{1}{r_{ab}} + \frac{1}{r_{12}} - \frac{1}{r_{a2}} - \frac{1}{r_{b1}} \right) \psi^2(r_{a1}) \psi^2(r_{b2}) dv_1 dv_2, \tag{1.26}$$

$$H_{12} = \iint e^2 \left(\frac{1}{r_{ab}} + \frac{1}{r_{12}} - \frac{1}{r_{a2}} - \frac{1}{r_{b1}} \right) \psi(r_{a1}) \psi(r_{a2}) \psi(r_{b1}) \psi(r_{b2}) dv_1 dv_2, \tag{1.27}$$

$$S_{12} = \iint \psi_{a1} \psi_{a2} \psi_{b1} \psi_{b2} dv_1 dv_2 = \int \psi_{a1} \psi_{b1} dv_1 \int \psi_{a2} \psi_{b2} dv_2 = S_{ab}^2 . \tag{1.28}$$

Here, the integrals $H_{11} = H_{22}$ are simply the electrostatic energy of inter-action of the two atoms. Integral H_{12} also describes the electrostatic inter-action of four particles, but it contains the products $e\psi(r_{a1})\psi(r_{a2})$ and $e\psi(r_{b1})\psi(r_{b2})$, which are the consequence of the fact that electrons 1 and 2 are indistinguishable. This integral cannot be interpreted on the basis of classical concepts. Integral H_{12} (1.27) is called the exchange integral; it describes the exchange part of the system energy, which basically determines the binding energy of the molecule. Similar exchange terms arise in calculating any multielectron system in the one-electron function approximation, including calculations of atoms by the self-consistent field method, described above. Integral S_{12} is called the overlapping integral.

The solution to (1.25) with a plus sign is called symmetric with an energy E_{symm}, the arrangement of the electron spins for it being antiparallel. The solution to (1.25) with a minus sign is called antisymmetric with an energy E_{ant}; it is antisymmetric in respect to the coordinates, and the arrangement of the spins is parallel. Figure 1.16 shows the curves for the dependence of energies E_{symm} and E_{ant} on the distance to the nucleus (1.25). It can be seen that E_{symm} has a minimum, i.e., it corresponds to the formation of a covalent

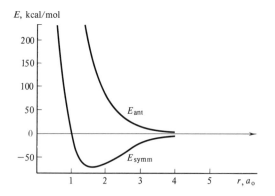

Fig. 1.16. Dependence of the energies E_{symm} and E_{ant} of two unexcited hydrogen atoms on the distance between nuclei

bond. E_{ant} is positive everywhere, i.e., it does not produce any bonds, although it determines a physically possible state at which the atoms repel each other. This calculation of the hydrogen molecule was approximate; later on a number of improvements were introduced into it, which resulted in excellent agreement of theory with the experimental values: bonding energy $E = 4.747$ eV $= E_{exp}$, and distance $d_{H-H} = 0.741$ Å $= d_{exp}$.

The above consideration of the H_2 molecule was the prototype of one of the methods [the *valence-bond (VB) method*] for calculating the chemical bond, in which the wave function of the whole system is composed of single-electron wave functions of the separate atoms, as in (1.24), with due consideration for all the permutations of the electrons; the solution of the Schrödinger equation with variation of the coefficients c helps to find the minimum of the system energy.

Thus, from the standpoint of the VB method, the nature of the covalent bond is explained quantum mechanically. The covalent bond is realized in the hydrogen molecule by a pair of electrons with antiparallel spins and is formed between multielectron atoms according to the same principles. One of the electrons forming a covalent bond can be contributed to the orbital, as we have seen, by each of the contacting atoms. But there is another possibility for the formation of a two-electron covalent bond, namely, when one of the combining atoms (or groupings) — a donor — has an excess of electrons not involved in any bond, while the other — an acceptor — has a free orbital unoccupied by electrons. Such a covalent bond in which there are ultimately two electrons per each valence is called a donor–acceptor bond.

1.2.4 Hybridization. Conjugation

None of the atomic orbitals, except s, have a spherical symmetry. Therefore, directional covalent bonds can be explained as a result of their combination with s orbitals and with each other. The formation of directional bonds in this way is called hybridization. For instance, sp hybridization, i.e., a combination of s and p orbitals, gives a directional sp orbital; sp^2 hybridization, i.e., a combination of one s and two p orbitals, produces trigonal bonds (Fig. 1.17).

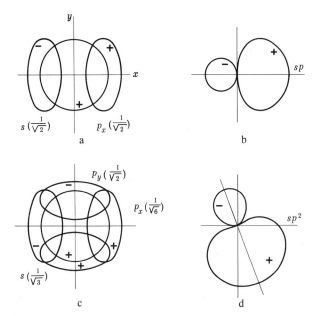

Fig. 1.17a – d. Generation of (**a, b**) hybridized atomic sp orbitals and (**c, d**) one of the sp^2 orbitals forming trigonal bonds of the carbon atom. (**a, c**) position of the s and p AO (the weights of the AO are indicated), (**b, d**) resultant hybridized orbitals, their positive regions correspond to superposition of the positive regions of the initial AO

The contribution of each AO to the resulting hybridized orbital is determined by its symmetry (antisymmetry) and by the condition of ortho-normalization of AO. The direction of the resulting bonds may not coincide with that of the density maximum of the initial AO, as seen from Fig. 1.17. Another example is the tetrahedral arrangement of single bonds of the carbon atom, such as in methane CH_4, diamond, or hydrocarbons. This bond is attributed to sp^3 hybridization with the formation of four equivalent orbitals of the type

$$\varphi_i = \frac{1}{2}(\psi_s \pm \psi_{px} \pm \psi_{py} \pm \psi_{pz}) . \qquad (1.29)$$

Here, ψ_s is spherically symmetric and ψ_p are elongated along three mutually perpendicular axes x, y, z (Fig. 1.18). But hybridization imparts tetrahedral direction to the resulting orbital; for instance, all the plus signs in (1.29) give an orbital $\varphi_{[111]}$ elongated in the direction [111]. Accordingly, the three other cases are $\varphi_{[1\bar{1}\bar{1}]}$, $\varphi_{[\bar{1}1\bar{1}]}$, and $\varphi_{[\bar{1}\bar{1}1]}$. Tetrahedral bonds, which are quite common, determine the structure of Ge, Si, and many semiconductor compounds.

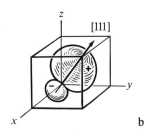

a

b

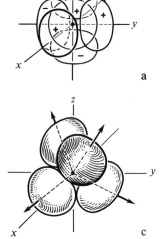

c

Fig. 1.18a–c. Scheme showing the origin of tetrahedral hybridization of the carbon atom. (a) superposition of one s and three p orbitals (positive regions p_x, p_y, and p_z are oriented along the positive directions of the x, y, z axes), (b) arising hybridized sp^3 orbital, (c) four such orbitals directed along the tetrahedron axes (only the positive regions are shown)

In addition to directional covalent bonds – linear, trigonal, and tetrahedral – other types of directional bonds are also known. Thus, the tetrahedral bonds of atoms of transition metals which have electrons in the s, p, and also d states are ascribed to the hybridization of d^3s states. The observed plane square configurations, as in ion $[PtCl_4]^{-2}$, are attributed to dsp^2 hybridization; octahedral, to sp^3d^2 hybridization; etc. (see Sect. 2.5).

The formation of hybrid orbitals is energetically favorable, because pairs of electrons electrostatically repelling one another arrange themselves at a maximum distance from one another.

The number of electron pairs depends on the participation of *s, p,* or also *d* electrons in the formation of the bond. The maximum spacing leads to their disposition at the vertices of the following configurations: 2 − line, 3 − triangle, 4 − tetrahedron, 5 − trigonal bipyramid, 6 − octahedron, 8 − square antiprism, etc., depending on the number of pairs.

Apart from the electron pairs involved in the bond, other electron pairs may remain on the free orbitals. Such pairs are called lone or unshared . For instance, the water molecule H:Ö: contains two lone pairs. If we interpret directional bonds from this point of view, the valence angles are deformed because the lone pair electrons repel the shared ones slightly more strongly than the latter repel one another. The "corner" structure of the H_2O molecules is thus ascribed to the directional *p* bonds and to the presence of two lone pairs (Fig. 1.19). The structure of the pyramidal molecule NH_3 is interpreted similarly. Theoretical calculations [1.13] and experimental investigations into the distribution of the electron density by the method of difference deformation syntheses[4] show maxima due to lone pairs of electrons (Fig. 1.20a − d).

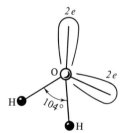

Fig. 1.19. Scheme of the structure of the water molecule; the arrangement of two lone electron pairs is indicated. The molecule has two positively charged projections (protons) and two negative ones, where the lone pairs are concentrated

If the covalent bond is formed by two or three electron pairs, then multiple (double or triple) bonds arise, which are stronger and shorter than single ones.

Not all the bonds in a number of compounds are of integral order; according to their characteristics, i.e., the energy and distance between the bonded atoms, they have an intermediate character, the bond order being fractional. Nevertheless, the compound formula can be written with classical valence dashes; for instance, $H_2C = CH − CH = CH_2$. With this alternation of bonds

---

[4] As indicated in [Ref. 1.6, Sect. 4.7.10] a map of deformation Fourier synthesis of electron density is the difference between the observed electron density of the crystal structure and the spherically symmetric electron densities of nonbonded atoms, whose centers occupy the same positions as in the crystal, i.e., it reveals precisely the deformation of the electron density caused by the chemical bond. Thermal vibrations of atoms are also taken into account.

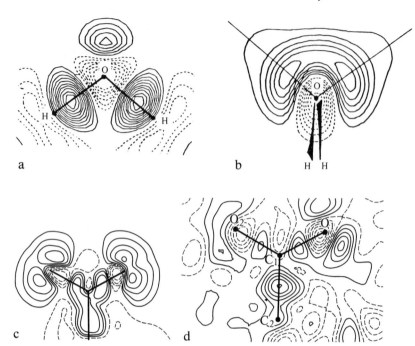

Fig. 1.20 a – d. Difference deformation electron density maps of some molecules, showing the charge of lone electron pairs and covalent bond bridges. (**a, b**) water molecule in $LiOH \cdot H_2O$. Contour interval $0.005 \; e \; \text{Å}^{-3}$. (**a**) the plane of $H-O-H$ atoms, (**b**) perpendicular plane [1.14], (**c, d**) group $C-C\begin{smallmatrix} O \\ O \end{smallmatrix}$ of the molecule of α-glycine, (**c**) theoretical calculation, (**d**) experimental data contour interval $0.008 \; e \; \text{Å}^{-3}$ [1.15]

of formally integral order their properties are actually partly or wholly equalized, becoming intermediate between those of "pure" double and single bonds. This is called "conjugation". Another classical example is benzene, for which it is possible to write two principal Kekulé valence schemes,

$$\text{(1.30)}$$

with single and double bonds. But in actuality all the bonds in benzene are equivalent, their order being $1\frac{1}{2}$. Equivalent bonds of the same nature (order $1\frac{1}{3}$) are inherent in the hexagonal net of carbon atoms in graphite (see Figs. 1.31 and 2.5). The higher the order of the bond, the shorter is the bond. The dependence of the interatomic distance on the bond order in various compounds is given in Fig. 1.46.

The physical significance of the conjugation phenomenon and of the formation of intermediate-order bonds lies in the fact that the electrons are actually not fixed on a definite bond, but belong to the molecule as a whole, and the condition of "multiplicity" of finding an electron pair on a bond is energetically advantageous in many cases, but not necessarily in all.

Interpretation of intermediate-order bonds in complex multiatomic molecules may involve the writing of classical valence schemes for them. Thus, the equivalence of bonds in benzene can formally be obtained by forming a "superposition" of two formulae (1.30). The structures, which are obtained by superposition of several structures corresponding to possible classical formulae, were named resonance structure by L. Pauling. The resonance concept helped to supply qualitative and semiquantitative [with calculation similar to (1.26–28)] descriptions of the structure of many molecules. But since the initial structures, between which the resonance occurs, do not exist as such, this concept is purely conventional.

A description of intermediate bonds more consistent with physical reality is given by the molecular-orbital method, which will be considered below.

Thus, the principal type of two-electron bond does not exhaust the whole diversity of covalent interactions between atoms. In addition to the above-mentioned compounds with intermediate bonds, so called electron-excessive and electron-deficient compounds are also known. In the former, the two-electron orbitals are all used up, but there are also electrons residing outside them. The bond in these compounds is comparatively weak. Electron-deficient compounds (typical examples are boranes) do not have a sufficient number of electrons to form two-electron orbitals, but here, too, covalent interaction is still possible.

1.2.5 Molecular-Orbital (MO) Method

This is the basic method in modern quantum chemistry. The valence-bond theory considers the bonds between any pair of atoms in terms of the interaction of pairs of electrons whose orbitals belong to mutually approaching atoms, i. e., are single centered. One can, however, proceed from a more general assumption and consider the orbitals of one electron in the field of all the nuclei making up the molecule (or crystal) – *multicenter single-electron functions* – and then take into account the interaction of such orbitals. Here, the Schrödinger equation is solved for fixed nuclei at rest (Born–Oppenheimer's adiabatic approximation).

It is further assumed that when an electron approaches one of the nuclei, its motion must be as if it were assigned by the corresponding AO. Thus, in the case of a diatomic molecule, when considering the motion of each electron in the field of both atoms simultaneously, as the initial approximation the MO are composed of the AO

$$\psi_m = c_a \psi_a \pm c_b \psi_b ,$$

and in the general case of multiatomic molecule

$$\psi_{mi} = \sum_p c_{ip} \psi_p , \tag{1.31}$$

where i is the number of MO, p are the number of AO, and c_{ip} are the coefficients determining the weight of the AO in MO. For instance, for a hydrogen molecule it is possible to use this method for constructing, from $1s$ orbitals, two MO with electron densities

$$|\psi_m|^2 \sim \psi_a^2 + \psi_b^2 \pm 2\psi_a\psi_b . \tag{1.32}$$

The degree of overlapping of the atomic orbitals is described by the integral of the overlapping

$$S = \int \psi_a \psi_b dv . \tag{1.33}$$

If the integral is positive, the electrons concentrate between the atoms, and a bonding MO[5] may arise, to which the plus sign in (1.32) corresponds. For a bonding MO (Fig. 1.21a), the electrons are between the nuclei of the mutually approaching atoms and thus draw them together. Figure 1.22a shows the electron density of the H_2 molecule calculated theoretically, and Fig. 1.22b gives the picture of the deformation difference density obtained by subtracting the spherically symmetric density of individual H atoms from the electron density of the molecule (Fig. 1.22a). The maximum between the H nuclei in Fig. 1.22b and the minima behind the nuclei show the electron redistribution due to the formation of a covalent bond; the maximum corresponds to the positivity of the integral of overlapping (1.33). If the signs of the AO are opposite when MO are being formed (Fig. 1.21b), then antibonding MO arise, the probability of finding electrons between nuclei is low, and, being behind both nuclei, the electrons of such orbitals weaken the bond. The lowest energy levels of the molecule are those which are in line with the formation of bonding MOs when the integral of overlapping (1.33) is positive, and which is equal to zero for antibonding MO. It should be emphasized that the positive or zero value of the integral of overlapping is simply determined from the symmetry and antisymmetry of the atomic orbitals (Fig. 1.23). The MO of a molecule with a definite symmetry are described by the irreducible representations of the corresponding point group G_0^3.

---

[5] For bonding and antibonding MO the corresponding German terms *gerade* and *ungerade* are also used.

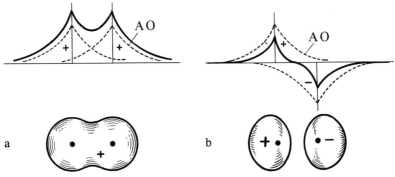

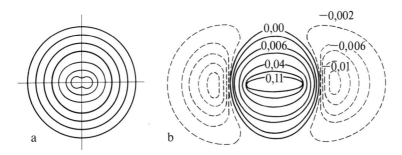

Fig. 1.21 a, b. Bonding and antibonding molecular orbitals of the H_2 molecule. **(a)** origin of bonding MO, **(b)** origin of antibonding MO

Fig. 1.22. (a) Electron density, **(b)** difference deformation density of the H_2 molecule

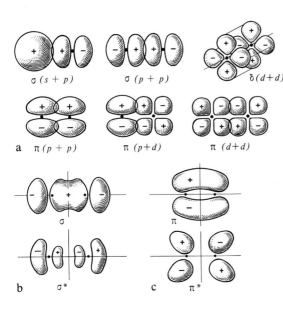

Fig. 1.23a–c. Origin of the most important molecular orbitals from atomic orbitals. AO symmetry, its signs, and symmetry of the superposition are essential, **(a)** AO combinations, **(b, c)** examples of bonding and antibonding MOs (the latter are marked with an asterisk). σ-MO is obtained from combination $(p + p)$-AO using the signs indicated in **(a)**; σ^*-MO is obtained on change of sign of the right p-AO. π- and π^*-MO are obtained similarly

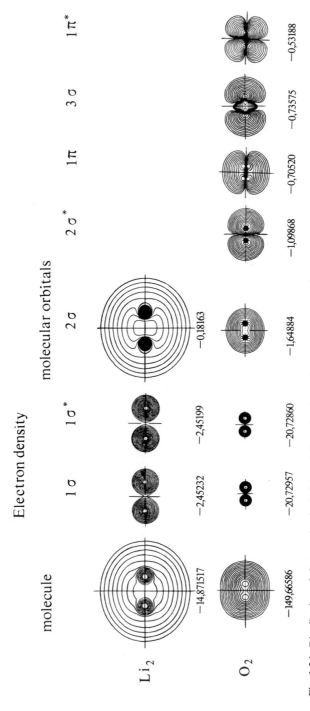

Fig. 1.24. Distribution of electron density in Li_2 and O_2 molecules and the electron density of individual molecular orbitals of these molecules. The symbols are the same as in Fig. 1.6, but three more outer contour lines are added. The bonding orbitals are 1σ, 2σ, 3σ, and 1π, the antibonding ones are $1\sigma^*$, $2\sigma^*$ and $1\pi^*$. The deep orbitals 1σ and $1\sigma^*$ of the inner shells differ very little from the initial AO [1.3]

Figure 1.23 shows the most important types of MO formed when AO combine. So-called σ bonds are formed when s AO combine or when s AO is combined with p AO, or also when two p AO directed along the line of the bond are combined. Another type of bond, namely a π bond, is formed when p AO combine normal to the line, or $p-d$ AO or $d-d$ AO; δ bonds arise from "parallel" d AO. Such MO, for instance, serve to form the electron distribution in the bonds of diatomic, second-period molecules – from Li_2 to F_2. The distribution of the electron density $\rho(r)$ in molecules such as Li_2 and O_2 is shown in Fig. 1.24. Linear combinations of the atoms involved in the bonds are constructed to explain and calculate their multiple (double, triple) bonds from the initial wave functions. Wave functions of different types (s, p, d, etc.) can be combined for this purpose. The electrons may be located on both bonding and antibonding MO. Generally, the bond order is defined as the difference between the numbers of electron pairs located on bonding and on antibonding MO. Figure 1.25a shows the scheme of the MO of the plane ethylene molecule $\begin{smallmatrix} H \\ H \end{smallmatrix}\!\!>\!C\!=\!C\!<\!\!\begin{smallmatrix} H \\ H \end{smallmatrix}$; sp^2 hybridization of C atoms on the basis of s, p_x, and p_y AO and a combination of these orbitals with the s AO of the H atom results in an MO with H atoms and a σ bond $C-C$. To the second valence dash corresponds the π MO produced from the p_z AO. From the standpoint of MO theory of the formation of intermediate-order bonds is interpreted as the result of MOs embracing the entire molecule. For instance, in benzene such a solid framework forms π MO (Fig. 1.25b).

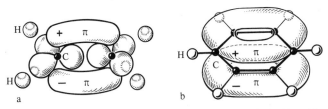

a   b

Fig. 1.25a, b. Chemical bond in molecules of ethylene C_2H_4 (**a**) and benzene C_6H_6 (**b**). Trigonal hybridized sp AO [shown only in (**a**)] form ordinary bonds of C atoms among themselves and $C-H$ bonds. The additional bond between the C atoms is due to the π MO. The π orbitals lie above and below the planes of the molecules, forming "bananas" in the case of ethylene and "bagels" in the case of benzene

Single σ bonds are cylindrically symmetric relative to the bond line. This means that the atomic groupings joined by these bonds can rotate about them to the extent permitted by the steric interaction of the other atoms contained in these groupings (see Fig. 2.61). The other covalent bonds – multiple and intermediate – are constructed from MOs of definite symmetry, which almost always determines the azimuthal arrangement of the groupings bonded by them, i.e., rotation about these bonds is not possible.

Calculations of multiatomic molecules and, all the more so, of crystals by the MO method are very complicated. The problem is to find the multielectron wave function of a given system by the self-consistent-field (SCF) method. Because of the complexity of these calculations they were impossible until the advent of modern electronic computers. The principal method used for solving such problems is the method of linear combinations of atomic orbitals (LCAO) and the molecular orbitals composed of them, the SCF–LCAO–MO method. The calculation is based on the choice of a "basis", i.e., a set of initial atomic functions describing the electron states of isolated atoms (or ions). One can also narrow down the basis by assuming the inner shells to be intact and taking into account the wave functions of only the outer shells, which are responsible for the chemical bond. Then, single-electron MO are formed as linear combinations of AO, while the full wave function $\Psi(1, 2, \ldots, N)$ of the molecule is determined from these MO,

$$\Psi = \frac{1}{\sqrt{N!}} \begin{vmatrix} \psi_1(1) & \psi_1(2) & \ldots & \psi_1(N) \\ \psi_2(1) & \psi_2(2) & \ldots & \psi_2(N) \\ \ldots\ldots\ldots\ldots\ldots\ldots\ldots \\ \ldots\ldots\ldots\ldots\ldots\ldots\ldots \\ \psi_N(1) & \psi_N(2) & \ldots & \psi_N(N) \end{vmatrix}. \tag{1.34}$$

The numbers in the arguments of orbitals ψ_i stand for three orbital and one spin coordinates of the electron. The Pauli principle is fulfilled for MO; each of them can carry not more than two electrons with differing spin coordinates. Functions ψ_i are orthonormalized, one of the results being the appearance of factor $1/\sqrt{N!}$. Function Ψ is antisymmetric. The larger the number N of the initial AO in the basis, the more accurate is the approximation of the molecular orbitals, but the calculations become much more complicated because the number of single-electron integrals of the type (1.26) is then approximately proportional to $\sim 1/2N^2$, and that of two-electron integrals of the type (1.27), to $\sim N^4$. Generally speaking, various functions can be chosen as the basis, provided they are close to the functions describing isolated atoms at small distances from nuclei. Of great importance in the choice of the basis is the simplification of the calculations. Use is made of numerical Hartree-Fock atomic orbitals or exponential hydrogenlike Slater orbitals of the type $r^{n-1} \exp(-\beta r) Y_{lm}$ which are more convenient for calculations, and also Gaussian approximations of $r^{n-1} \exp(-\alpha r^2) Y_{lm}$. Carrying out complete, non-empirical, ab initio calculations takes into acount all the electrons of the system; positions may be assigned for the nuclei, but, in a more general sense the nuclei are not fixed. These calculations are extremely complicated; therefore, it is customary to make semiempirical calculations only for valence electrons or some of them, fixing the nuclei and the other electrons. In calculations of comparatively simple molecules, such as CO_2, bases of several tens of Slater or Gaussian functions are usually employed, and the number of integrals is of the order of $10^5 - 10^6$, which is time consuming. To simplify the solution of such problems, various approximations are used.

1.2.6 Covalent Bond in Crystals

The problem of calculating the chemical bond in crystals is solved by the universal method: by finding the solution of the Schrödinger equation $H\Psi = E\Psi$ for a crystal. As the number N of atoms in crystals is enormous, it may seem impossible to find the solution. However, the most important simplifying factor is the presence of translational periodicity in a crystal. The crystal potential determining Hamiltonian H is periodic, $v(r) = v(r + t)$, where t is any lattice translation. On the other hand, for the same reason wave function Ψ can be represented as a sum of Bloch functions

$$\psi_i = \chi(r)\exp(2\pi i kr), \tag{1.35}$$

where k is the wave vector of the electron in the reciprocal lattice. The solution of the Schrödinger equation is expressed precisely by Bloch functions. This consideration leads to the band theory of electrons in a crystal, which is covered in Chap. 3. The band theory predicts the allowed energies of free electrons in the crystal $E(k)$. It is a general approach applicable not only to the covalent, but also to the metallic and ionic bond.

The interaction of electrons with the crystal lattice is due to their wave nature. Similar phenomena arise in electron diffraction by crystals [Ref. 1.6, Sect. 4.8]. Proper free electrons of a crystal are also reflected from the lattice planes, which leads to a fundamental conclusion that not all the values of energies $E(k) = h^2 k^2/2m$ are possible. The permissible values of vector k are limited in the reciprocal space by polyhedra, i.e., Brillouin zones (Fig. 1.26). As we shall see in Chap. 3, inside these zones k (and hence E) takes on a practically continuous multitude of values (actually, there are N values, where N is the number of unit cells in the volume of a crystal). Within a zone the energy is a continuous function of k. The mathematical methods used for solving the Schrödinger equation for a crystal strongly depend on the bond in

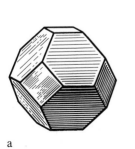

a

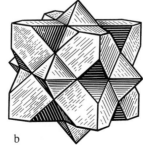

b

Fig. 1.26. (a) First and (b) second Brillouin zones for the structure of copper

the crystal concerned. If the electron – atom bond is strong, as in ionic and covalent crystals, these methods are closer to the above-discussed methods for analyzing the covalent bond in molecules. On the other hand, in metals where the outer electrons of atoms are collective, it is also possible to apply other approaches (see Sect. 1.2.8). In all the cases the lattice potential can be represented as a sum of atomic potentials $V(r) = \sum v_a(r + t)$, and the mutual overlapping of v_a ultimately defines the distribution of the coupling electrons among the atoms. Then, the potential is also expanded in a Fourier series $V(r) = \sum_H v_H [-\exp(2\pi i Hr)]$, which yields the set of equations for functions ψ_i (1.35), the Fourier coefficients of potential v_H, and energies $E(k)$. The calculation is greatly simplified because the inner shells (atomic cores) remain unaltered and can be excluded from consideration; it suffices, therefore, to take into account the potential on the periphery of the atoms and between them, which changes comparatively slowly.

The band structure and wave functions of the electrons for covalent crystals are calculated according to the above-described schemes. Since the covalent bond is realized by electrons localized on MOs between pairs of neighboring atoms, it can be described very well on the basis of the SCF – LCAO – MO method with expansion of the atomic functions in spherical harmonics of the type (1.2). A set of electrons of N crystal atoms is regarded as distributed over $2N$ two-center orbitals of the type (1.31). Indeed, the characteristics of the bond in a single tetrahedral bond $C - C$ in an organic molecule or in diamond are very similar. Analogous tetrahedral bonds arise in crystals of Si and Ge, and in $A^{III}B^V$ compounds; in the latter, they are already of a donor – acceptor nature. Other methods – those of so-called ortho-gonalized plane waves and of the pseudopotential (see Sect. 1.2.8) – also help to calculate the characteristics of the covalent bond in crystals.

For covalent crystals, the energy-level structure is such that the energy necessary for transferring an electron from a completely filled $1s$ band (valence band) to the vacant $2s$ band (conduction band) is high. This energy (the width of the forbidden band) is equal to 5.4 eV for diamond, and $\sim 1 - 3$ eV (cf. Fig. 1.11) for typical semiconductors. Here, certain states of the atoms (s, p, etc.) can be correlated with the corresponding energy bands of the crystal.

As an example, we consider the calculation of the electron density of covalent crystals of the diamond type (Ge, Si, GaAs, ZnSe, etc.) carried out by the pseudopotentials method [1.16]. After solving the Schrödinger equation, one finds the wave functions $\psi_{n,k}(r)$ of the valence bands n; the electron density of each state is proportional to $|\psi_{n,k}|^2$, and the total electron density is equal to $e \sum |\psi_{n,k}|^2$ [see (1.39)]. For $n = 1$ ($n = 2$) the valence density is similar to the s states of the free atoms (Fig. 1.27a), and for $n = 3$ ($n = 4$) (Fig. 1.27b), to the p states. The summary distribution (Fig. 1.27c) reflects the tetrahedral sp^3 state with an increase in density of the valence electrons between the atoms. For GaAs and, to a greater extent, for ZnSe, the ionic component of the bond is prominent. Thus, for ZnSe the first band –

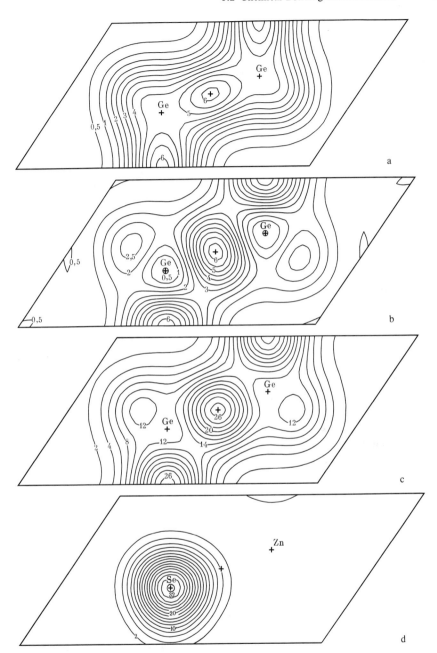

Fig. 1.27a – d. Theoretically calculated valence electron density contour map in the $(1\bar{1}0)$ planes of Ge (**a, b, c**) and ZnSe (**d**) crystals. (**a**) electron density of the first valence band of ρ_1 (*s* type) of Ge, (**b**) electron density of the third band of ρ_3 (*p* type) of Ge, (**c**) total electron density (sp^3 type) of Ge, (**d**) electron density of the first band of ZnSe showing a shift of the charge towards Se (ionic component of the bond) [1.16]

the state with $n = 1$ (Fig. 1.27 d) – clearly describes the ionic component of the bond: the corresponding charge is concentrated around Se. The summary distribution of the valence density of ZnSe also has a covalent "bridge" on the bond, as in Ge.

An example of the application of these methods to the ion structure is the calculation of $\rho(r)$ for a MgO crystal using the pseudopotentials method

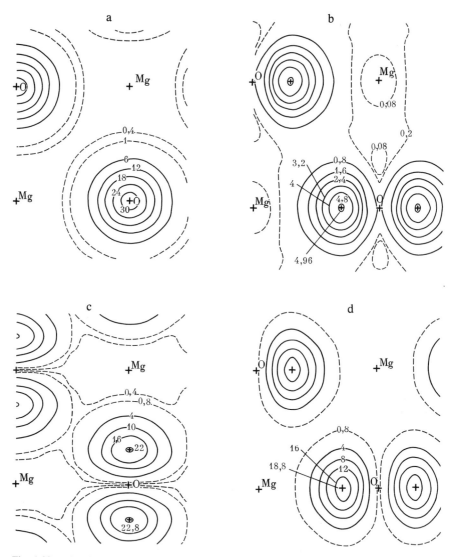

Fig. 1.28a – d. Theoretically calculated band components of the valence electron density in the (100) plane of MgO crystal passing through the centers of the atoms. (a) ρ_1 (s type); (b) ρ_2, (c) ρ_3, (d) ρ_4 (all p type) [1.17]

[1.17]. To find Bloch functions and distributions of $\rho_n(r)$ corresponding to the four valence bands, use was made of the pseudopotentials found from the experimental energy dependence of the imaginary part of the dielectric constant. As can be seen from Fig. 1.28 a – c, the $\rho(r)$ of the valence bands are very similar to the charge distribution in the relevant atomic states. According to calculations, the Mg atom is practically fully ionized, and the O states correspond to the s state and to three p states (recall that the electron density of the inner filled shell is excluded). The charge of O is $0.9\ e$.

In estimating the energy and the force constants of the covalent bonds a sufficiently good approximation is provided by the curves of the potential interaction energy $u(r)$ of the type shown in Fig. 1.12 obtained semi-empirically or by approximating the exact calculations. A satisfactory description of attraction forces is given by an expression of the type ar^{-m} ($m = 4$). As r decreases and a minimum is passed, the short-range repulsive forces increase sharply; they are due to the electrostatic repulsion of nuclei and electron shells of various atoms which exceeds the attraction of the nuclei to the electrons and the exchange energy [see (1.27)]. Repulsion can be approximated by the expression br^{-n}, and then equations of the type (1.16, 17) ($m = 4$ and $n = 6-9$) are applicable for covalent forces. Quantum-mechanical calculations of the repulsive potential give the exponential dependence; for the covalent forces

$$u(r) = -ar^{-m} + c\exp(-\alpha r), \quad m = 4. \tag{1.36}$$

From the condition $du/dr = 0$, which determines the equilibrium state of r_p and the energy $u_p = u(r_p)$, we find

$$u(r) = \frac{u_p}{m - \alpha r}\left\{\frac{\alpha r_p^{m+1}}{r^m} + m\exp[\alpha(r - r_p)]\right\} \tag{1.37}$$

The energy of covalent bonds depends on their order; triple and double bonds have the highest energy of all types of bonds, for instance, 225 kcal/mol in the triple bond in N_2 and 150 kcal/mol in the double bond $C=C$. In single bonds (or per unit bond order) it ranges from $60-70$ kcal/mol (for the strongest bonds) to $30-40$ kcal/mol or less (for instance, 36 kcal/mol for F_2). A lower r_e corresponds to a higher binding energy. The constant α can be calculated via the second derivative d^2u/dr^2 at the point of minimum. For a covalent bond this derivative is large and the bond is rigid; its characteristics change only slightly (including the value of r) on the transition from the molecule to a crystal.

It is noteworthy that there is a certain small component of the covalent bond even in the most typical ionic compounds. Similarly, with the exception of such genuinely covalent crystals as diamond, other covalent compounds composed of different sorts of atoms have the ionic component of the bond.

For instance, in borazon BN (a structural analogue of diamond) a charge transfer from boron to nitrogen already takes place; the same is true of another compound with this type of structure ZnS, where the effective charges are estimated at $0.5 - 0.8$ e.

Such a partly ionic (i), and partly covalent (c) bond can be described within the framework of the VB method by the wave function

$$\Psi = a_i \psi_i + a_c \psi_c \qquad (1.38)$$

where the degree of ionicity $\varepsilon = a_i^2/(a_i^2 + a_c^2)$. The degree of ionicity can also be estimated by the MO – LCAO theory, leading to (1.20).

It is easy to take into account the ionic component by adding to (1.36) the first term ar^{-1} from (1.15, 16), i.e., electrostatic attraction coefficient a is determined by the effective charges.

Semiempirical equations of the type (1.36, 37) only permit estimating the energy as a function of the interatomic distance between the bound atoms. At the same time the covalent bond is also directional, with the corresponding energy of deformation of valence angles. Rotations of groupings about a single bond are also possible, which defines the fraction of the so-called torsional energy. These contributions to the energy are covered by the appropriate equations, which will be treated in Sect. 2.6.

1.2.7 Electron Density in a Covalent Bond

The electron density of any system – an atom, a molecule, or a crystal – is described by the square of its wave function $|\Psi|^2$. While described by orthogonal orbitals ψ_i of the type (1.7, 34) the electron density has the form

$$\rho(r) = e|\Psi|^2 = \sum_{i=1}^{N} |\psi_i|^2. \qquad (1.39)$$

Each atom is a highly concentrated cloud of electron density; the overwhelming number of its electrons are within a peak of comparatively small radius (Fig. 1.9). The electrons realizing the valence bond are associated with a certain region of the wave function Ψ, and the relevant increase in the values of the electron density between atoms corresponds to the overlapping integral (1.33).

Let us consider, as an example, the structure of an element, for instance diamond, C. By analogy with the above construction for NaCl in Fig. 1.14, let us plot the radial electron density functions $D(r)$ [see (1.8)] of the initial atoms towards each other from the ends of the segment connecting the nearest atoms. The maxima of these functions for the outer orbitals of both p shells will be found at approximately the same place (Fig. 1.29a). Recall again that this construction is conventional; it shows the place where the orbitals

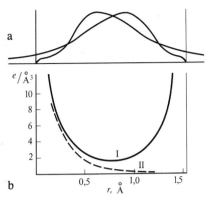

Fig. 1.29a, b. Electron density of a single C – C bond. (a) overlapping of the radial density distribution maxima of the valence orbitals of the C atoms, (b) experimentally found distribution of the electron density along the line of the covalent C – C bond (*curve I*) and in the direction "between the bonds" (*curve II*)

overlap, but does not describe the actual electron density $\rho(r)$. In the distribution of the electron density of the crystal there is no maximum of $\rho(r)$ on the bond line. It can only be said that in the case of the formation of a covalent bond the summary electron density along this bond, i. e., on the straight line joining the bound atoms, is higher than that at the same distance from the atom, but in a direction not corresponding to the bond. It is also higher than the practically near-zero electron density between the atoms in the case of a purely ionic bond (Fig. 1.13). At the same time, if we eliminate the electron density of the inner shells by subtraction, such a maximum is observed for the distribution of the *valence* electrons. This is indicated both by theoretical (Fig. 1.27b, c) and experimental data obtained from precision x-ray investigations with calculation of $\rho(r)$ by a Fourier synthesis. Figure 1.30a shows the electron density map of diamond (a classic example of a crystal having a covalent bond) on a plane passing through one of the series of atoms

forming the structure. The "bridges" of increased density can be

can be seen between the C atoms. An increase in ρ is determined by the maxima of the difference deformation density (Fig. 1.30b).

Similar, very accurate data for Si have been obtained by measuring x-ray structure factors (not intensities) on the basis of the pendulum solution of dynamic theory [Ref. 1.6, Sect. 4.3]. (The standard deviation of ρ is 0.007 e $\mathring{A}^{-3}$.) Figure 1.30c, d gives difference maps of the valence electron density of Si (the inner-shell electron density has been subtracted). The valence density peaks are elongated along the bond line and are approximately cylindrically symmetric relative to it. Their height of 0.69 e $\mathring{A}^{-3}$ is in good agreement with the theoretically calculated value, 0.65 e $\mathring{A}^{-3}$. Figure 1.30e, f shows the deformation difference density (the total spherically symmetric electron distribution of nonbonded atoms has been subtracted)[6]. According to these data the deformation difference electron density of the Si – Si bond is slightly lower

---

[6] See footnote 4, p. 31.

than that of C–C in diamond. According to these and other data the deformation density $\rho_{def.}$ of a single bond at a maximum is $0.3-0.4$ $e\text{Å}^{-3}$. According to theoretical data the summary charge of the valence-electron peak on the bond is about 0.1 e.

Fig. 1.30a–f. Experimental electron density maps of diamond and silicon. (**a**) cross section $\rho(xxz)$ of diamond through the centers of C atoms (peak height 174 $el/\text{Å}^3$), (**b**) corresponding difference deformation density [1.10], (**c, d**) electron density of valence electrons ρ_{val} in a Si crystal along the plane of Si–Si bonds (**c**) and perpendicular to the Si–Si bond in its center (**d**). The lines are drawn at intervals of 0.1 $e/\text{Å}^3$; (**e, f**) the same for the deformation difference density ρ_{def}, the contours are drawn at intervals of 0.05 $e/\text{Å}^3$, negative values are shown by dashed lines [1.18]

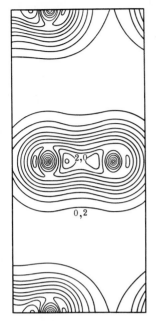

Fig. 1.31. Electron density of valence electrons in the C – C bond in graphite (vertical section of the unit cell) [1.19]

In the case of multiple bonds the electron density is naturally higher. Figure 1.31 presents the valence density of the bonds between the carbon atoms in graphite (the $1s^2$ electrons are excluded). These bonds are hybridized, their order being $1\frac{1}{3}$.

Similar information was obtained for the structure of organic molecules in crystals by combining x-ray and neutron-diffraction data (recall that the latter supply accurate positional parameters of the nuclei and the parameters of the anisotropic thermal vibrations). Examples are given in Fig. 1.32 (see also Fig. 1.20). Thus, in investigating cyanuric acid (Fig. 1.32a) the difference deformation synthesis clearly reveals maxima due to the covalent bond, unshared electrons of the O atoms, and the electron outflow (negative electron density) from the external part of H atoms. The same features are prominent on difference valence (Fig. 1.32b) and deformation (Fig. 1.32c) syntheses of deutero $-\alpha-$ glycylglycine (a deuterated crystal is more suitable for neutron diffraction studies).

Figure 1.32d, e represents cross sections of difference deformation syntheses of the butatriene grouping for tetraphenylbutatriene, a representative of the cumulenes. The experiment was carried out at 100 K. The characteristics of the double bonds C $=$ C are of interest here. The values of $\rho_{\mathrm{def.}}$ at their peaks are $0.9\,e\,\text{\AA}^{-3}$ for the internal and $0.75\,e\,\text{\AA}^{-3}$ for the external bond, which naturally exceeds the values characteristic of single bonds. The sections perpendicular to the bonds (drawn through the bond midpoint) are elliptical, i.e., these bonds (as distinct from single ones) do not show cylindrical symmetry. The charge in the external C $=$ C bond is elongated along the

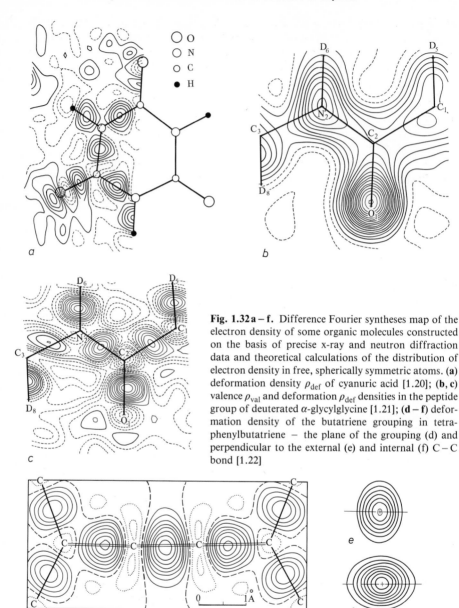

Fig. 1.32a – f. Difference Fourier syntheses map of the electron density of some organic molecules constructed on the basis of precise x-ray and neutron diffraction data and theoretical calculations of the distribution of electron density in free, spherically symmetric atoms. (**a**) deformation density ρ_{def} of cyanuric acid [1.20]; (**b, c**) valence ρ_{val} and deformation ρ_{def} densities in the peptide group of deuterated α-glycylglycine [1.21]; (**d – f**) deformation density of the butatriene grouping in tetraphenylbutatriene – the plane of the grouping (d) and perpendicular to the external (e) and internal (f) C – C bond [1.22]

normal to the butatriene plane, and in the internal bond, in this plane. This agrees with the predictions of the classic theory of π electrons in cumulene systems (cf. Fig. 1.25a, see also Fig. 2.75).

Deformation and valence difference syntheses also enable one to establish the charge at the peak by integrating over the positive or negative electron density peak at the site of the subtracted atom. It turns out that there is almost always some charge redistribution from atom to atom in a molecule. Thus, in the peptide group of the molecule of deutero $-\alpha-$ glycylglycine $O_2CCD_2NDCOCD_2ND_3$ the O atoms have a negative charge of about $-0.5\ e$, and N, about $-0.4\ e$; atoms of C bound with O have a positive charge of $+0.3$ to $+0.4\ e$; and those bound with D, a weak negative charge of $-0.1\ e$; all the D atoms are charged positively, $+0.1$ to $+0.3\ e$. The summary charge of the bond electrons can be similarly established by integrating over the corresponding peaks.

Methods for a parametric description of experimentally obtained electron distribution of valence-bound atoms in various systems have also been developed. One such approach consists in expanding ρ with respect to a set of basis functions φ

$$\rho(xyz) = \sum_{\mu} \sum_{\nu} P_{\mu\nu} \varphi_\mu \varphi_\nu^* \qquad (1.40)$$

and finding coefficients $P_{\mu\nu}$ [1.23]. Slater's or Gaussian orthonormalized orbitals (multipoles) can be chosen as basis functions as in (1.34) [1.24 – 26]; the anisotropic thermal motion is also taken into account. On the basis of such a description and taking into account the symmetry of the system it is possible to calculate the electron population density $q(A)$ of atoms and $q(AB)$ of bonds between atoms, and to compare it with the data obtained with theoretical predictions. It is essential to take into account the quantum properties of the electron distribution – the Pauli principle – in many calculations. The matrix of $P_{\mu\nu}$ values should then satisfy condition $P^2=P$ [1.27].

Most of the results of the calculations of $q(A)$ and $q(AB)$ agree with the theoretical predictions, some others diverge, for instance as regards the shape and distribution of the electron density regions corresponding to lone pairs.

1.2.8 Metallic Bond

In a covalent bond, superposition and some redistribution of electrons of the outer shells of atoms take place, the orbitals of these electrons being mainly localized between the pairs of neighboring atoms. The discrete energy levels of the atoms change and form a quasicontinuous filled band in the crystal.

The nature of a metallic bond is the same as that of covalent bond, namely, sharing the outer electrons; the localization of these electrons, however, is different. In metallic atoms the outer orbitals is filled with a small number of electrons, and the ionization energy I^+ necessary for the detachment of outer electrons, which can be considered a measure of the stability of the orbital (see Table 1.2), is small. In most cases this is a spherically symmetric s orbital, which is comparatively wide. When metallic atoms draw close together and form crystals of metals or alloys, these orbitals overlap with a great number of identical neighboring orbitals; for instance, in metals with a face-centered cubic lattice they overlap with 12 orbitals (Fig. 1.33). Therefore the concept of localization of outer electrons near a given atom or between pairs of atoms loses its meaning, and their entire system, described by the wave function, Ψ, common for the whole crystal, is characterized by its approximately uniform value in the space between the atoms. This also corresponds to the classical concepts of the electron theory about the presence of a "gas" of free electrons in metals.

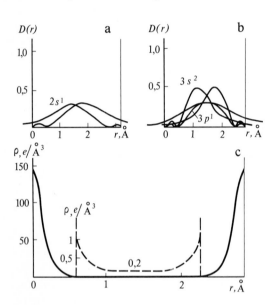

Fig. 1.33a–c. Electron density distribution in the structure of metals. Overlapping of the radial distribution functions of outer electrons in Li (a) and Al (b). Experimentally determined electron density in aluminium (c) along the Al–Al line (the dashed curve referring to the atom periphery is scaled up)

As mentioned in Sect. 1.2.6, the methods for calculating the metallic bond are based on the band theory. In metals, the electrons can freely shift to vacant levels immediately adjoining the filled levels, which actually explains the metallic properties. Since the band is not filled completely, the surface of vectors $\mathbf{k}$, corresponding to the maximum energy and called the Fermi surface, does not touch the faces of the Brillouin zone or forms "channels" in the periodic reciprocal space (see Chap. 3 for details). The methods for calculating the structure of metals use various models. Since only the outer electrons are common for the atoms of metals, the problem can be solved, for instance for alkali metals, by representing the lattice potential as a sum of the potentials of the ionic core with the shell of the noble gas atom in whose field the conduction electrons are moving (one-electron approximation). If we divide the whole volume of the lattice into equal Dirichlet polyhedra (see [Ref. 1.6, Fig. 2.89] surrounding the given atom (such a construction for metals is called a division into Wigner – Zeitz cells, see Sect. 2.8), it can be assumed that within the filled shell of the core the potential is spherically symmetric, and outside the shells it changes only slightly, being practically constant. By fitting the solutions together at the boundaries of shells and polyhedra one can find the general solution.

In the one-electron approximation, the interaction due to Coulomb repulsion between the free electrons is neglected. There are several methods which take into account these effects, and also exchange interaction and spin effects. One can ultimately obtain, in terms of one metallic atom, simplified expressions for the energy of the type

$$u(r) = -\frac{a}{r} + \frac{b_1}{r^2} + \frac{b_2}{r^3} + b_3, \tag{1.41}$$

where the constants are calculated theoretically or semiempirically from the compressibility data, etc. Expression (1.41) with the first two terms can also be obtained from the classical concepts about the interaction of a negatively charged gas with the positive ionic cores. The energy curves for alkali metals, obtained with the aid of quantum-mechanical calculations, are given in Fig. 1.34. It is important to emphasize that, in view of the cooperative nature of the interaction, these curves are only valid for a crystal; they do not describe the interaction between a pair of atoms, for instance in a Me_2 "molecule". The physical meaning of the attraction forces in the metal structure is that the potential distribution in it enables the outer electrons to occupy lower energy levels than in free atoms.

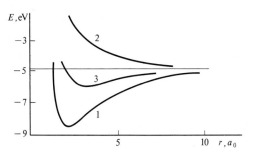

Fig. 1.34. Energy curves for the lattice of metallic sodium. (1) interaction energy of a free electron with Na^+; (2) kinetic energy of electrons; (3) total energy

In metals with several electrons in the outer shell the interaction is more complicated because not all these electrons are free, i.e., not all of them shift to the conduction band. Therefore, the covalent interaction also contributes a certain component to the attraction forces. A purely metallic bond is non-directional, i.e., it is spherically symmetric, therefore, many structures, for instance those of the metals Na, K, and many others, are close-packed cubic. At the same time, for many metals the effects of bond orientation become apparent, when, for instance, body-centered cubic or hexagonal structures are formed. The quantum-mechanical calculations have only been performed in several cases. In such calculations, the pseudopotential method is used for multivalent metals. In this method the pseudowave function is calculated, instead of the true wave function; the latter must be orthogonal to all the wave functions of electrons in the inner shells and, therefore, has a complex oscillating form within the ionic cores of the metal. The pseudowave function coincides with the true wave function outside the ionic cores, but remains smooth inside them. The equation defining the pseudowave function is obtained from the conventional Schrödinger equation by replacing the local true potential with the nonlocal pseudopotential. This includes, in addition to the term corresponding to attraction, a term describing the effective repulsion of conduction electrons from the ionic cores. In approximate calculations of the interaction energy and other characteristics of metals, equations of the type (1.41) can also be used.

X-ray investigations positively confirm the presence of a continuously distributed constant electron density in the metal structure and the ionization of the outer atomic shell. The electron density of an ionized metallic atom is spherically symmetric [1.8, 28, 29]. The mean electron density of the inter-atomic space is constant and equals about $0.15 - 0.20 \, e \, \text{Å}^{-3}$ (Fig. 1.33c, cf. Fig. 1.14b); the obtained values of the number of electrons in atoms for some metals are as follows: 10,4 for Mg, 10.2 for Al, 23.0 for Fe, and 20.0 for Cr, which approximately corresponds to ionization equal to the valency of the metal.

Note that the above-discussed three basic types of "strong" bonds — ionic, covalent, and metallic — are close in energy and characteristic interatomic distances. Below (in Sect. 1.4) we shall revert to the causes of this similarity. We recall once more that in many compounds the bond is inter-mediate in nature. This refers, for instance, to such an important class of compounds as semiconductors, which are formed by such elements as P, S, Ge, Si, Ga, As, Se, Sb, Te, etc., as well as to some intermetallic compounds. The bond in them cannot be allocated to a single definite type; is has features of both a covalent and a metallic bond, and sometimes of an ionic bond. The ionization of atoms and the existence of a small valence bridge electron density are detected in such compounds by experimental x-ray investigations (cf. Fig. 1.27a–d, see also Sect. 2.4).

It should always be kept in mind that a chemical bond in a compound is a quite definite state of the outer electrons of a given multiatomic system. Its division into certain components is, to a large extent, conventional.

It should also be emphasized that many substances which under normal conditions are insulators or semiconductors experience phase transitions and acquire metallic properties with an increase in pressure; the bond in them also acquires a metallic character. This is natural because forced mutual approach of atoms under external pressure increases the overlapping of outer shells and, hence, the number of shared outer electrons; the energy spectrum also changes — the bands merge together. Thus, Te becomes a metal at ~ 40 kbar, Ge at 160 kbar, and InSb at 20 kbar. The problem of metallic hydrogen is of interest. According to theoretical estimates ordinary molecular hydrogen H_2

can be transformed into metallic hydrogen at a pressure of about 2 megabar. It has been suggested that this phase may be metastable, i.e., it may remain metallic after the pressure is lifted, and will also be superconducting. Some hypotheses assume that depending on the particular structure of the energy spectrum (for instance, for Ni), a pressure increase may, contrariwise, result in a loss of metallic properties.

1.2.9 Weak (van der Waals) Bonds

Noble gases, whose atoms have completed shells, crystallize into highly symmetric structures at low temperatures. Organic molecules having a system of strong, completely saturated (covalent) bonds also form crystals. A number of physical characteristics (melting point, mechanical properties, etc.) show that the bonds between the particles in all such crystals are weak. According to x-ray data, the shortest distances between non-valence-bonded atoms, i.e., atoms of "contacting" neighboring molecules or atoms of noble gases, considerably (by 50 or 100%) exceed the "short" distances of strong bonds. The attraction forces operating here are commonly called van der Waals forces, because they explain the correction for molecular attraction in the van der Waals equation for the state of gases.

If molecules possess a constant electric moment μ, one of the components of these forces is the classical dipole – dipole interaction. It is reduced by the thermal motion, which disturbs the dipole orientation. The energy of such interaction is defined by

$$u_1(r) = -\frac{2}{3}\mu^4 r^{-6}\frac{1}{kT}. \tag{1.42}$$

This is called the orientation effect. Naturally, its fraction in intermolecular interactions is large only for molecules with a high μ, for instance, for H_2O and NH_3.

Some contribution to the molecular interaction is also made by another, so-called induction effect, which takes into account the possibility of the polarization of molecules by one another, i.e., the possibility of inducing dipoles. The corresponding energy is also proportional to r^{-6},

$$u_2(r) = -2\alpha\mu^2 r^{-6}, \tag{1.43}$$

where α is the polarization.

The principal component of the intermolecular forces is the so-called dispersion interaction between neutral atoms or molecules, which completely explains, in particular, attraction between the atoms of inert elements. This interaction is due to the presence, in the atoms, of instantaneous dipoles induced by the moving electrons of neighboring atoms and can be considered quantum mechanically. According to F. London, while finding it one must take into account both the ground, ψ_0 and φ_0, and excited, ψ_n and φ_n, states of the mutually approaching atoms. Since the atoms are spaced wide apart and only remote regions of the wave functions overlap, the exchange can be neglected. It is possible to deduce, on the basis of the second approximation of perturbation theory, that

$$u_3(r) = -Kr^{-6}, \quad K = \frac{3h}{2} \frac{v_1 v_2}{v_1 + v_2} \alpha_1 \alpha_2. \tag{1.44}$$

Here, K is expressed in terms of polarization α_1, α_2, and the characteristic frequencies v_1, v_2 of atom excitation are the same as those responsible for the light dispersion. If we take into account not only the dipole, but also the multipole interaction, terms with r^{-8} and r^{-10} also appear. The zero energy is taken into consideration as well. The potential of repulsion of molecules or ions is expressed by the exponential function. As a result, the following equation holds true for the description of interaction of the atoms of neighboring molecules:

$$u(r) = u_1 + u_2 + u_3 + c\exp(-\alpha r) = -ar^{-6} + c\exp(-\alpha r). \tag{1.45}$$

This equation is the same as (1.35), but with $m = 6$ and with different a, c, and α. For instance, for interactions of C atoms $a = 358$ kcal/mol, $c = 4.2 \times 10^4$ kcal/mol, and $\alpha = 3.58$ Å^{-1}. The fraction of the three components in the intermolecular bond energy depends on the dipole moment μ and the polarization α of the molecules. For instance, for H_2O $u_1 = 190$, $u_2 = 10.0$, and $u_3 = 93.0$ (erg $\cdot$ 10^{-60}); and for CO $u_1 = 0.003$, $u_2 = 0.05$, and $u_3 = 67.5$. Thus, for molecules with a small or zero dipole moment (these constitute the vast majority of the molecules of organic compounds) the molecular inter- action energy is practically entirely due to the dispersion forces. Equation (1.45) can be rewritten in the form (1.37) by introducing the equilibrium distance r_e and energy u_e. The characteristic distance r_e for the van der Waals forces is $3-4$ Å. At the same time, the exponential repulsive term in (1.45) produces a sufficiently rapidly increasing left-hand branch of the interaction energy curve (Fig. 1.13, Curve II), so that the mutual approach of non- valence-bound atoms is also sharply limited. This permits introducing the concept of intermolecular radii (see Sect. 1.4).

The van der Waals forces of molecular interaction are much weaker than the forces of the covalent, ionic, and metallic bonds. They may be called forces of weak interaction in contrast to the three types of strong bonds. These forces rapidly diminish with distance; the minimum of (1.45) is shallow and less prominent than for strong interactions (Fig. 1.12). Therefore the distances between non-valence-bounded atoms in crystal structures have (for a given pairs of atoms) a slightly greater spread than for strong bonds (see Fig. 1.50).

1.2.10 Hydrogen Bonds

One more variety of binding forces is known, the so-called hydrogen bond. It is formed between H atoms included in the groupings NH or OH and electronegative atoms of N, O, F, Cl, or S, which is denoted schematically as $AH \ldots B$.

The localization of H atoms is sufficiently achieved by all three diffraction methods (Figs. 1.32, 35 – 37). The maximum of the potential, which corresponds to the position of the nucleus, is recorded by the electron diffraction technique. The value of the potential, which increases with reducing electron density of the shell, points to some positive ionization of the H atom in the H bond (Fig. 1.35). The position of the proton (or deuterium) and its thermal motion are determined by the neutron diffraction method (Fig. 1.36). The position of the electron cloud of H is determined by x-ray diffraction using difference syntheses of electron density. If the peaks of all the atoms, except H, are subtracted (Fig. 1.37, see also Fig. 2.64), it is usually found that the maximum of the electron density ρ_H does not coincide with the position of the proton but is shifted towards the atom with which H forms a covalent bond.

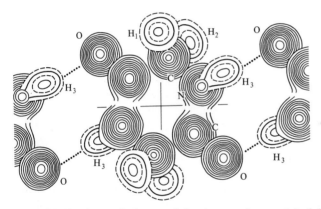

Fig. 1.35. Fourier synthesis map of the electrostatic potential of diketopiperazine constructed from electron diffraction data. Chains of molecules linked by hydrogen bonds in the crsytal are clearly seen. Solid contours are drawn at 15 V, and dashed ones, at 7.5 V intervals. The potentials of the H_1 and H_2 atoms of the CH_2 group are 32 and 33 V, and that of the H_3 atom of the NH group involved in the NH – H hydrogen bond is 36 V, which points to its ionization [1.30]

Thus, according to spectroscopic and neutron diffraction data the C – H distance is 1.09 Å and the N – H distance about 1.00 Å, while the maximum of the electron density peak is located 0.1 – 0.2 Å closer to C or N atom. The shift of the maximum of ρ_H is attributed to the anisotropy and anharmonicity of thermal motion and the ionization of the H atom, especially when it is involved in the hydrogen bond. The "external" part of the H atom actually has a decreased electron density. The same is revealed by Fourier deformation difference syntheses with subtraction of the spherically symmetric electron

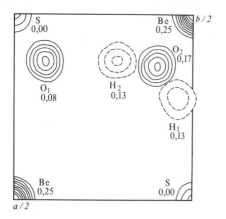

Fig. 1.36. Fourier synthesis of the nuclear density of $BeSO_4 \cdot 4H_2O$. The contours are drawn at intervals of 0.46 cm $\cdot$ 10^{-12} Å^{-3}, the z coordinates of the nuclei are indicated [1.31]

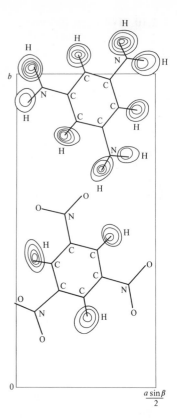

Fig. 1.37. Difference electron density map of the ▷ 1:1 complex of s-trinitrobenzene and s-triaminobenzene; the electron density of the hydrogen atom is shown [1.32]

density of the H atom (see Figs. 1.20, 33). This is manifested in the deformation synthesis: the outer region has negative values of difference density.

The H bonds of molecules in liquids and gases are often responsible for the formation of dimers and, in crystals, also of chains (Fig. 1.35) or two- and three-dimensional nets; the latter are observed, for instance, in the structure of ice (Fig. 1.38). The hydrogen bond is directional, the atom B lies approximately on the continuation of the covalent bond $A - H$ and is spaced not more than 20° from this straight line.

The distances in NH...N and NH...O bonds vary from 2.7 to 3.0 Å. The OH...O bonds are divided into short (2.45 − 2.6 Å) and long ones (up to 2.9 Å). According to x-ray data the O − H distances in hydrates of metal fluorides lie in the range of 0.7 − 1.0 Å, and the OH − F distance, 2.50 − 2.90 Å [1.33]. It can be seen that these distances are, as a rule, shorter than in van der Waals interactions between A and B atoms and are a priori shorter than in van der Waals H and B contacts. The energy of hydrogen bonds is slightly higher (i. e., they are stronger) than that of weak molecular interaction; it equals about 5 − 10 kcal/mol. The principal features of the hydrogen bond A H...B can be explained by assuming that the H atom in the A H grouping is partially ionized; according to various data, including x-ray and electron diffraction, it

By now, over a thousand magnetic structures have been investigated. A more detailed discussion of the relationship between ordering and the domain structure of magnetic materials and the macroscopic magnetic properties of crystals will be given in [1.7].

1.3 Energy of the Crystal Lattice

1.3.1 Experimental Determination of the Crystal Energy

The free energy F of a system of atoms forming a crystal structure consists of the potential energy of the chemical bond U between the atoms (cohesion energy) and the free energy of thermal motion F_T

$$F = U + F_T. \tag{1.46}$$

To decompose a crystal into atoms at the temperature of absolute zero and separate them to an infinite distance from each other requires work equal to the potential energy of the chemical bond taken with the opposite sign $-U$. This energy in terms of 1 mole (or gram atom) of the substance is precisely the energy of the crystal (the energy of the complete dissociation of the crystal — i.e., of its "atomization"), which is by tradition called the "lattice energy." For elements, metals and alloys, and covalent structures the crystal energy is equal to the heat of sublimation S at absolute zero. The heat of sublimation, in turn, is equal to the sum of the heat of evaporation E and the heat of fusion F, plus the heat of dissociation of the molecules D, provided they arise in evaporation or sublimation,

$$-U = S + D = E + F + D, \tag{1.47}$$

For molecular crystals, it is natural to use molecules, rather than atoms, as structural units; accordingly, the energy of breaking the intermolecular bonds should be used for the lattice energy; it is equal to the heat of sublimation S or to the sum $E + F$ at absolute zero. Thus, theoretical calculations of the lattice energy can be compared with the experimental data.

Historically, the first calculations were made for ionic crystals; their lattice energies implied the energy U_i of the dissociation of such a crystal into ions, rather than atoms. It is obvious that U_i differs from U (1.47) by the ionization energies of cations I^+ and the affinity I^- of anions for electrons

$$-U_i = S + D + I^+ - I^-. \tag{1.48}$$

Experimental values of U_i for ionic crystals can be obtained by means of the so-called Born–Haber circular process, which is shown in Fig. 1.42 for a

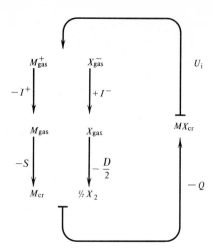

Fig. 1.42. Born – Haber cycle

compound of the type MX, where M and X are single-charged ions (for instance, NaCl). The energy values in all the stages of the process (except I^+ and I^-) reflect the change in the heat content in corresponding reactions at 298 K. From the figure it follows that

$$-U_i = Q + S + I^+ + \frac{D}{2} - I^- . \tag{1.49}$$

Molecular crystals have the lowest lattice energies (on decomposition to molecules), $U \sim 1-5$ kcal/mol. The lattice energy U of elements ranges from several tens to a hundred kilocalories per gram atom. For instance, it is equal to 130 kcal/g at. for B, 26 for Na, 170 for C, 42 for Ca, 101 for Co, 26 for I, and 160 for Os. The lattice energy of ionic and covalent crystals is of the order of hundreds of kilocalories per mole.

1.3.2 Calculation of the Potential Energy

Finding the free energy of a crystal is a key factor in determining all the thermodynamic functions and other characteristics of the crystal structure, in particular, its lattice parameters, various physical constants, etc.

If we vary F (1.46) for an arbitrary arrangement of atoms, corresponding then to the equilibrium crystal structure, there is the deepest minimum of F, at which

$$\delta F = \delta U + \delta F_T = 0 . \tag{1.50}$$

The principal contribution to (1.46) is made by lattice energy U; the condition $\delta U = 0$ determines the arrangement of atoms in a given lattice at absolute zero. Finding the equilibrium arrangement of atoms in a crystal is a complicated task; in calculating the lattice energy it is usually assumed that the atomic coordinates are known, although it is possible to solve the more complex problems as well. It is also clear that if we neglect the free energy of thermal motion, we cannot solve the problem of phase transitions and poly-morphism in solids. We shall elaborate on this in Chap. 4; meanwhile, we shall consider the possibilities of potential energy computations.

Quantum-mechanical calculations of the crystal energy are rather intricate, but calculations of U become reasonably straightforward if pairwise interactions are assumed and semiempirical expressions for energies of the type (1.16, 40) or their variations are used.

Thus, assuming that the interatomic forces are central and the atom-atom interaction potential energy has the form

$$u_{ik} = u_{ik}(r_{ik}) , \qquad (1.51)$$

i.e., that they are independent of each other, the potential energy of any atomic association in a crystal can be represented by a sum of such expressions. It is a good approximation for all types of bonds, except the metallic. Therefore, for pairwise interactions (for instance, for electrostatic ones) the total potential energy is the sum of interatomic potentials over all the possible pairs

$$U = \frac{1}{2} \sum_{i,k} u_{ik}(r_{ik}) . \qquad (1.52)$$

For n atoms, the sum (1.52) contains $n(n-1)/2$ terms, the factor $1/2$ arising since each function u_{ik} refers to a pair of atoms. Functions (1.51) decrease rapidly and make an appreciable contribution only at r_{ik} not exceeding $10-20$ Å, which reduces the number of terms in the sum, nevertheless leaving it rather considerable.

In the case of a covalent bond an atom interacts almost exclusively with its nearest neighbors, and therefore only the terms $u_{ik}(r_{ik})$ of the interaction between such covalently bound pairs of atoms can be taken into consideration in sum (1.52). For a metallic bond, the pairwise interactions are not considered when the calculations are made in terms of one atom, for instance by (1.41), and the whole lattice is actually taken into account.

Energy calculations are simplest if the compound is homodesmic, i.e., if it has one type of bond, and if its chemical composition is not complicated. Then the number of terms in u_{ik} (1.51) is not large. As the crystal structure is periodic, summation (1.52) can be performed over the atomic arrangement within a unit cell, and then over the cells. These sums are called lattice sums.

The first, and now classical calculations of the ionic energy of lattices U_i were carried out by M. Born and his associates. Let us consider such a calculation for a NaCl-type lattice. We find lattice sum (1.52) for the first — electrostatic — term. The environment of any Na or Cl atom is the same. Each of them is surrounded by six neighbors of opposite signs at a distance of $r = d(AB)$, then by 12 neighbors of the same sign at a distance of $r\sqrt{2}$, then by eight neighbors of opposite signs at a distance of $r\sqrt{3}$, and so on (Fig. 1.13). Consequently, the electrostatic energy of interaction of one atom of a NaCl-type structure with the other atoms of the lattice is

$$ U = \frac{e^2}{r}\left(6 - \frac{12}{\sqrt{2}} + \frac{8}{\sqrt{3}} - \frac{6}{\sqrt{4}} + \frac{24}{\sqrt{5}} - \cdots\right) = \frac{e^2}{r}M. \tag{1.53}$$

The series in (1.53) converges rapidly; the value M is called Madelung's constant. Its value in (1.53) for a NaCl-type structure is 1.748; for other structural types, the values of M are as follows: 1.76 for CsCl, 5.04 for CaF_2, 1.64 for ZnS, 4.38 for CdI_2, and 24.24 for Al_2O_3. (In the case of non-univalent ions, M includes, as a factor, the product of the formal charges of cation and anion.) In addition to the electrostatic energy of attraction and repulsion, which is covered by the sum (1.53), we should take into account the contribution of the repulsive forces between the electron shells — the second term in (1.15) of the type r^{-n}. Similarly to M in (1.53), factor M' will arise, but since the repulsive forces decay rapidly with distance, M depends almost only on the nearest neighbors of the given atom. In the final analysis, taking into consideration that in the case under review all the atoms make an identical contribution to the total lattice energy and also assuming that there are N such atoms in a gram molecule, we get

$$ -U_i(r) = N(Me^2 r^{-1} - M'br^{-n}). \tag{1.54}$$

The second term in (1.54) is small compared with the first. Thus, Born's equation (1.54) helps to calculate the lattice energy in high-symmetry ionic crystals. A comparison of the calculated values with the experimental data is given in Table 1.3.

Table 1.3. Comparison of the experimental and theoretical values of ionic energy of some crystal lattices (in kcal/mol)

U_i	LiF	NaCl	RbI	CaF_2	MgO	$PbCl_2$	Al_2O_3	ZnS	Cu_2O	AgI
Experiment U_i	242	183	145	625	950	521	3618	852	788	214
Theory U_i	244	185	149	617	923	534	3708	818	644	190

In the case of single-charged ions the agreement is good, which supports the theoretical conclusions. When the ion charge increases, the agreement becomes worse, but is still satisfactory. In the light of the available data, however, it cannot be regarded as a confirmation of Born's theory. Indeed, the ion charge in such compounds, for instance, in Al_2O_3, does not coincide with the formal valency and is nonintegral. On the other hand, determination of U_i^e for such compounds according to the Born – Haber cycle (1.49) is also inconsistent, because it includes the electron "affinity" of a multicharge ion, which in fact does not exist. Besides, for the last compounds given in Table 1.3 there is a considerable fraction of covalent bond.

If we assume that the equation for the potential energy of pairwise interactions has the form (1.16) or (1.41), its parameters can be found from macroscopic measurements of the crystal. Thus, substitution of (1.16) into (1.52) yields

$$U = \frac{a}{2} \sum_{ik} r_{ik}^{-m} + \frac{b}{2} \sum_{ik} r_{ik}^{-n} , \tag{1.55}$$

and the four parameters a, b, m, and n can be found from the experimental values of the lattice energy, the molar volume, the compressibility, and the thermal expansion of the crystal.

The Born equation (1.53) was simplified by A. F. Kapustinsky who used the following considerations. The value of M for different structures can be replaced by a near-constant coefficient if it is referred to the sum of ions Σ in the formula unit. Then

$$U = 256 \frac{Z_1 Z_2 \gamma}{d(AB)} \text{ [kcal/mol] .} \tag{1.56}$$

This equation deviates from expression (1.54) by about $1\% – 3\%$. The energy of interaction of a definite ion with the lattice as a whole, which can be calculated by (1.56) in the simplest cases, should not depend strongly on the particular structure. Therefore A. E. Fersman suggested that this energy should be regarded approximately as a constant increment of a given ion, and named that increment the "energy constant" (EC). The values of EC for some ions are as follows:

K^+	Na^+	Li^+	Cu^+	Ba^{2+}	Fe^{2+}	Mg^{2+}	Al^{3+}	F^-	Cl^-	Br^-	O^{2-}	S^{2-}	N^{3-}
0.36	0.45	0.55	0.70	1.35	2.12	2.15	4.95	0.37	0.25	0.22	1.55	1.15	3.60

In this approximation the lattice energy is simply equal to $U = 256 \Sigma EC$. This equation, as well as (1.56) or its slightly more complicated versions, is approximate because it is based on the concept of a pure ionic bond in all cases and overstates the share of energy contributed by multicharge ions. Yet it enables one to estimate the energy of complicated structures, for instance, minerals, which is essential to geochemistry.

In the years following the development of M. Born's classic theory, calculations of the crystal lattice energy were carried out for many crystals on the basis of more accurate approximations of the interaction potentials, mainly for the repulsive term, and also on the basis of quantum-mechanical considerations. In these calculations the repulsive potential has a more accurate form, $\exp(-\alpha r)$ (1.36), due consideration being given to van der Waals, dipole – dipole (1.44), and multipole interactions, to the zero vibrations of the lattice, and, finally, to the so-called many-particle interactions, i.e., the over-lapping of remote regions of the wave functions of atoms not only with the nearest, but also with the next nearest neighbors.

Thus, for NaCl the electrostatic term (1.53) is 205.6 kcal/mole, van der Waals attraction, 5.7, repulsion energy, 24.9, zero energy, 1.4, and the total ionic energy of the lattice $U_i = 185.2$ kcal/mole at 298 K, which practically agrees with experiment.

Quantum-mechanical calculations of crystal energy require solving the Schrödinger equation and can be carried out, as mentioned above, for a lattice on the basis of the MO – LCAO method with expansion of Bloch's functions (1.35).

For ionic crystals, the first electrostatic term (1.52) appears, in a further quantum-mechanical consideration, in the same form, while the other components are the energy due to consideration of the extended (as opposed to the pointlike) distribution of the ion charge and the covalent exchange energy corresponding to (1.27). The most comprehensive calculations with an allowance for the various corrections, not only give the values of the lattice energy, but also allow one to calculate the equilibrium interatomic distances r, i.e., ultimately the parameters of the lattice, This can be done by varying the value of r in the expressions for the energy and finding its minimum. Thus, for LiF it was found that $d(\text{LiF}) = 2.01$ Å ($d_e = 2.00$ Å). Since the terms appearing in the ultimate expression are large and have different signs, the theoretical and experimental energy calculations sometimes disagree.

Energies U_a (atomization) for some crystals (including the ionic ones) are given in Table 1.4. Recall that U_a is always less than U_i by $-I^+ + I^-$ (1.48). For instance, U_i^e is equal to 242 kcal/mol for LiF and 950 kcal/mol for MgO (see Table 1.3).

Table 1.4. Comparison of experimental and theoretical values of atomization energy for certain crystals (in kcal/mol)

Substance	LiF	NaCl	KI	MgO	CaF$_2$	AgI	Al$_2$O$_3$	SiO$_2$
Experiment U_a	199	150	122	239	374	108	730	445
Theory U_a	202	152	125	262	428	116	695	416

For covalent crystals, exact energy calculations are naturally possible only on the basis of quantum-mechanical considerations. A simplified calculation can be made in the pairwise-interaction approximation (1.52), where the terms corresponding to the nearest neighbors remain, and the summation can be reduced to the interaction of neighbors in a single unit cell. The energies of covalent crystals are high, for instance, $U = 170$ kcal/g at for diamond.

For metals, quantum-mechanical calculation actually yields the lattice energy per atom. The values of coefficients a, b_1, and b_2 in equation (1.41) are obtained by summation of several terms of equal degrees but with different signs; therefore, they are not reliable enough, and the theoretical and the experimental values of U do not always agree sufficiently well (Table 1.5).

Table 1.5. Comparison of experimental and theoretical values of lattice energy for certain metals (in kcal/mol)

Substance	Li	Na	K	Cu	Be
Experiment	39	26	23	81	75
Theory	36	24	16	33	36 − 53

Using lithium as an example we can illustrate the above-mentioned difference between the interaction energy of a pair of atoms characterized by the potential curve $u(r)$ and that of identical atoms in the lattice, when the equilibrium distance is determined by condition $dU/dr = 0$, rather than by $du/dr = 0$. In the Li_2 molecule, the binding energy is equal to 1.14 eV, and distance r_e is about 2.7 Å. In a lithium crystal $U = 1.7$ eV, and the interatomic distance is greater, 3.03 Å. Although the energy of a metal lattice cannot be represented as a sum of pairwise interactions, the formal division of 1.7 eV by 12 yields 0.14 eV (each Li atom has 12 neighbors), and thus helps to estimate the weakening of "individual" bonds, with an ultimate gain in the crystal lattice energy compared with the molecule. A similar effect of an increase in interatomic distances and energy gain compared with molecules is also observed for ionic crystals. At the same time rigid covalent bonds only slightly change their characteristics in molecules and crystals.

1.3.3 Organic Structures

For the simplest crystals with van der Waals bonds the calculation of the lattice energy by (1.45 − 52) gives satisfactory agreement with experiment (Table 1.6).

Table 1.6. Comparison of experimental and theoretical values of lattice energy for some substances (in kcal/mol)

U	Ne	Ar	O_2	CH_4	Cl_2
Experiment	0.52	1.77	0.74	2.40	6.00
Theory	0.47	1.48	1.48	2.70	7.18

Since the van der Waals forces rapidly decrease with distance, it is sufficient to sum (1.52) over the atomic pairs within a sphere of radius 10 − 15 Å.

For complicated molecules, a good approximation of the potential of the forces acting between atoms of different molecules is (1.45), the "6 exp" potential. Thus, to compute hydrocarbon structures it will suffice to find the constants in (1.45) for three types of interaction, C and C, C and H, and H and H, proceeding from the experimental data for several typical structures. Then, assuming these potentials to be universal, it is possible to use them for analyzing all the known or new structures of this type. If the molecules have a dipole or quadrupole moment, one can also compute the corresponding electrostatic interaction; the hydrogen bonds can also be accounted for with the aid of an potential curve.

With this approach [1.36], a sufficiently good physical model of an organic crystal can be obtained and its structure and properties described. The simplest problem is to calculate the potential energy, i.e., the heat of sublimation for a given structure. Here, we obtain satisfactory agreement with experiment, for instance, for benzene $U_{exp} = 11.0$ kcal/mol, and $U_{theor} = 11.7$ kcal/mol. In general, when the structure's energy is described by (1.52) on the basis of atom-atom potentials of the type (1.45), U is found to be a multidimensional function of the periods a, b, c and angles α, β, γ of the unit cell, the coordinates of the centers of gravity of the molecules in the cell x_i, y_i, z_i, and the Eulerian angles θ, φ, and ψ, which assign the orientation of the molecules,

$$U = U(a, b, c, \alpha, \beta, \gamma, x_1, y_1, z_1, \theta_1, \varphi_1, \psi_1, x_2, y_2, z_2, \theta_2, \varphi_2, \psi_2, \ldots). \qquad (1.57)$$

The structure at absolute zero corresponds to the minimum of this multidimensional function. It is difficult to solve the problem in the general form, and one can investigate the shape and minima of the energy surface only by changing some variables and assuming the others to be fixed. For instance,

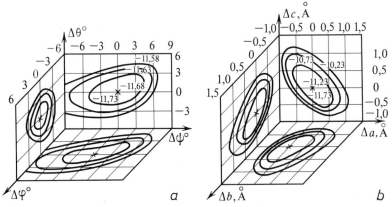

Fig. 1.43a, b. Cross sections of $u(\theta, \varphi, \psi)$ (**a**) and $u(x, y, z)$ (**b**) at the minimum of the energetic surface of benzene. The energy values are indicated in kcal/mol [1.36]

knowing the unit cell and the space group of a crystal, one can seek the orientation of the molecules. Thus, for benzene the minimum actually gives agreement between the calculated and observed orientation with an accuracy to $1° - 3°$. It is also possible to find the parameters of the unit cell with a given orientation of the molecules (Fig. 1.43) or both the unit cell and the orientation with a given space group and the number of molecules in the cell, and thus one can assign in (1.57) $x, y, z, \theta, \varphi, \psi$ of only one molecule, while for the others the same values depend on the symmetry elements.

Proceeding from an analysis of the shape of the minimum we can compute the physical characteristics of a crystal, for instance, $\partial^2 U/\partial x_i^2$ defines the elasticity coefficients and the shape of the surfaces of compression along different directions. It is possible to identify the characteristics of the spectrum of intermolecular vibrations, and so on. By stating the problem as in (1.57) the existence of the crystal structure is postulated beforehand by introducing the unit cell parameters. In principle, one can try to simulate the very origin of three-dimensional periodicity, i. e., of the crystal, so that the cell and symmetry are obtained automatically, by calculating U with assigned interaction of a sufficiently large number of atoms or molecules.

Recall that the minimum of U corresponds to the structure at the temperature of absolute zero. The structure of a given phase may remain unaltered (experiencing only thermal expansion) up to the melting point. In other cases, a change in structure − a phase transition − may take place. This is due to the contribution of the free energy of the thermal motion of atoms or molecules to the general expression (1.46) (for more details see Sect. 4.3).

1.4 Crystallochemical Radii Systems

1.4.1 Interatomic Distances

Entering into a chemical bond and forming a crystal, atoms arrange themselves at definite distances from each other. The experimental data on crystal structures prove that interatomic distance $d(AB)$ between a given pair of atoms A and B in one type of a chemical bond remains constant with an accuracy to about $0.05 - 0.1$ Å in all structures, i.e., it is practically independent of the given structure. This distance corresponds to the minimum of the curve for the potential interaction energy (Fig. 1.12) of a given pair of atoms in the crystal.

Each atom has its own spatial distribution of electrons, which, even for outer electrons, changes comparatively little with the formation of a strong chemical bond, to say nothing of the weak van der Waals bond. Therefore it is possible, as a first approximation, to assign certain "sizes", i.e., certain constant "radii" to atoms, depending on the type of bond, so that the distances between different pairs of atoms are sums of these values. This is called the additivity of the crystallochemical radii. Once established, crystallo-chemical radii are well maintained in newly discovered structures, i.e., they possess the power of prediction. Thus, the huge amount of experimental data on interatomic distances accumulated in structural investigations is generalized in the radii systems. At the same time, as we shall see below, the very concept and the concrete values of crystallochemical radii are, in a sense, conventional.

The concept of crystallochemical radii has been developed by many crystallographers and geochemists; the trail was blazed by *Bragg*, who proposed the first system of radii in 1920 [1.37]. A considerable contribution to this area was made by *Goldschmidt* [1.38] and later by other researchers, who compiled various tables of radii. New improvements in radii systems have been introduced in recent years.

1.4.2 Atomic Radii

By considering the interatomic distances in the structures of elements it is easy to construct a system of the atomic radii of the elements r_{at}. This quantity is equal to half the shortest interatomic distance (Fig. 1.44) $r_{at} = 1/2\, d(AA)$. In the structures of elements (Sect. 2.1) the atoms are bound by a metallic or covalent bond; therefore, the set of atomic radii can be subdivided into r_m and r_c according to the type of bond. Experiments confirm the additivity of such radii also for the structures of compounds with the corresponding type of bond. For instance, the $C - C$ distance in diamond is equal to 1.54 Å, i.e., the covalent radius (of a single bond) of carbon is 0.77 Å. The $Si - Si$ distance in silicon is equal to 2.34 Å, whence the corresponding radius is 1.17 Å. The observed $Si - C$ distance in silicon carbide, 1.89 Å, is in good agreement with the sum $r(C) + r(Si) = 1.94$ Å. Thousands of such examples can be quoted. The same is true for metallic radii. For instance, $r(Nb) = 1.45$ Å, $r(Pt) = 1.38$ Å, $\sum r = 2.83$ Å, and the distance between these atoms in the compound is 2.85 Å. When constructing the system of metallic radii, the data on the distances between different metallic atoms in intermetallic compounds are also taken into consideration. Thus,

$$d(AB) \simeq r_{at}(A) + r_{at}(B)\,. \tag{1.58}$$

The system of atomic radii is presented in Table 1.7 and Fig. 1.45. Inter-atomic distances with a given type of bond vary to some extent, depending on the coordination. They decrease with decreasing coordination number (c.n.), each bond becoming stronger and, hence, shorter. For metals, the r values in Table 1.7 are given for c.n. 12; the decrease in distance for other c.n. is as follows: c.n. 8 by 2%, c.n. 6 by 4%, and c.n. 4 by 12%.

The covalent bond is directional. The lengths of these bonds and the relevant resulting coordinations depend on the bond multiplicities. For C, N,

Table 1.7. Atomic radii [Å]

Period	Ia	IIa	IIIa	IVa	Va	VIa	VIIa	VIIIa			Ib	IIb	IIIb	IVb	Vb	VIb	VIIb	VIIIb
1																	H · 0.46	He 1.22
2	Li 1.55	Be 1.13											B 0.91	C 0.77	N 0.71	O	F	Ne 1.60
3	Na 1.89	Mg 1.60											Al 1.43	Si 1.34	P 1.3	S	Cl	Ar 1.92
4	K 2.36	Ca 1.97	Sc 1.64	Ti 1.46	V 1.34	Cr 1.27	Mn 1.30	Fe 1.26	Co 1.25	Ni 1.24	Cu 1.28	Zn 1.39	Ga 1.39	Ge 1.39	As 1.48	Se 1.6	Br	Kr 1.98
5	Rb 2.48	Sr 2.15	Y 1.81	Zr 1.60	Nb 1.45	Mo 1.39	Tc 1.36	Ru 1.34	Rh 1.34	Pd 1.37	Ag 1.44	C 1.56	In 1.66	Sn 1.58	Sb 1.61	Te 1.7	I	Xe 2.18
6	Cs 2.68	Ba 2.21	La 1.87	Hf 1.59	Ta 1.46	W 1.40	Re 1.37	Os 1.35	Ir 1.35	Pt 1.38	Au 1.44	Hg 1.60	Tl 1.71	Pb 1.75	Bi 1.82	Po	At	Rn
7	Er 2.80	Ra 2.35	Ac 2.03															

Subgroup

	IVa	Va	VIa	VIIa	VIIIa								
Lanthanides	Ce 1.83	Pr 1.82	Nd 1.82	Pm	Sm 1.81	Eu 2.02	Gd 1.79	Tb 1.77	Dy 1.77	Ho 1.76	Er 1.75	Tu 1.74	Yb 1.93	Lu 1.74
Actinides	Th 1.80	Pa 1.62	U 1.53	Np 1.50	Pu 1.62	Am	Cm	Bk	Cf	Es	Fm	Md	(No)	Lr

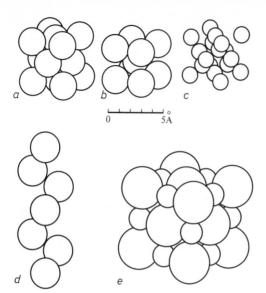

Fig. 1.44a – e. Simplest structures depicted as contacting spheres. (a) Cu, (b) α-Fe, (c) diamond, (d) α-Se, (e) NaCl. In the structures of elements the sphere radius r_{at} is defined as half the interatomic distance. The radii of spheres in structures of different atoms (NaCl) are determined on the basis of the additivity principle in series of structures and other data

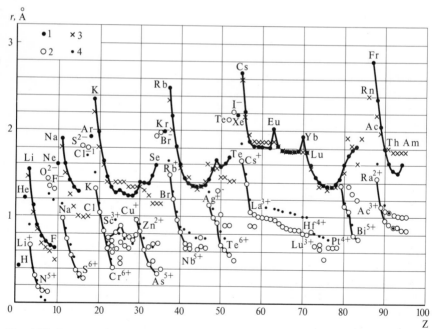

Fig. 1.45. Systems of crystallochemical radii. (*1*) atomic radii r_{at}; (*2*) ionic radii r_i; (*3*) atomic – ionc radii r_{ai}; (*4*) physical ionic radii r_{ph} (in the graph they are given for the maximum valence and c.n. 6 or for the maximum c.n. for the given element if it is less than 6)

O, and S the shortening of r_c, as compared with the single bond, equals 12% – 14% for a double bond, and 20% – 22% for a triple bond. The dependence of the length of the C – C bond on order (including intermediate-order bonds) is presented in Fig. 1.46. Tetrahedral single covalent bonds are very common. The corresponding system of radii is given in Table 1.13.

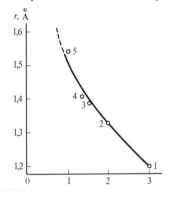

Fig. 1.46. Dependence of the bond length between carbon atoms on the order of the bond; (*1*) acetylene; (*2*) ethylene; (*3*) benzene; (*4*) graphite; (*5*) diamond

As has already been mentioned, the equilibrium interatomic distance for a covalent and metallic bond corresponds to a strong overlapping of the outer shells, which is illustrated by the scheme of overlapping radial density functions in Fig. 1.29 and 1.33. Therefore the atomic radii (1.58) are approximately equal to the orbital radii (1.13) of the outer shells

$$r_{at} \simeq r_o, \quad d(AB) \simeq r_o(A) + r_o(B). \tag{1.59}$$

The curves of the atomic radii versus the atomic number are defined by the structure of the electron shells of the atom, and the graph of Fig. 1.45 is close to that of the orbital radii in Fig. 1.10. Both graphs reflect the rules for the filling of electron shells. The appearance of a new shell (the beginning of a period) increases r_{at}; then, with increasing Z inside the period, r_{at} is reduced because the Coulomb forces of attraction of the electrons to the nucleus become stronger. Towards the end of the large periods the increase in the number of electrons nonetheless gradually increases r_{at}. The filling of inner shells either slightly affects r_{at} or, in lanthanides and actinides, reduces them.

As the orbital radii r_o (1.12) characterize the free atoms, and the over-lapping of the outer shells of nonexcited atoms is only an approximation for calculating the equilibrium interatomic distance, from which r_{at} is found, the agreement between the graphs of r_{at} in Fig. 1.45 and r_o in Fig. 1.10 is not complete. A comparison of Tables 1.1 and 1.7 and of Figs. 1.10 and 1.45 shows that the orbital radii r_o approximately coincide with the atomic radii r_{at} for the beginning and the middle of the periods, respectively. At the end of each period, r_o continues to fall off with increasing Z, while the atomic radii r_{at} of period III decrease slower than the r_o and the r_{at} in the larger periods IV,

V, and VI even begin to increase. For some elements (for instance, Ag, Sb, Te, Hg, Ta, Pb, Bi) the divergences may reach several tenths of an Angstrom. This can evidently be explained by the insufficiency of the pure AO model for these elements; thus it is necessary to take into account the interactions of their electron shells in crystals, and the corresponding changes in energy levels.

1.4.3 Ionic Radii

By analogy with the system of atomic radii, it is possible to construct a system of ionic radii r_i for ionic compounds. The procedure of establishing them from the interatomic distances is, however, ambiguous. To find r_i, use is made of the cation–anion distances in a series of isomorphous (i.e., identically built) structures. Classical series of this kind are those of cubic structures of halides of alkali metals and also of some oxides. Figure 1.44 depicts the structure of NaCl, and Fig. 1.47, a series of face-centered structures isomorphous to it (three structures – CsCl, CsBr, and CsI – are built differently; the anion atom centers the cube, not its face). From the observed cation–anion $(A-B)$ distances

$$d(A_I B) = r_i(A_I) + r_i(B) , \qquad d(A_{II} B) = r_i(A_{II}) + r_i(B)$$

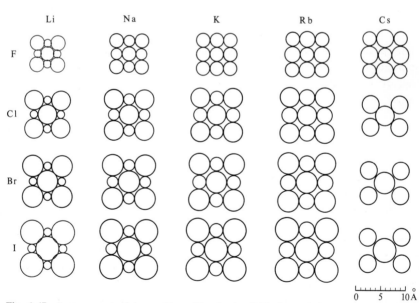

Fig. 1.47. Arrangement of ions with radii r_i in the (100) face of the cubic unit cell of the alkali–halide structures. For the structures of CsCl, CsBr, and CsI, the arrangement of the ions in the diagonal (100) plane is shown

one finds the differences in cation sizes

$$r_i(A_I) - r_i(A_{II}) = d(A_I B) - d(A_{II} B) \qquad (1.60)$$

and, similarly, from $d(AB_I)$ and $d(AB_{II})$, those of the anions

$$r_i(B_I) - r_i(B_{II}) = d(AB_I) - d(AB_{II}) . \qquad (1.61)$$

It is clear that in order to establish a definite system one must either have r_i of some atoms as a "reference" or employ some other data. Indeed, if a certain system of r_i helps to calculate the interatomic distances

$$d(AB) = r_{cat} + r_{an} , \qquad (1.62)$$

then the system of radii

$$r'_{cat} = r_{cat} \pm \delta , \qquad r'_{an} = \sigma_{an} \mp \delta , \qquad (1.63)$$

would give the same distances

$$d(AB) = r'_{cat} + r'_{an} . \qquad (1.64)$$

Proceeding from the data on molar refraction, *Goldschmidt* [1.38] assumed that the ionic radius of the fluoride anion $r(F^-) = 1.33$ Å and that of oxygen $r_i(O^{2-}) = 1.32$ Å. *Pauling* [1.39, 40] adopted crystals of NaF, KCl, RbBr, and CsI as standards, assuming the ratio r_{cat}/r_{an} in them to be about 0.75, and established a somewhat different system, in which $r_i(F^-) = 1.36$ Å, $r_i(O^{2-}) = 1.40$ Å. One more criterion helping to eliminate the uncertainty in the choice of δ (1.63) consists in using the distances in structures with large anions and assuming that these (identical) anions are in contact [1.41]. In the isomorphous series of alkali halides (Fig. 1.43) LiCl and LiBr are such structures, and the "touching" of anions is also observed in a number of other structures. In such a case

$$d(BB) \simeq 2r_{an} . \qquad (1.65)$$

The "classical" systems of effective ionic radii are built so that r_{an} sufficiently describes both the cation – anion (1.62) and anion – anion (1.65) distances in the case of "contacting" anions, although some deviations are observed for large anions. For instance, $r_i(I^-) = 2.20$ Å, $2r_i(I^-) = 4.40$ Å, and the $I-I$ distance in LiI is 4.26 Å.

The ionic radii corresponding to Goldschmidt's system, as corrected by *Belov* and *Boky* [1.42], who assumed $r_i(O^{2-}) = 1.36$ Å, are given in Fig. 1.45 and Table 1.8a. In this table, the anion radii of each period of Mendeleyev's table are larger than the cation radii in the same period. For instance, $r(Li^+) = 0.68$ Å and $r(F^-) = 1.33$ Å; both types of radii increase when going down the subgroup $r(Na^+) = 0.98$ Å, $r(Cl^-) = 1.81$ Å, etc. According to the system of r_i, cations never touch each other.

The distances between ions observed in simple structures usually agree with the sum of the corresponding radii with an accuracy up to about $1\% - 3\%$. In complicated structures with asymmetric coordination and also in structures with large anions the deviations from additivity may reach a few per cent. The coordination can be taken into account by a corresponding correction. The tabulated r_i refer to a cation coordination number equal to 6. On transition to a larger coordination number the radii r_i somewhat increase; for instance, for a c.n. 8 they increase by 3% and for 12, by 12%. The deviations from additivity observed in structures with large ions were attributed to the comparative softness and "polarizability" of their electron shell. For instance, in AgBr, $\sum r_i = 3.09$ Å, and $d(\text{AgBr}) = 2.88$ Å.

The above-considered ionic radii are called effective, because their system corresponds fairly well to the purpose of describing and predicting interatomic distances. Many regularities in crystal chemistry, including the coordination of atoms in a number of compounds, phenomena of isomorphous substitutions, etc., were interpreted on the basis of the "classical" system, but some inconsistencies arose in its practical use.

At the time of the introduction of the system, the knowledge of the electron distribution in atoms and crystals was very vague, and the values of ionic radii did not correlate with it. If we refer, for example, to Fig. 1.14 and draw the boundary corresponding to the effective ionic radii, its course will not be related to any real characteristic of the electron distribution in atoms or to the parameters of the interaction curve.

The present-day chemical bond theory and experimental x-ray data on the electron density distribution in crystals provide physical substantiation for the concepts of crystallochemical radii, in general, and ionic radii, in particular.

In the case of a pure ionic bond the electron density peaks of cations and ions in crystals are isolated, and a reasonable criterion for determining the ionic radii would be the distance from the peaks to the electron density minimum along the straight line joining their centers [1.43 – 45]. The minimum for NaCl is shown in Fig. 1.14. Such ionic radii found on the basis of experimental x-ray data on the electron density distribution in the structures of alkali – metal halides, which may be called "x-ray" or "physical" (r_{ph}) radii, have the following values [Å]:

Li^+	Na^+	K^+	Rb^+	Cs^+	F^-	Cl^-	Br^-	I^-
0.94	1.17	1.49	1.63	1.86	1.16	1.64	1.80	2.05

This minimum of electron density lies in the zone of the decrease of clashing "tails" of the orbitals of the ionized metallic atom and anion. According to (1.23) this zone has a width Δ, and the minimum is found to lie near $\Delta/2$, i.e., the cation radius

$$r_{ph}(A^+) = r_0(A^+) + k\Delta, \quad k \approx 0.5. \tag{1.66}$$

Table 1.8a. Ionic radii

Subgroup

Period	Ia	IIa	IIIa	IVa	Va	VIa	VIIa	VIIIa	Ib	IIb	IIIb	IVb	Vb	VIb	VIIb	VIIIb
1															H 1^- 1.36, 1^+ 0.00	He
2	Li 1^+ 0.68	Be 2^+ 0.34									B 3^+ 0.20	C 4^+ 0.2, 4^+ 0.15, 4^- 2.60	N 3^+, 5^+ 0.15, 3^- 1.48	O 2^- 1.36	F 1^- 1.33	Ne
3	Na 1^+ 0.98	Mg 2^+ 0.74									Al 3^+ 0.57	Si 4^+ 0.39	P 3^+, 5^+ 0.35, 3^- 1.86	S 2^- 1.82, 6^+ 0.29	Cl 1^- 1.81, 7^+ 0.26	Ar
4	K 1^+ 1.33	Ca 2^+ 1.04	Sc 3^+ 0.83	Ti 2^+ 0.78, 3^+ 0.69, 4^+ 0.64	V 2^+ 0.72, 3^+ 0.67, 4^+ 0.61, 5^+ 0.4	Cr 2^+ 0.83, 3^+ 0.64, 6^+ 0.35	Mn 2^+ 0.91, 3^+ 0.70, 4^+ 0.52, 7^+ 0.46	Fe 2^+ 0.80, 3^+ 0.67; Co 2^+ 0.78, 3^+ 0.64; Ni 2^+ 0.74	Cu 1^+ 0.98, 2^+ 0.80	Zn 2^+ 0.83	Ga 3^+ 0.62	Ge 2^+ 0.65, 4^+ 0.44	As 3^+ 0.69, 5^+ 0.47, 3^- 1.91	Se 2^- 1.93, 4^+ 0.69, 6^+ 0.35	Br 1^- 1.96, 7^+ 0.39	Kr
5	Rb 1^+ 1.49	Sr 2^+ 1.20	Y 3^+ 0.97	Zr 4^+ 0.82	Nb 4^+ 0.67, 5^+ 0.66	Mo 4^+ 0.68, 6^+ 0.65	Tc	Ru 4^+ 0.62; Rh 3^+ 0.75, 4^+ 0.65; Pd 4^+ 0.64	Ag 1^+ 1.13	Cd 2^+ 0.99	In 1^+ 1.30, 3^+ 0.92	Sn 2^+ 1.02, 4^+ 0.67	Sb 3^+ 0.90, 5^+ 0.62, 3^- 2.08	Te 2^- 2.11, 4^+ 0.89, 6^+ 0.56	J 1^- 2.20, 7^+ 0.50	Xe
6	Cs 1^+ 1.65	Ba 2^+ 1.38	La 3^+ 0.82, 4^+ 0.90	Hf 4^+ 0.82	Ta 5^+ 0.66	W 4^+ 0.68, 6^+ 0.65	Re 6^+ 0.52	Os 4^+ 0.65; Ir 4^+ 0.65; Pt 4^+ 0.64	Au 1^+ 1.37	Hg 2^+ 1.12	Tl 1^+ 1.36, 3^+ 1.05	Pb 2^+ 1.26, 4^+ 0.76	Bi 3^+ 1.20, 5^+ 0.74, 3^- 2.13	Po	At	Rn
7	Fr	Ra 2^+ 1.44	Ac 3^+ 1.11													

Lanthanides	Ce 3^+ 1.02, 4^+ 0.88	Pr 3^+ 1.00	Nd 3^+ 0.99	Pm 3^+ 0.98	Sm 3^+ 0.97	Eu 3^+ 0.97	Gd 3^+ 0.94	Tb 3^+ 0.89	Dy 3^+ 0.88	Ho 3^+ 0.86	Er 3^+ 0.85	Tu 3^+ 0.85	Yb 3^+ 0.81
													Lu 3^+ 0.80
Actinides	Th 3^+ 1.08, 4^+ 0.95	Pa 3^+ 1.06, 4^+ 0.91	U 3^+ 1.04, 4^+ 0.89	Np 3^+ 1.02, 4^+ 0.88	Pu 3^+ 1.01, 4^+ 0.86	Am 3^+ 1.00, 4^+ 0.85	Cm	Bk	Cf	Es	Fm	Md	No
													Lr

Table 1.8b. Ionic radii

Ion	ec	c.n.	sp	r_{ph}	r_i
Ac³⁺	$6p^6$	6		1.26	1.12
Ag¹⁺	$4d^{10}$	2		0.81	0.67
		4		1.14	1.00
		4sq		1.16	1.02
		5		1.23	1.09
		6		1.29	1.15
		7		1.36	1.22
		8		1.42	1.28
Ag²⁺	$4d^9$	4sq		0.93	0.79
		6		1.08	0.94
Ag³⁺	$4d^8$	4sq		0.81	0.67
		6		0.89	0.75
Al³⁺	$2p^6$	4		0.53	0.39
		5		0.62	0.48
		6		0.675	0.535
Am²⁺	$5f^7$	7		1.35	1.21
		8		1.40	1.26
		9		1.45	1.31
Am³⁺	$5f^6$	6		1.115	0.975
		8		1.23	1.09
Am⁴⁺	$5f^5$	6		0.99	0.85
		8		1.09	0.95
As³⁺	$4s^2$	6		0.72	0.58
As⁵⁺	$3d^{10}$	4		0.475	0.335
		6		0.60	0.46
At⁷⁺	$5d^{10}$	6		0.76	0.62
Au¹⁺	$5d^{10}$	6		1.51	1.37
Au³⁺	$5d^8$	4sq		0.82	0.68
		6		0.99	0.85
Au⁵⁺	$5d^6$	6		0.71	0.57
B³⁺	$1s^2$	3		0.15	0.01
		4		0.25	0.11
		6		0.41	0.27
Ba²⁺	$5p^6$	6		1.49	1.35
		7		1.52	1.38
		8		1.56	1.42
		10		1.61	1.47
		11		1.66	1.52
		12		1.71	1.57
Be²⁺	$1s^2$	3		0.30	0.16
Be²⁺	$1s^2$	4		0.41	0.27
		6		0.59	0.45
Bi³⁺	$6s^2$	5		1.10	0.96
		6		1.17	1.03
		8		1.31	1.17
Bi⁵⁺	$5d^{10}$	6		0.90	0.76
Bk³⁺	$5f^8$	6		1.10	0.96
Bk⁴⁺	$5f^7$	6		0.97	0.83
		8		1.07	0.93
Br¹⁻	$4p^6$	6		1.82	1.96
Br³⁺	$4p^2$	4sq		0.73	0.59
Br⁵⁺	$4s^2$	3py		0.45	0.31
Br⁷⁺	$3d^{10}$	4		0.39	0.25
		6		0.53	0.39
C⁴⁺	$1s^2$	3		0.06	−0.08
		4		0.29	0.15
		6		0.30	0.16
Ca²⁺	$3p^6$	6		1.14	1.00
		7		1.20	1.06
		8		1.26	1.12
		9		1.32	1.18
		10		1.37	1.23
		12		1.48	1.34
Cd²⁺	$4d^{10}$	4		0.92	0.78
		5		1.01	0.87
		6		1.09	0.95
		7		1.17	1.03
		8		1.24	1.10
		12		1.45	1.31
Ce³⁺	$6s^1$	6		1.15	1.01
		7		1.21	1.07
		8		1.283	1.143
		9		1.336	1.196
		10		1.39	1.25
		12		1.48	1.34
Ce⁴⁺	$5p^6$	6		1.01	0.87
		8		1.11	0.97
		10		1.21	1.07
		12		1.28	1.14
Cf³⁺	$6d^1$	6		1.09	0.95
Cf⁴⁺	$5f^8$	6		0.961	0.821
Cf⁴⁺	$5f^8$	8		1.06	0.92
Cl¹⁻	$3p^6$	6		1.67	1.81
Cl⁵⁺	$3s^2$	3py		0.26	0.12
Cl⁷⁺	$2p^6$	4		0.22	0.08
		6		0.41	0.27
Cm³⁺	$5f^7$	6		1.11	0.97
Cm⁴⁺	$5f^6$	6		0.99	0.85
		8		1.09	0.95
Co²⁺	$3d^7$	4		0.72	0.58
		5		0.81	0.67
		6	ls	0.79	0.65
		6	hs	0.885	0.745
		8		1.04	0.90
Co³⁺	$3d^6$	6	ls	0.685	0.545
		6	hs	0.75	0.61
Co⁴⁺	$3d^5$	4		0.54	0.40
		6	hs	0.67	0.53
Cr²⁺	$3d^4$	6	ls	0.87	0.73
		6	hs	0.94	0.80
Cr³⁺	$3d^3$	6		0.755	0.615
Cr⁴⁺	$3d^2$	4		0.55	0.41
		6		0.69	0.55
Cr⁵⁺	$3d^1$	4		0.485	0.345
		6		0.63	0.49
		8		0.71	0.57
Cr⁶⁺	$3p^6$	4		0.40	0.26
		6		0.58	0.44
Cs¹⁺	$5p^6$	6		1.81	1.67
		8		1.88	1.74
		9		1.92	1.78
		10		1.95	1.81
		11		1.99	1.85
		12		2.02	1.88
Cu¹⁺	$3d^{10}$	2		0.60	0.46
		4		0.74	0.60
		6		0.91	0.77
Cu²⁺	$3d^9$	4		0.71	0.57
		4sq		0.71	0.57
		5		0.79	0.65
		6		0.87	0.73
Cu³⁺	$3d^8$	6	ls	0.68	0.54

Ion	Config.	CN	Spin	R_1	R_2
D^{1+}	$1s^0$	2		0.04	−0.10
Dy^{2+}	$4f^{10}$	6		1.21	1.07
		7		1.27	1.13
		8		1.33	1.19
Dy^{3+}	$4f^9$	6		1.052	0.912
		7		1.11	0.97
		8		1.167	1.027
		9		1.223	1.083
Er^{3+}	$4f^{11}$	6		1.030	0.890
		7		1.085	0.945
		8		1.144	1.004
		9		1.202	1.062
Eu^{2+}	$4f^7$	6		1.31	1.17
		7		1.34	1.20
		8		1.39	1.25
		9		1.44	1.30
		10		1.49	1.35
Eu^{3+}	$4f^6$	6		1.087	0.947
		7		1.15	1.01
		8		1.206	1.066
		9		1.260	1.120
F^{1-}	$2p^6$	2		1.145	1.285
		3		1.16	1.30
		4		1.17	1.31
		6		1.19	1.33
F^{7+}	$1s^2$	6		0.22	0.08
Fe^{2+}	$3d^6$	4	hs	0.77	0.63
		4sq	hs	0.78	0.64
		6	ls	0.75	0.61
		6	hs	0.920	0.780
		8	hs	1.06	0.92
Fe^{3+}	$3d^5$	4	hs	0.63	0.49
		5		0.72	0.58
		6	ls	0.69	0.55
		6	hs	0.785	0.645
		8	hs	0.92	0.78
Fe^{4+}	$3d^4$	6		0.725	0.585
Fe^{6+}	$3d^2$	4		0.39	0.25
Fr^{1+}	$6p^6$	6		1.94	1.80
Ga^{3+}	$3d^{10}$	4		0.61	0.47
		5		0.69	0.55
		6		0.760	0.620
Gd^{3+}	$4f^7$	6		1.078	0.938
		7		1.14	1.00
Gd^{3+}	$4f^7$	8		1.193	1.053
		9		1.247	1.107
Ge^{2+}	$4s^2$	6		0.87	0.73
Ge^{4+}	$3d^{10}$	4		0.530	0.390
		6		0.670	0.530
H^{1+}	$1s^0$	1		−0.24	−0.38
		2		−0.04	−0.18
Hf^{4+}	$4f^{14}$	4		0.72	0.58
		6		0.85	0.71
		7		0.90	0.76
		8		0.97	0.83
Hg^{1+}	$6s^1$	3		1.11	0.97
		6		1.33	1.19
Hg^{2+}	$5d^{10}$	2		0.83	0.69
		4		1.10	0.96
		6		1.16	1.02
		8		1.28	1.14
Ho^{3+}	$4f^{10}$	6		1.041	0.901
		8		1.155	1.015
		9		1.212	1.072
		10		1.26	1.12
I^{1-}	$5p^6$	6		2.06	2.20
I^{5+}	$5s^2$	3py		0.58	0.44
		6		1.09	0.95
I^{7+}	$4d^{10}$	4		0.56	0.42
		6		0.67	0.53
In^{3+}	$4d^{10}$	4		0.76	0.62
		6		0.940	0.800
		8		1.06	0.92
Ir^{3+}	$5d^6$	6		0.82	0.68
Ir^{4+}	$5d^5$	6		0.765	0.625
Ir^{5+}	$5d^4$	6		0.71	0.57
K^{1+}	$3p^6$	4		1.51	1.37
		6		1.52	1.38
		7		1.60	1.46
		8		1.65	1.51
		9		1.69	1.55
		10		1.73	1.59
		12		1.78	1.64
La^{3+}	$4d^{10}$	6		1.172	1.032
		7		1.24	1.10
		8		1.300	1.160
		9		1.356	1.216
		10		1.41	1.27
La^{3+}		12		1.50	1.36
Li^{1+}	$1s^2$	4		0.730	0.590
		6		0.90	0.76
		8		1.06	0.92
Lu^{3+}	$4f^{14}$	6		1.001	0.861
		8		1.117	0.977
		9		1.172	1.032
Mg^{2+}	$2p^6$	4		0.71	0.57
		5		0.80	0.66
		6		0.860	0.720
		8		1.03	0.89
Mn^{2+}	$3d^5$	4	hs	0.80	0.66
		5	hs	0.89	0.75
		6	ls	0.81	0.67
		6	hs	0.970	0.830
		7	hs	1.04	0.90
		8		1.10	0.96
Mn^{3+}	$3d^4$	5		0.72	0.58
		6	ls	0.72	0.58
		6	hs	0.785	0.645
Mn^{4+}	$3d^3$	4		0.53	0.39
		6		0.670	0.530
Mn^{5+}	$3d^2$	4		0.47	0.33
Mn^{6+}	$3d^1$	4		0.395	0.255
Mn^{7+}	$3p^6$	4		0.39	0.25
		6		0.60	0.46
Mo^{3+}	$4d^3$	6		0.83	0.69
Mo^{4+}	$4d^2$	6		0.790	0.650
Mo^{5+}	$4d^1$	4		0.60	0.46
		6		0.75	0.61
Mo^{6+}	$4p^6$	4		0.55	0.41
		5		0.64	0.50
		6		0.73	0.59
		7		0.87	0.73
N^{3-}	$2p^6$	4		1.32	1.46
N^{3+}	$2s^2$	6		0.30	0.16
N^{5+}	$1s^2$	3		0.044	−0.104
		6		0.27	0.13
Na^{1+}	$2p^6$	4		1.13	0.99
		5		1.14	1.00
		6		1.16	1.02
		7		1.26	1.12
		8		1.32	1.18
		9		1.38	1.24

Table 1.8b (continued)

Ion	ec	c.n.	sp	r_{ph}	r_i
Na^{1+}	$2p^6$	12		1.53	1.39
Nb^{3+}	$4d^2$	6		0.86	0.72
Nb^{4+}	$4d^1$	6		0.82	0.68
Nb^{5+}	$4p^6$	8		0.93	0.79
		4		0.62	0.48
		6		0.78	0.64
		7		0.83	0.69
Nd^{2+}	$4f^4$	8		0.88	0.74
		8		1.43	1.29
		9		1.49	1.35
Nd^{3+}	$4f^3$	6		1.123	0.983
		8		1.249	1.109
		9		1.303	1.163
		12		1.41	1.27
Ni^{2+}	$3d^8$	4		0.69	0.55
		4sq	ls	0.63	0.49
		5	hs	0.77	0.63
		6	ls	0.830	0.690
Ni^{3+}	$3d^7$	6		0.70	0.56
		6		0.74	0.60
NH4$^+$	$3d^6$	6		0.62	0.48
No^{2+}	$5f^{14}$	6		1.24	1.1
Np^{2+}	$5f^5$	6		1.24	1.10
Np^{3+}	$5f^4$	6		1.15	1.01
Np^{4+}	$5f^3$	6		1.01	0.87
		8		1.12	0.98
Np^{5+}	$5f^2$	6		0.89	0.75
Np^{6+}	$5f^1$	6		0.86	0.72
Np^{7+}	$6p^6$	6		0.85	0.71
O^{2-}	$2p^6$	2		1.21	1.35
		3		1.22	1.36
		4		1.24	1.38
		6		1.26	1.40
		8		1.28	1.42
OH^{1-}		2		1.18	1.32
		3		1.20	1.34
		4		1.21	1.35
		6		1.23	1.37
Os^{4+}	$5d^4$	6		0.770	0.630
Os^{5+}	$5d^3$	6		0.715	0.575
Os^{6+}	$5d^2$	5		0.63	0.49
Os^{6+}	$5d^2$	6		0.685	0.545
Os^{7+}	$5d^1$	6		0.665	0.525
Os^{8+}	$5p^6$	4		0.53	0.39
P^{3+}	$3s^2$	6		0.58	0.44
P^{5+}	$2p^6$	4		0.31	0.17
		5		0.43	0.29
		6		0.52	0.38
Pa^{3+}	$5f^2$	6		1.18	1.04
Pa^{4+}	$6d^1$	6		1.04	0.90
		8		1.15	1.01
Pa^{5+}	$6p^6$	6		0.92	0.78
		8		1.05	0.91
		9		1.09	0.95
Pb^{2+}	$6s^2$	4py		1.12	0.98
		6		1.33	1.19
		7		1.37	1.23
		8		1.43	1.29
		9		1.49	1.35
		10		1.54	1.40
		11		1.59	1.45
		12		1.63	1.49
Pb^{4+}	$5d^{10}$	4		0.79	0.65
		5		0.87	0.73
		6		0.915	0.775
		8		1.08	0.94
Pd^{1+}	$4d^9$	2		0.73	0.59
Pd^{2+}	$4d^8$	4sq		0.78	0.64
		6		1.00	0.86
Pd^{3+}	$4d^7$	6		0.90	0.76
Pd^{4+}	$4d^6$	6		0.755	0.615
Pm^{3+}	$4f^4$	8		1.11	0.97
		9		1.233	1.093
		6		1.284	1.144
Po^{4+}	$6s^2$	8		1.08	0.94
Po^{6+}	$5d^{10}$	6		1.22	1.08
Pr^{3+}	$4f^2$	6		0.81	0.67
		8		1.13	0.99
		9		1.266	1.126
		6		1.319	1.179
Pr^{4+}	$4f^1$	6		0.99	0.85
		8		1.10	0.96
Pt^{2+}	$5d^8$	4sq		0.74	0.60
		6		0.94	0.80
Pt^{4+}	$5d^6$	6		0.765	0.625
Pt^{5+}	$5d^5$	6		0.71	0.57
Pu^{3+}	$5f^5$	6		1.14	1.00
Pu^{4+}	$5f^4$	6		1.00	0.86
		8		1.10	0.96
Pu^{5+}	$5f^3$	6		0.88	0.74
Pu^{6+}	$5f^2$	6		0.85	0.71
Ra^{2+}	$6p^6$	8		1.62	1.48
		12		1.84	1.70
Rb^{1+}	$4p^6$	6		1.66	1.52
		7		1.70	1.56
		8		1.75	1.61
		9		1.77	1.63
		10		1.80	1.66
		11		1.83	1.69
		12		1.86	1.72
		14		1.97	1.83
Re^{4+}	$5d^3$	6		0.77	0.63
Re^{5+}	$5d^2$	6		0.72	0.58
Re^{6+}	$5d^1$	6		0.69	0.55
Re^{7+}	$5p^6$	4		0.52	0.38
		6		0.67	0.53
Rh^{3+}	$4d^6$	6		0.805	0.665
Rh^{4+}	$4d^5$	6		0.74	0.60
Rh^{5+}	$4d^4$	6		0.69	0.55
Ru^{3+}	$4d^5$	6		0.82	0.68
Ru^{4+}	$4d^4$	6		0.760	0.620
Ru^{5+}	$4d^3$	6		0.705	0.565
Ru^{7+}	$4d^1$	4		0.52	0.38
Ru^{8+}	$4p^6$	4		0.50	0.36
S^{2-}	$3p^6$	6		1.70	1.84
S^{4+}	$3s^2$	6		0.51	0.37
S^{6+}	$2p^6$	4		0.26	0.12
		6		0.43	0.29
Sb^{3+}	$5s^2$	4py		0.90	0.76
		5		0.94	0.80
		6		0.90	0.76
Sb^{5+}	$4d^{10}$	6		0.74	0.60
Sc^{3+}	$3p^6$	6		0.885	0.745

$$r_m = \tfrac{1}{2}d(AA), \quad d(AB) \simeq r_m(A) + r_m(B) .\qquad (1.71)$$

These radii are presented in Table 1.10. Recently, proceeding from the statistical treatment of distances in a large number of organic structures, values of r_m for some atoms were found, differing slightly from those given in Table 1.10, namely for N 1.50; O 1.29; S 1.84; Cl 1.90; Br 1.95; I 2.10 Å. Intermolecular radii r_m are always larger than strong-bond radii. As the potential function for weak bonds has a very gradual slope (Curve II in Fig. 1.12), the additivity condition (1.71) permits deviations up to tenths of Å units. Figure 1.50 shows a histogram of deviations of van der Waals contact distances, found experimentally, from standard ones [1.51]. According to the system of r_m a molecule formed by covalently bound atoms should be represented, with respect to the intermolecular contacts in the crystal, as if it were wrapped up in a "coat" with the shape of spheres of indicated radii (Fig. 1.51).

Table 1.10. Intermolecular (van der Waals) radii

			H	He
			1.17	1.40
C	N	O	F	Ne
1.70	1.58	1.52	1.47	1.54
Si	P	S	Cl	Ar
2.10	1.80	1.80	1.78	1.88
	As	Se	Br	Kr
	1.85	1.90	1.85	2.02
		Te	I	Xe
		2.06	1.96	2.16

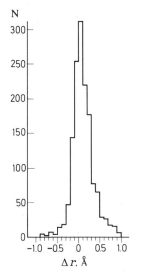

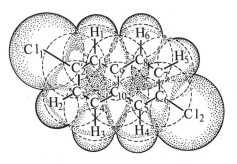

Fig. 1.50 Histogram of the distribution of deviations of interatomic distances at van der Waals contacts from standard values

Fig. 1.51. Molecule as a geometric body framed in a "coat" of van der Waals radii

1.4.6 Weak- and Strong-Bond Radii

In molecular and noble gas crystals the equilibrium distances between non-covalently bound atoms are defined by the repulsive branch of the weak inter-action curve (Fig. 1.12); the system of intermolecular radii serves as a geometric model of this repulsion.

As we have already said, a similar interaction takes place between anions in ionic crystals. Vivid examples are layer lattices, which are flat packets of anions with cations inside (Fig. 1.52). The packets contact each other only by means of anions. These crystals exhibit cleavage due to weak bonds between layers. As in intermolecular contacts, the strong bonds inside some of the structures are saturated, while the contacting anions have the filled shell of noble gas atoms. The distances between them are nearly the same as in molecular crystals; for instance, the $Cl-Cl$ distance in $CdCl_2$ is 3.76 Å. The tabulated values of the effective ionic radius of Cl^- and the van der Waals radius of Cl practically coincide, $r_i(Cl) = 1.81$ Å and $r_m(Cl) = 1.78$ Å.

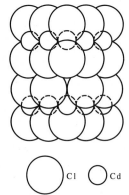

○ Cl   ○ Cd   **Fig. 1.52.** Layer structure of $CdCl_2$. Contact of sheets having formula $CdCl_2$ is due to weak $Cl-Cl$ bonds

Another important example is the distance between oxygen atoms. The distance between non-valence-bound anions O^{2-} in ionic structures – oxides, silicates, and inorganic salts – ranges from 2.5 to 3.2 Å, which corresponds to the effective ionic radius of oxygen, taken to be $r_i(O^{2-}) = 1.38$ Å. The tabulated values of the intermolecular radius $r_m(O)$ are $1.36-1.52$ Å, i.e., it is the same value.

Anions in ionic structures do not attract each other (if we neglect the weak van der Waals forces); they experience Coulomb repulsion, but are united in crystal structures owing to their interaction (attraction) with cations. The distances between the anions correspond to the equilibrium of all these forces. Thus, the similarity between the effective ionic radii of anions and the inter-molecular radii of the corresponding atoms

$$r_{an} \simeq r_i \tag{1.72}$$

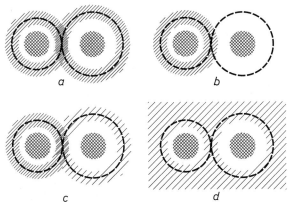

Fig. 1.53a – d. Schematic illustration of various types of chemical bonds. Only two bound atoms are shown conventionally. (*Hatching*) outer orbitals; (*cross-hatching*) inner atomic shells; (*dotted circles*) orbital radius.

(a) Covalent bond. At the site of the superposition of valence orbitals the electron density is increased; it exceeds the sum of the electron densities of the valence orbitals of isolated atoms. **(b)** Ionic bond. The cation (on the right) has lost an electron from the valence orbital, which was incorporated into the anion shell. The anion is so positioned, however, that its shell is at the same distance from the inner cation shell at which the now-ionized electron of the cation was located. **(c)** Covalent bond with a fraction of ionicity − a case intermediate between (a) and (b). The cation shell is depleted and the anion shell enriched in electrons; at the site of the overlapping the electron density exceeds the sum of the electron densities of the valence orbitals of the isolated atoms. **(d)** Metallic bond. The electrons of the valence orbitals are now distributed uniformly throughout the crystal lattice space (except in the core shells); the interatomic electron density is relatively low

is not accidental − it reflects the identical physical nature of the interaction between atoms whose strong chemical bonds are already saturated. These radii may be called weak-bond radii. They can be interpreted physically as quantities determining the minimum distance of "touching" of non-valence-bound atoms (Figs. 1.49f, 50, 51). The physical model corresponding to a strong bond (covalent, metallic, or anion − cation) is the interpenetration of the outer orbitals of the contacting atoms, although it acquires a specific form for each of the main types of bond (Fig. 1.53). Thus, the interpenetration is most pronounced in the covalent bond; the cation electrons are fully or partly incorporated in an anion shell in the case of ionic bond; the valence electrons are collective in metals. From this standpoint, ionic structures, as well as molecular ones, can be described by two systems of radii, namely by the radii of the strong cation − anion bond, and (if there are anion − anion contacts) by weak-bond radii (Fig. 1.49c, f).

In conclusion we reiterate the most essential elements in the construction of the various systems of crystallochemical radii. The covalent, metallic, and "atomic − ionic" radii are approximately equal to the orbital radii of the outer atomic electrons, which are responsible for the formation of a strong chemical bond. The radii of weak bonds describe the geometric picture of the contacts in weak van der Waals atomic interactions. The system of physical ionic radii gives the distances to the minimum of the electron density at cation – ion contacts. The formal scheme of effective ionic radii also describes the interatomic distances, since the additivity conditions are also fulfilled.

1.5 Geometric Regularities in the Atomic Structure of Crystals

1.5.1 The Physical and the Geometric Model of a Crystal

The theory of crystal structure, i.e., the theory of its formation from an assembly of interacting atoms, is based on the general principles of thermodynamics, solid-state physics, and quantum mechanics. At the same time, the result of this interaction is geometrically amazingly simple; in the unit cell of a three-dimensionally periodic structure the atoms occupy fixed positions at definite distances from each other.

The very consideration of this geometry, irrespective of its physical causes, using certain relatively simple physical or chemical data which are also geometrized, promotes the understanding of many regularities in the atomic structure of crystals.

In the geometric model of a crystal, the arrangements of the crystal's structural units, i.e., atoms or molecules, the distances between them as well as their coordination are considered. Proceeding from the systems of crystallochemical radii, atoms can be modeled as hard spheres, and molecules as solids of more complicated shape. Then packings of such spheres or bodies can be analyzed. The formally geometric consideration is supplemented by taking into account the nature of the chemical bond between the atoms; stable groupings of atoms – coordination polyhedra, complexes, molecules, etc. – are studied, as are their shapes and symmetries and their relationship with the space symmetry of the crystal.

The geometric model of a crystal is the most simplified version of its physical model. The geometric approach was the starting point for the development of concepts of the atomic structure of crystals. It is naturally limited and cannot claim to explain crystal structures in all their details. It does help, however, to formulate and describe a number of regularities in the structure of crystals in a simple and pictorial form.

1.5.2 Structural Units of a Crystal

The very expression "the atomic structure of a crystal" shows that the ultimate structural units with respect to which the structure of a crystal is considered (at any rate, at the geometric level) are the atoms. In many cases, however, even before the formation of a crystal or in the course of its formation, the atoms, owing to their chemical nature, draw together into certain stable groupings which are preserved in the crystal as an entity and can conveniently and legitimately be regarded as structural units of crystals. The isolation of structural units according to their crystallochemical features makes possible their definite geometric and symmetric description.

In dividing the crystals into structural units according to the types of association of the atoms, one should consider whether the chemical bond forces acting between all the atoms are the same or different. In the former case the crystals are called *homodesmic*. Since all their interatomic bonds are the same type, the interatomic distances differ but little. Examples of homodesmic structures are those of metals and alloys; covalent structures, as well as many ionic ones, are also homodesmic. The structural units of such crystals are the atoms themselves, which form a three-dimensional network of approximately equivalent bonds; at the same time, definite structural groupings can sometimes be singled out among them.

Stable, isolated, finite groupings or complexes of atoms in a crystal can form if the types of bonds are different. Such crystals are called *heterodesmic*. More often than not, the bonds inside such a structural grouping are fully or partly covalent. A typical example are organic molecules; strong covalent bonds act within the atoms, but weak van der Waals bonds operate between unit molecules in the crystal. Examples of structural groupings in inorganic crystals are the complex anions CO_3^{2-}, SO_4^{2-}, NO_2^-, water molecules H_2O, the complexes $[PtCl_4]^{2-}$ and $[Co(NH_3)_6]^{3+}$, intermetallic complexes $MoAl_{12}$ (Fig. 1.54), etc. These groupings are finite in all three dimensions; sometimes they are called "insular" groupings. But there may also be one- or two-dimensionally extended structural groupings, chains and layers, which will be discussed below.

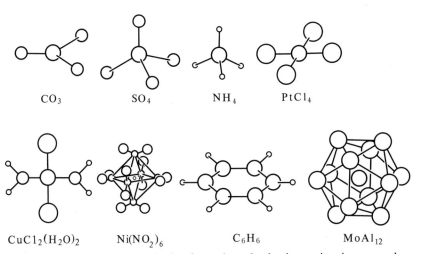

CO_3 SO_4 NH_4 $PtCl_4$

$CuCl_2(H_2O)_2$ $Ni(NO_2)_6$ C_6H_6 $MoAl_{12}$

Fig. 1.54. Examples of structural units of crystals: molecules, inorganic anions, complexes, the intermetallic complex $MoAl_{12}$

1.5.3 Maximum-Filling Principle

In considering the formation of a crystal out of structural units from the geometric standpoint, one should take into account their own shape and symmetry, and the nature of interaction among them.

If we first refer to crystals whose structural units are atoms, they can be divided into two subtypes as regards the intrinsic symmetry of atoms. If the forces acting among atoms are central or approximately so, as in metallic and ionic crystals as well as in crystals of solidified noble gases, one can speak of an intrinsic, spherical or approximately spherical point symmetry of atoms $\infty/\infty\ m$. The other subtype is given by the covalent crystals. The atoms in them have directed bonds and can be regarded as having a definite and nonspherical point symmetry. We shall treat them later. If a crystal has limited or any other kind of structural groupings, the interaction between them can also be reduced to the central interactions among the atoms of neighboring groups despite the great diversity in their forms and symmetry.

The principal contribution to the free energy of a crystal F (1.46) is made by the potential energy of interaction U, i.e., the lattice energy. All expressions for the energy of the central interaction forces are dependent on the interatomic distance (1.51); the energy minimum is attained at the equilibrium interatomic distances which are typical of given atoms and of the type of bond between them. These distances can be expressed as the sums of the corresponding crystallochemical radii. The more atoms approaching equilibrium interatomic distances, i.e., the more terms in (1.52) having r_{ik} corresponding to the largest $u(r_{ik})$, the larger is U (1.52). However, the repulsive branch of the interaction curve, which can be interpreted geometrically as the finite size of atoms (Fig. 1.12), limits the number of atoms reaching the equilibrium interatomic distances. All these circumstances can be expressed as the geometric principle of *maximum filling*. It consists of the following: that, subject to the action of central or near-central forces of attraction, the atoms or more complex structural units of a crystal always tend to approach each other so that the number of permissible shortest contacts is maximal. This can also be expressed as a tendency towards a maximum number n of atoms (or structural units) per unit volume Ω at distances r_{ij}, not less than the standard permissible distances r_{st},

$$\frac{n}{v_{(r_{ij} \geqslant r_{st})}} \to \max . \tag{1.73}$$

The most appropriate formulation is based on the concept of crystallochemical radii r. If there is a "radius", then naturally a crystal model consists of atoms — balls having these radii and volumes $V_{at} = 4/3\,\pi r^3$ are in contact with each other (Figs. 1.44a, b; 47). If we introduce the ratio of the sum of volumes occupied by atoms to the volume of the cell Ω, i.e., the so-called packing coefficient,

$$\sum V_{at}/\Omega = q \to \max \qquad (1.74)$$

then the maximum filling principle can be described as that of the packing coefficient maximum. It is most clearly expressed in structures built according to the method of the closest packing of identical balls ($q = 74.05\%$). But it is also used when considering the geometric interpretation of structures made up of different atoms as the packing of balls of different radii. The closest packings will be specially considered below. In general, however, according to this principle a structure should not have any vacant sites where atoms (balls) of the largest radii could be accommodated, and there must be as few sites as possible which are not occupied by other atoms (balls) having smaller radii.

Similarly, if we assign some volume V_i to multiatomic structural groupings or molecules, wrapping them in a "coat" of weak-bond radii (as shown in Figs. 1.49, 50), then the maximum-filling principle will be expressed analogously (1.74)

$$\sum V_i/\Omega = q \to \max . \qquad (1.75)$$

Being the geometrization of the concepts of the crystal structure, the maximum-filling principle, or the same idea expressed as the closest-packing principle, is naturally qualitative, because it explains the main tendency in the formation of structures consisting of mutually attracted particles, but not the individual features of the specific structures. Yet owing to its simplicity and generality it plays an important part in crystal chemistry and sometimes aids in drawing quantitative conclusions, for instance in considering the structure of some ionic and molecular crystals (see Sect. 2.3.6).

1.5.4 Relationship Between the Symmetry of Structural Units and Crystal Symmetry

As has already been noted, the atoms in a structure − when they are its structural units − can be regarded either as spherically symmetric or as having a different point symmetry corresponding to the orientation of the covalent bonds (cf. Fig. 1.18c). Multiatomic finite structural groupings also have a definite point symmetry, most often crystallographic (Fig. 1.54), but there may be groupings with a noncrystallographic symmetry as well. The force field around such groupings, i.e., the field of the potential interaction energy, is anisotropic and corresponds to the symmetry of the grouping itself, which presupposes, to some extent, the possibility of its contacts with other atoms or with similar or different groupings. But this anisotropy is not very large, and for geometric consideration it is often possible to assume that the forces of interaction between charged groupings are near central. For instance, the structure of Na_2SO_4 can be considered as built up of mutually attracted structural units of Na^+ and SO_4^{2-} (Fig. 1.55).

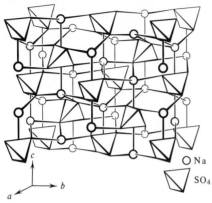

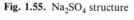

Fig. 1.55. Na$_2$SO$_4$ structure

○ Na

▽ SO$_4$

Inorganic and organic molecules are the most diversified as regards their structure and symmetry. They include asymmetric (symmetry I), centrosymmetric ($\bar{I}$), and higher symmetry molecules (see Fig. 2.72) up to giant virus molecules with an icosahedral pseudospherical symmetry (see Figs. 2.169, 174).

Let us now see whether the space symmetry Φ of the structure arising from the given structural units is related (and if so, then how) to the point symmetry G_0^3 of these units. When forming a crystal, atoms or more complex structural units occupy definite positions in the unit cell, and arranging themselves in accordance with one or several regular point systems of the general (with symmetry 1) or particular (with symmetry K) position of group Φ. Consider whether there is any connection between the intrinsic symmetry G_0^3 of the structural unit and the symmetry K of its position in the crystal.

On the one hand, many examples of such a relationship are known. For instance, the spherically symmetric atoms in the structures of metal elements occupy all positions with the maximum symmetry $m\bar{3}m$ in the cubic, high-symmetry space group $Fm\bar{3}m$; the same symmetry is observed for Na and Cl atoms, with central forces acting between them in the structure of NaCl. The tetrahedral carbon atom in the diamond structure occupies a position with a tetrahedral symmetry $\bar{4}3m$, or, more precisely, predetermines the symmetry of this position. The trigonal complex anion CO_3^{2-} "imposes" the rhombohedral symmetry on the calcite structure (see Fig. 2.19), etc. In such cases it can be said that the symmetry of position K, being a subgroup of the symmetry of structural unit G_0^3, is maximally close to, or coincides with it: $G_0^3 \supseteq K$. This corresponds to the fulfilment of the so-called Curie principle of interaction of symmetries [1.7]. The field of all the particles of the structure, surrounding a given particle with the symmetry G_0^3, interacts with its own field. This interaction ultimately determines the symmetry K.

On the other hand, in many structures symmetric atoms or molecules occupy positions K, which are less symmetric than their proper symmetry. In this case, too, the Curie principle holds, i.e., the point group of positions is a subgroup $K \subset G_0^3$ of the point group of the structural unit, but the anti-symmetrization may be quite strong so that sometimes the positions may even be asymmetric, $K = 1$. For instance, benzene molecules with a high symmetry $6/mm$ are packed in an orthorhombic structure, and the symmetry of the positions of the molecule center is $\bar{1}$. The tetrahedral grouping SiO_4 in different crystals occupies both the positions corresponding to this symmetry (cristobalite) and those whose symmetry is lower, i.e., 2, m, or 1 in many silicates.

The answer to our question is simple. The determining principle of the formation of a structure is the principle of energy minimum, which for un-directed interaction forces is expressed geometrically as the maximum-filling principle. If the symmetry of a structural unit during the formation of a crystal is consistent with or helpful in attaining this minimum, then a structure is formed with such a space group Φ and such positioning of the structural units within it which are closest to their proper symmetry. But if the minimum is attained when the structural units occupy low-symmetry positions, their proper symmetry either does not play any role or is used only partly, i.e., it does not completely coincide with the symmetry of positions.

When speaking of the relationship between the symmetry of positions K and that of structural unit G_0^3, one may distinguish two cases. In one, the symmetry G_0^3 of an isolated structural unit in the crystal taken by itself is preserved $G_0^3 = G_{0(cr)}^3$. In the other, the effect of the field of the surrounding particles (the lattice field) is such that the structure of a molecule changes and its symmetry is reduced, $G_0^3 \supset G_{0(cr)}^3 \supseteq K$.

It is worth mentioning one more circumstance, which is associated with the simplicity or complexity of the chemical formula of a given substance. The number of most symmetric positions in any space group Φ is limited. For structures with a simple formula this may not be meaningful. For instance, in ionic structures having simple formulae of the type AX, AX_2, etc., central forces operate between the atoms which can occupy high-symmetry positions, and the structures usually show a high symmetry. But if there are many sorts of atoms, the number of such positions is simply insufficient, and the symmetry of the structure is reduced. Therefore, usually the more complex the chemical formula of an inorganic compound, the lower is the symmetry of its crystal structure.

If speaking of the relationship between the point symmetry of a structural unit and its position in the crystal, one must also mention those cases where the latter is higher than the former. This is naturally impossible for molecules "at rest", but it can be achieved statistically, either by averaging over all the unit cells of the crystal or by thermal reorientation of the molecules or of their rotation, $K = G_{0(statist.)}^3 \supset G_0^3$. Thus, cases are known where asymmetric (with symmetry 1) molecules, which are, however, approximately centrosymmetric in shape, form a centrosymmteric crystal, in which they statistically occupy a position with symmetry $\bar{1}$. Owing to thermal reorientations the grouping NH_4 with a proper symmetry $\bar{4}3m$ occupies positions with the symmetry $m\bar{3}m$ in some crystals, while in certain compounds it is in the state of complete spherical rotation.

We have considered the relationship between the point symmtery of structural units (with nondirectional binding forces between them) and the point symmetry of their positions in the crystal structure. What is more important, however, the space symmetry Φ also regulates the mutual arrangement of the structural units, not only by means of point symmetry operations, but also by translations and symmetry operations with a translational component. Let us take a look at Fig. 1.56, which illustrates the role of symmetry operations when particles of arbitrary shape contact each other. It is seen that, from the standpoint of the principle of maximum mutual approach of neighboring particles, point symmetry operations (Fig. 1.56a) are actually inappropriate since they impose the condition of pairwise arrangement of the centers of the atoms along lines (or in planes) perpendicular to the corresponding symmetry elements, and this precludes the most economical utilization of space. On the contrary, symmetry elements with a translational component (Fig. 1.56b) let some particles enter the spaces between others; more precisely, these elements are virtually formed with such packing. Therefore, the important consequences of the maximum-filling principle are, the rules concerning, firstly, the utilization of symmetry elements having a translational component, and, secondly, the implications of the appearance of point symmetry elements between neighboring atoms; if such elements exist, the atoms or structural units should be arranged on them.

If the binding forces are directional, as in covalent crystals, the close-packing principle contradicts the physical model picturing the attainment of the potential energy minimum. Then, representing the structure by means of contacting balls also is not adequate any more. The energy minimum is

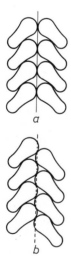

Fig. 1.56a, b. Packing of structural units in a crystal. Point symmetry operations (in this case mirror plane m) are not appropriate because the projections of the structural units confront each other (a). Close packing is promoted by symmetry operations which a translational component (b)

attained precisely upon saturation of the directional covalent bonds; in this case the principle of a proper point symmetry of atoms is always in the foreground. Therefore, the covalent structures in which atoms have relatively few neighbors (1.3, 4, 7) in accordance with their symmetry, are "lacier" than the maximum-filling structures. Examples are diamond (Fig. 1.44c) and other tetrahedral structures, octahedral structures of the NiAs type (Fig. 1.57). The symmetry of these structures is defined by the symmetry of the constituent atoms. Although covalent structures are "lacy", the volumes per atom in structures with distinct types of bonds differ only slightly, on the average, since the covalent bonds are usually slightly shorter than the ionic or metallic.

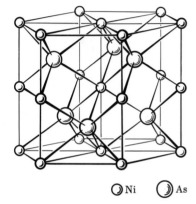

$\bigcirc$ Ni $\bigcirc$ As

Fig. 1.57. The structure of NiAs with octahedral Ni bonds

Concerning the relationship of the point symmetry of structural units G_0^3, their positions in the crystal K, and the space symmetry of structure G_3^3, one must also mention the so-called local, noncrystallographic symmetry. In some molecular (Sect. 2.6.2) and biological (Sect. 2.9.4) structures such packing of the structural units is observed when they are related symmetrically, for instance, by axes 2; these symmetry elements do not belong to group G_0^3 (they are said to be noncrystallographic), but act locally, say, between definite (not all) pairs of particles (Fig. 2.150). The so-called noncrystallographic screw axes, which link chains of particles running along different directions in the crystal, are also known (see Fig. 2.75). Later (Sect. 2.6) we shall see how all these phenomena can be explained from the standpoint of the packing energy minimum principle and from the structural features of the units packed.

1.5.5 Statistics of the Occurrence of Space Groups

The above considerations are confirmed by the statistics of the distribution of the investigated structures among the space groups. Such data were obtained by processing the results for 5,600 inorganic and 3,200 organic structures (there are 8,800 structures in all) [1.52, 53]. They are listed in Table 1.11, the groups being arranged in the order of increasing symmetry.

Thus, 40% of the inorganic structures belong to only nine groups, 60% of the organic structures to six groups, and half of all the structures to the 12 groups Φ listed in Table 1.11.

Table 1.11. Distribution of crystal structures among space groups Φ [%]

Φ	Inorganic structures	Organic structures	All structures
$C_i^1 - P\bar{1}$	1	5	3
$C_2^2 - P2_1$	–	8	3
$C_{2h}^5 - P2_1/c$	5	26	13
$C_{2h}^6 - C2/c$	4	7	5
$D_2^4 - P2_12_12_1$	–	13	5
$D_{2h}^{15} - Pbca$	–	3	1
$D_{2h}^{16} - Pnma$	7	–	5
$D_{3d}^5 - R\bar{3}m$	4	–	2
$D_{6h}^4 - P6_3/mmc$	4	–	3
$O_h^1 - Pm\bar{3}m$	4	–	3
$O_h^5 - Fm\bar{3}m$	9	–	6
$O_h^7 - Fd\bar{3}m$	5	–	3

The occurrence of the listed groups is due to the above consideration concerning the advantage of the presence of "packing" symmetry elements with a translational component, which is typical of practically all the groups of Table 1.11 and is most clearly manifested in inorganic structures having complicated formula, as well as in organic structures. If a structure has a relatively simple formula, its atoms can arrange themselves on point symmetry elements. This mainly explains the lower part of the table; all the groups listed in it belong to the structures of elements and simple inorganic compounds.

Only 40 groups Φ (including the just listed 12) are relatively widespread. Representatives of 197 groups Φ (out of 219) have been found. An analysis of the frequency of occurrence of groups Φ shows, however, that, with the increase of the number of investigated crystals the representatives of groups not yet observed may be found.

The center of symmetry $\bar{1}$ is present in most of the structures investigated – in 82% of the inorganic and 60% of organic, or altogether in 74% of the substances[8].

---

[8] In recent years, due to the development of the structure analysis technique, the percentage of investigated noncentrosymmetric structures has increased.

Table 1.12. Relationship between coordination n and values of g and g'

Coordination	n	g	g'
Tetrahedron	4	1.225	0.225
Octahedron	6	1.414	0.414
Thomson (wringed) cube	7	1.592	0.592
	8	1.645	0.645
Cube	8	1.732	0.732
Cubooctahedra	12	2.000	1.000

If g is intermediate between the "nodal" values, for instance, $1.225 \leqslant g < 1.414$, then only four anions can settle around the cation (not necessarily contacting each other) at a given $d(AB)$. As g increases, a transition to larger coordination numbers and to the corresponding polyhedra may (and must, from the standpoint of close packing) take place. If, on the contrary, there is contact of all the outer spheres, then on realization of coordination n and at g less than the corresponding nodal value (e. g., for the octahedron $n = 6$ and $g < 1.414$), the cation will "dangle" among the surrounding anions. Such a configuration is unstable from the geometric standpoint. Thus, at a given g, the geometrically most probable coordination is realized when g is less than the nearest nodal value, but lower coordinations are also possible. Thus, although at $g = 1.645$ the formation of a Thomson cube with $n = 8$ is already possible, at these and higher values a stable grouping is often an octahedron with $n = 6$ (this is true, for example, of KF and RbCl). The Magnus rules hold for a number of ionic crystals. So, for many halides of alkali metals (see Fig. 1.47) the values of g correspond to the observed octahedral coordination. The structures of CsCl, CsBr, and CsI have a cubic coordination in conformity with the values of g (1.75, 1.84, and 1.91). At the same time, these rules are not fulfilled in many cases. Thus, the structures of LiCl, LiBr, and LiF have $g < 1.415$, i. e., they would be expected to be tetrahedral (the octahedron is "impossible" for them), but in fact they crystallize as NaCl, with an octahedral coordination. This is due to the conventional nature of the geometric model.

1.5.10 Closest Packings

A number of the structures of elements, alloys, and ionic crystals are built on the close-packing principle. Consider the geometry and symmetry of the closest packings of spheres and some related questions.

One two-dimensional close-packed layer of identical spheres is depicted in Fig. 1.61 together with its intrinsic symmetry elements. Axes 6 pass through the centers of the spheres, and axes 3, through the interstices (holes) between spheres. There are twice as many interstices as spheres. The next, identical layer will be packed most closely if its spheres are placed over the voids, between the spheres of the lower layer (Fig. 1.62). The common elements for

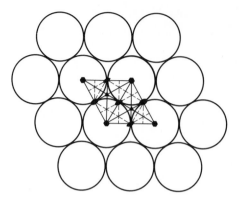

Fig. 1.61. Close-packed layer of spheres and its symmetry

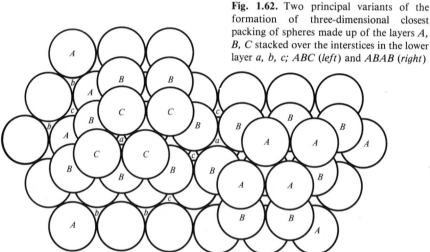

Fig. 1.62. Two principal variants of the formation of three-dimensional closest packing of spheres made up of the layers A, B, C stacked over the interstices in the lower layer a, b, c; ABC (*left*) and $ABAB$ (*right*)

two, or any number of layers superimposed in this manner will be the axes 3 and planes m, as before. Hence, the space groups of the closest packings are all the groups which have subgroup $P3m1$; these are (besides the foregoing) $R3m$, $P\bar{3}m1$, $R\bar{3}m$, $P\bar{6}m2$, $P6_3mc$, $P6_3/mmc$, and $Fm\bar{3}m$; in all, there are eight such groups [1.55].

If we denote (as in Fig. 1.62) the initial layer of spheres by A, the possible ways of placing the next layer, either B or C, over it are b and c. The symbol is irrelevant until the third layer is determined (in Fig. 1.62 it is B) which may be placed over the interstices of layer B. Thus, any combinations of arrangements of new layers can be described by a sequence of letters A, B, and C, and no identical letters can stand side by side, since that would mean placing one sphere over another. The packing coefficient for any closest packing is 74.05%. The two-layer packing ... $ABAB$ ... (right-hand side of Fig. 1.62 and Fig. 1.63a) is called the closest hexagonal and has the symmetry $P6_3/mmc$. A remarkable arrangement is the three-layer packing ... $ABCABC$

... (left-hand side of Fig. 1.62, and Fig. 1.63b). The stacking of the initial three close-packed two-dimensional layers produces an arrangement such that precisely identical layers in it (and only in it) can be singled out along three other directions, as seen from Fig. 1.63b. This packing is cubic face centered, its symmetry being $Fm\bar{3}m$.

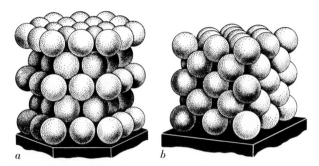

a   *b*

Fig. 1.63a, b. Hexagonal (**a**) and cubic (**b**) closest packing. In the cubic packing the closest-packed layers can be singled out in four planes perpendicular to the body diagonals of the cube

Further, we can consider the stacking of layers with a periodicity n, greater than for three layers. Before we do that, we should note that the obvious three-letter notation is redundant. This follows even from the fact that the sequences ... $ABAB$ ..., $ACAC$ ..., and $BCBC$ ... denote one and the same hexagonal packing. It would suffice to label a given layer according to the way the two nearest layers adjoin it from below and from above, namely by the hexagonal (h) or cubic (c) law. Then the two principal packings will be written as follows (the symmetry is also indicated):

$$
n = 2 \begin{array}{c} ABAB \\ hhhh \end{array} D_{6h}^4 = P6_3/mmc, \qquad n = 3 \begin{array}{c} ABCABC \\ ccc\dots \end{array} O_h^5 = Fm\bar{3}m, \tag{1.78}
$$

where n is the number of layers in the packing. Figure 1.64 shows the coordination polyhedra of spheres of type h and c, i.e., cubooctahedra, in packing c, a regular, and in h, a "wrung" one (hexagonal, an analog of the cubooctahedron). Packings with $n = 4, 5$ are as follows:

$$
n = 4 \begin{array}{c} ABAC \\ chch \end{array} D_{6h}^4 = P6_3/mmc, \qquad n = 5 \begin{array}{c} ABCAB \\ hcch \end{array} D_{3d}^3 = P\bar{3}m1.
$$

There are two packings with $n = 6$,

$$
n = 6 \begin{array}{c} ABCACB \\ hcchcc \end{array} D_{6h}^4 = P6_3/mmc, \qquad \begin{array}{c} ABABAC \\ chhhch \end{array} D_{3h}^1 = P6m2.
$$

Figure 1.65 shows some packings with different numbers of layers.

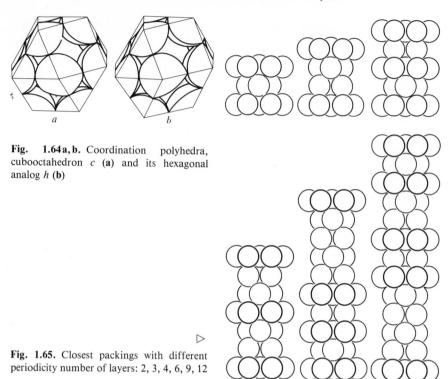

Fig.   1.64a, b. Coordination   polyhedra, cubooctahedron c **(a)** and its hexagonal analog h **(b)**

▷

Fig. 1.65. Closest packings with different periodicity number of layers: 2, 3, 4, 6, 9, 12

A complete system of the closest packings of spheres was proposed by *Belov* [1.55]. With a fixed n, the number of different packings is naturally limited, but it increases with n,

number of layers   2 3 4 5 6 7 8 9 10 11 12 ...

number of packings   1 1 1 1 2 3 6 7 16 21 43 ...

An example of 8-layer packing is $h\,c\,c\,c\,h\,c\,c\,c$, and of 12-layer, $h\,h\,c\,c\,h\,h\,c\,c\,h$ $h\,c\,c$. It is significant that while at small n all the spheres are symmetrically equal, i. e., they occupy a single regular point system, an increase in n renders this impossible. This reflects the above-described competition of the principles of closest packing and preservation of proper symmetry of particles in the lattice, which arises when the number of particles in the unit cell is large.

The structures of many elements − metals [Na, Al, Cu, Fe, Au, etc. (c), Mg, Be, α Ni, Cd, Zn, etc. (h)] and solid inert elements − are built according to the principle of cubic face-centered (fcc) and hexagonal closest packing (hcp). In some cases the structure deviates from the ideal, examples being the rhombohedral or tetragonal distortion of cubic packing, and deviation from the "ideal" value of $c/a = 1.633$ in hexagonal packing (Fig. 1.66).

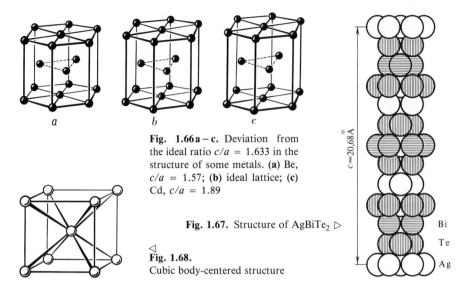

Fig. 1.66a–c. Deviation from the ideal ratio $c/a = 1.633$ in the structure of some metals. (**a**) Be, $c/a = 1.57$; (**b**) ideal lattice; (**c**) Cd, $c/a = 1.89$

Fig. 1.67. Structure of $AgBiTe_2$ ▷

◁
Fig. 1.68.
Cubic body-centered structure

$c = 20,68 \overset{\circ}{A}$

Bi
Te
Ag

Closest packings are also realized in structures made up of different atoms, provided these structures are homodesmic and the atomic radii are approximately equal. A good example is the structure of alloys, such as Cu_3Au (see Fig. 2.9b) and semiconductor alloys containing Bi, Sb, S, Te, Ge, and Ag with coordination 12, all of whose atoms have radii of $1.6 - 1.8$ Å. These alloys form multilayer packings, sometimes with a statistical population of more than one position; an example of such a compound is presented in Fig. 1.67.

Body-centered cubic packing (bcc) is common in the structures of elements and a number of compounds (Fig. 1.68). If we regard it formally as a packing of equal spheres, the packing factor will be equal to 68.01%; it is less than the factor of 74.05% for closest packings, but is still large. Transitions between such two types of cubic structures (for instance, α-γ-δ-Fe: bcc-fcc-bcc) are very interesting. From the geometric standpoint one can say that the stability of bcc structures is due to the fact that the principle of high symmetry of the atoms is well maintained, and that their coordination (8 + 6) is very high. But a comprehensive explanation of this type of structure can be achieved only by taking into account all the features of atomic interaction and lattice dynamics.

1.5.11 Structures of Compounds Based on Close Packing of Spheres

Between closest-packed spheres there are interstices, or holes of two types. Having covered the structure of the densest layer with the next one (Fig. 1.62), we shall now see that holes of the lower layer which are covered by spheres of the upper are surrounded by four spheres, the centers of which form a tetra-

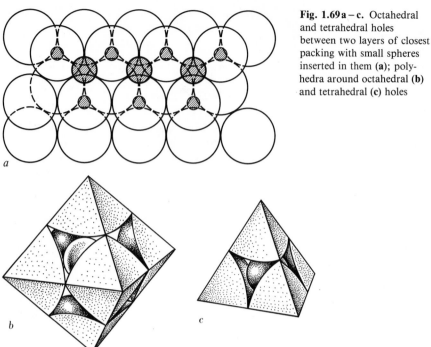

Fig. 1.69 a – c. Octahedral and tetrahedral holes between two layers of closest packing with small spheres inserted in them (**a**); polyhedra around octahedral (**b**) and tetrahedral (**c**) holes

hedron (the hole a is covered by a sphere A, or b is covered by B, and c, by C). They are called tetrahedral holes (Fig. 1.69a). There are three spheres above and below holes of the second type; they are octahedral [a has nonmatching spheres B and C above and below, and so on (Fig. 1.69a)]. In any three-dimensional closest packing the number of tetrahedral holes is twice that of the spheres, while the number of octahedral holes is equal to that of the spheres.

The holes could be filled with smaller spheres contacting the principal spheres of radius R. The radius of such a small sphere for a tetrahedral hole is equal to $0.225\,R$, and for an octahedral hole $0.415\,R$ (Fig. 1.69b, c). Using the ionic radii systems we can consider the geometric possibilities of "populating" the holes in an anion packing with cations. These possibilities depend on the formal values of g or g' according to Table 1.12, which correspond to octahedral and tetrahedral coordinations.

However, in the vast majority of crystals built according to this scheme the "close-packed" spheres of anions no longer contact each other, since the cations "spread out" the packing irrespective of the fact whether we regard it from the standpoint of the system of effective or physical ionic radii or, all the more so, from the standpoint of the atomic-ionic radii of strong bonds. It nevertheless turns out that a number of structures can be described with the aid of such models. It can, therefore, be said that the system of anion – cation bonds, the presence of covalent interactions directed along a tetrahedron or an octahedron, and the mutual repulsion of like-charged ions result in an arrangement where the centers of atoms, in particular those of anions, occupy positions corresponding to the closest packing of spheres (but of a larger diameter). In other words, the positions corresponding to a close packing are energetically favorable. This can no longer be explained from the standpoint of the geometric theory of closest packings, because the main condition, i.e., the mutual contact of the principal spheres, is violated. On the other hand, the principles of maximum filling and symmetry are nearly always maintained.

Proceeding from the versions of closest packings of anions and from the different possibilities of populating voids of one type or another with cations, we can describe, following *Belov* [1.55], various structures. Such a description is equivalent to representing these structures as combinations of vacant or filled octahedra and tetrahedra connected in various ways.

For instance, populating all the octahedral holes in the closest cubic packing produces the structural type NaCl, and populating all of them in hexagonal packing, the NiAs type (Fig. 1.70, cf. Fig. 1.57). Note that the centers of the atoms residing in octahedral holes are arranged according to the close-packing law, and that the entire structure can just as well be regarded as one having the holes populated by anions in a cation packing. The centers of both sorts of atoms form identical lattices, but are shifted relative to one another.

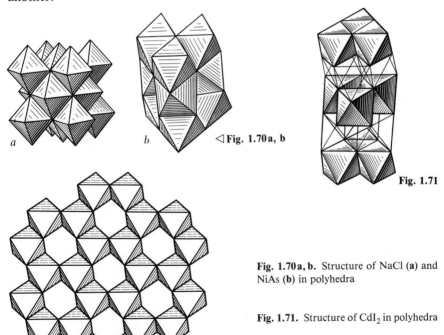

◁ Fig. 1.70 a, b

Fig. 1.71

Fig. 1.70 a, b. Structure of NaCl (a) and NiAs (b) in polyhedra

Fig. 1.71. Structure of CdI$_2$ in polyhedra

◁ **Fig. 1.72.** One layer of Al$_2$O$_3$ structure

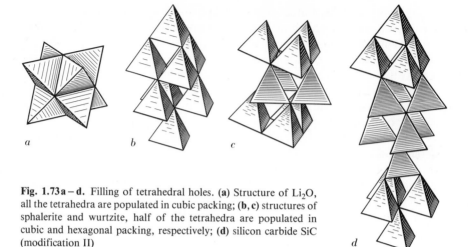

Fig. 1.73a–d. Filling of tetrahedral holes. **(a)** Structure of Li$_2$O, all the tetrahedra are populated in cubic packing; **(b, c)** structures of sphalerite and wurtzite, half of the tetrahedra are populated in cubic and hexagonal packing, respectively; **(d)** silicon carbide SiC (modification II)

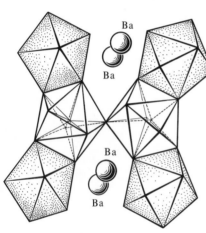

◁
Fig. 1.74. Eight-vertex polyhedra around Lu atoms in the structure of BaLu$_2$F$_8$, Ba is surrounded by 11 or 12 atoms of F

Fig. 1.75a, b. Structures with the cube as a coordination polyhedron. **(a)** Cube in CsCl, **(b)** packing of cubes in CaF$_2$. Structure of CsCl – solid packing of cubes, and of CaF$_2$ – chess-like packing
▽

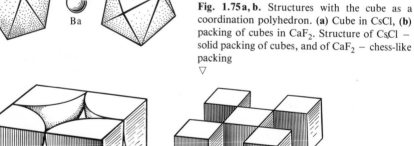

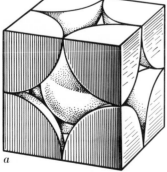

If half of the octahedral holes in a hexagonal packing are occupied, we obtain type CdI_2 (Fig. 1.71); the filling of two-thirds of the holes according to the corundum motif results in Al_2O_3 type (Fig. 1.72). The filling of all the tetrahedral holes in the closest cubic packing yields the Li_2O type of structure (Fig. 1.73a).

As indicated above, by using the formal scheme for the accommodation of atoms and holes in close packing the structures can be described by directional, tetrahedral or octahedral, covalent bonds. Thus, filling half of the tetrahedral holes in cubic packing we obtain the structure of sphalerite, ZnS (Fig. 1.73b), and in a hexagonal, wurtzite (Fig. 1.73c). Various multilayer modifications of SiC, in which tetrahedral nets alternate in different sequences of h and c, are built similarly (Fig. 1.73d).

If the $d(AB)/r_{an}$ ratio increases, then, according to Table 1.12, the coordination number must also increase, and the structures cannot be fitted into the close-packing scheme. High coordination numbers 7, 8, and 11 are indeed observed for the large cations of Ca, Sr, Ba, etc. The coordination polyhedra may be very complex for them (Fig. 1.74). Classical examples of regular polyhedra – cubes – are represented by the structures CsCl and CaF_2 (Fig. 1.75a, b); in the latter, half of the cubes are vacant.

1.5.12 Insular, Chain and Layer Structures

If the bonds between the atoms in a structure are of distinct types (heterodesmic structures), such structures are coordination unequal, and it is possible to single out groupings of atoms with $m = 0, 1, 2$, which are bound together by short (strong) bonds. The atoms of such distinct groupings are bound by weak (long) bonds. The configurations of the atoms within a grouping depend on the mutual arrangement of the groupings as a whole, i. e., the groupings are stable units of the structure. Since strong bonds are satured within the groupings, the mutual packing of groupings as stable units is largely defined by weak forces.

The insular structures ($k = 0$) with finite "zero-dimensional" groupings of atoms "islands", include all the molecular compounds except high polymers (Fig. 1.76), structures containing finite complexes of metals with inorganic or organic ligands (Fig. 1.77), and some others. These structures will be considered in detail in Sect. 2.5, 6.

Sometimes it is difficult to draw a clear-cut boundary between zero-dimensional and equal-coordination structures. For instance, such stable groupings as NH_4 and SO_4 are finite, but they are not separated from the other surrounding atoms by large distances.

Structures with one-dimensional groupings ($k = 1$) are called chain structures. Typical examples are crystalline polymers built up of infinitely long molecules, and the inorganic structures such as selenium (Fig. 1.44d), semiconductor-ferroelectric SbSI (Fig. 1.78), the complex compound $PdCl_2$ (Fig. 1.79), and chain silicates of the asbestos type.

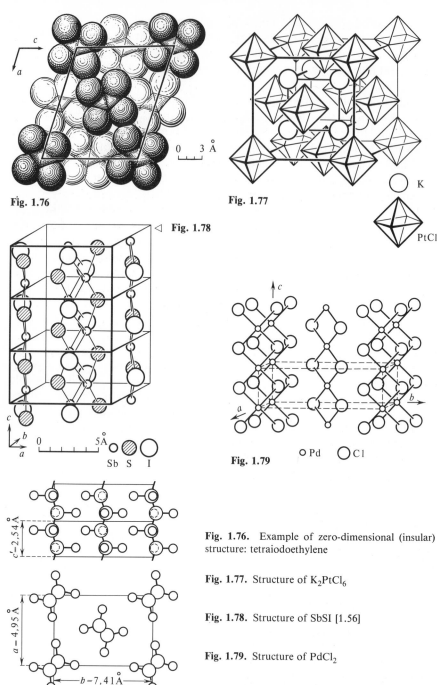

Fig. 1.76

Fig. 1.77

◁ **Fig. 1.78**

Fig. 1.79

○ Pd   ◯ Cl

○ K

PtCl$_6$

Sb S I

○ C   ◯ H

Fig. 1.76. Example of zero-dimensional (insular) structure: tetraiodoethylene

Fig. 1.77. Structure of K_2PtCl_6

Fig. 1.78. Structure of SbSI [1.56]

Fig. 1.79. Structure of $PdCl_2$

◁ **Fig. 1.80.** Structure of paraffin C_nH_{2n+2} in two projections

The chain itself is described by one of the symmetry groups G_1^3. In a crystal this symmetry can only decrease, according to the same principles as for insular structures. An example is the structure of paraffins C_nH_{2n+2} (Fig. 1.80). The symmetry of the chains themselves is mmc, but when they are packed in a lattice, only the "horizontal" planes m and axes 2 remain, and "packing" symmetry elements n and a arise; the space group of the structure is $Pnam$. Such a rigid chain retains its configuration in the crystal. Another version is also possible, where the bonds between some rigid chain units, or links, allow different mutual orientations of these elements with the preservation of chain continuity. In this case the conformation of the chains when packed in a crystal structure is, to a large extent, due to the weak bonds between the chains, which are arranged parallel to each other in chain structures.

Layer structures ($k = 2$) have groupings of atoms bounded by strong bonds; these groupings extend infinitely in two dimensions. Good representatives of such structures are graphite (Fig. 2.5), layer structures of the CdI_2 type (Fig. 1.71), and layer silicates (Fig. 1.81), which can also be regarded as a two-dimensional system of coupled stable coordination polyhedra. Structures which are composed of thick "multistory" layers are sometimes called packets. The stacking of layers according to the close-contact principle can sometimes be done in ways, which differ very little energywise, and therefore various modifications of such structures often arise. For instance, CdI_2 has many modifications, and there are numerous modifications of various clay minerals, differing in the stacking of the packets or in some features of the structure of the packets themselves.

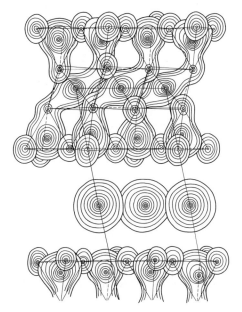

Fig. 1.81. One layer of the mica structure; three-dimensional Fourier synthesis of the lattice potential [1.57]

Concluding this section we emphasize once again that the allocation of some structure to the coordination-equal, the chain, or the layer type sometimes cannot be made rigorously enough. In coordination-equal structures it is often possible to single out chains or layers which explicitly play an independent structural role, but are in close contact with other atoms of the structure, etc. Some structures are made up of units of two types, for instance, finite complexes and layers or chains, and so on.

1.6 Solid Solutions and Isomorphism

1.6.1 Isostructural Crystals

Many crystals have an identical atomic structure, i.e., are isostructural. This means that their space groups are identical and the atoms are located over the same regular point systems (RPS); for instance, atoms A of one structure and atoms A' of another occupy the same RPS, B and B' occupy another, and so on. It is obvious that isostructural substances are also "isoformular", i.e., their formulae are identical as regards the number of the corresponding atoms. Isostructural crystals may be of different complexity, beginning with simple substances and ending with complicated compounds. For instance, face-centered cubic metals and crystals of inert elements are isostructural. Alkali – halide compounds of the type NaCl (Fig. 1.46), a number of oxides such as MgO, and many alloys such as TiN may serve as other examples. There are large series of isostructural compounds with formulae AB_2, AB_3, ABX_2, etc. Each isostructural series is named after one of its most common (or first-discovered) representatives, for instance, the structural types α-Fe, NaCl, CsCl, K_2PtCl_4, etc.

The concept of isostructural crystals is formally geometric. The same structural type may cover crystals with different types of bonds, for instance, ionic and metallic. Yet geometric similarity indicates that the symmetry of the binding forces must be the same; for example, these forces, referring to corresponding atoms, must be, for instance, spherically symmetric or identically oriented.

1.6.2 Isomorphism

If crystals are isostructural and have the same type of bond, they are called isomorphous. The parameters of the unit cells of such crystals are close. Their similarity is also manifested macroscopically: owing to the above reasons their external forms are very similar (hence the term isomorphism). This is exactly how isomorphism was discovered at the macroscopic level by E. Mitcherlich in 1819 when observing crystals of KH_2PO_4, KH_2AsO_4, and $NH_4H_2PO_4$. Goniometric measurements showed that these tetragonal crystals have identical simple forms, and that the angles between the corresponding faces are similar. Another classical isomorphous group is rhombohedral carbonates MCO_3, M = Ca, Cd, Mg, Zn, Fe, or Mn, for which the vertex angles of the rhombohedrons differ by not more than $1° - 2°$. The physical properties of isomorphous substances are similar as well. The present-day investigations of isomorphism are mainly based on x-ray and other diffraction data, which have made it posible to elaborate and extend the initial concepts of isomorphism.

1.6.3 Substitutional Solid Solutions

X-ray studies have revealed a close relationship between isomorphism and solubility in the solid state – the phenomena referring to the formation of so-called solid solutions. Isomorphism of crystals is often associated with the possibility of the formation of a series of homogeneous solid solutions of iso-morphous substances with a phase diagram of the type shown in Fig. 1.82a.

The most common type of solid solution is a substitutional solution, in which atoms of one component substitute for atoms of another in a given regular point system (Fig. 1.83). The probability of finding a substituting atom at these points is a constant value depending on the composition of the solid solution. If, for instance, atoms of sort A in phase AB can be replaced

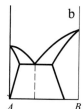
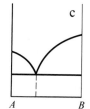

Fig. 1.82a – c. Various types of phase diagrams.
(a) The components form continuous solid solutions; (b) limited solubility case; (c) the components do not form solid solutions

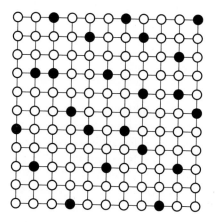

Fig. 1.83. Structure of a substitutional solid solution. Only one regular point system according to which substitution occurs is shown conventionally, there may be other regular point systems according to which no substitution occurs

by atoms of sort A', whose atomic fraction is equal to x, then such a solid solution will be described by the formula $A_{1-x}A'_xB$. In this case the probability of detecting the atom A at a point belonging to a given RPS (or in a given sublattice, as it is sometimes called) is equal to $1 - x$ and that of detecting the atom A', x.

Consequently crystals with a formula $A_{1-x}A'_xB$ $(0 < x < 1)$ for any x are isostructural to each other and with the extreme terms of the series, i.e., compounds AB and $A'B$ (Figs. 1.84, and 2.9a).

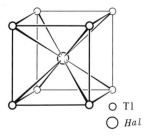

O Tl

O Hal

Fig. 1.84. Unit cell of a solid solution $TlHal'_x Hal''_{1-x}$. At the center of the cell the halogen atoms may replace each other statistically in any concentration

Thus, the term "isomorphism" is actually used to denote two similar but not quite identical concepts. The first implies similarity of the structure and the shape of crystals of different (but related) chemical compositions, and the second, mutual substitution of atoms or other structural units in crystalline phases of variable composition.

Investigations into the atomic structure of isomorphous substances and their solid solutions have helped to establish not only the geometric similarity of their structures, but also the geometric restrictions on the sizes of the atoms, which can replace one another. For instance, KBr and LiCl are isostructural and have an identical type of bond, but they do not form a series of homogeneous solid solutions. This is due to the significant difference in the size of their ions. An analysis carried out by V. M. Goldschmidt and corroborated by numerous subsequent investigations showed that in ionic compounds the radii of ions substituting for each other usually do not differ by more than $10\% - 15\%$; about the same range of difference between atomic radii exists in the isomorphous structures of covalent and metallic compounds. Therefore, the sizes of the unit cells of isomorphous substances, the interatomic distances in them, and the coordinates of generally situated atoms differ but slightly.

The dimensions of the unit cells of solid solutions of isomorphous substances are approximately linearly dependent on the concentration; this is Vegard's law (Fig. 1.85). Often deviations from this law are observed. The curve of the concentration dependence of the periods may slightly buckle upwards or downwards, or have an S shape. Since the atoms of the solute and the solvent generally differ in size, the introduction of impurity atoms into the solvent lattice results in two effects: a macroscopically uniform strain of the

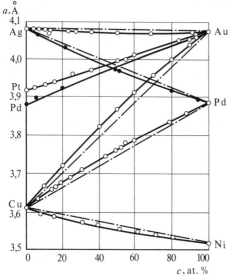

Fig. 1.85. Vegard's law: the periods of unit cells *a* of solid solutions of isomorphous substances are linearly dependent on the concentration *c*

crystal lattice of the solvent and local displacements due to each impurity atom.

Thus, the unit cell parameters of solid solutions, observed by x-ray diffraction, and the interatomic distances are averaged over all the cells of the crystal. Uniform strain results in a concentration dependence of the periods of the lattice of a solid solution, which is observed by the x-ray method. Because of the local displacements produced by solute atoms, each cell is distorted to some extent; its size and shape vary depending on the substituting atoms and the neighboring cells (Fig. 1.86). The local displacements are responsible for the weakening of the intensity of the x-ray reflections and the appearance of diffuse scattering concentrated near them. On the average, however, the long-range order is preserved despite local distortions; all the atoms deviate statistically from certain mean positions, which correspond to the ideal three-

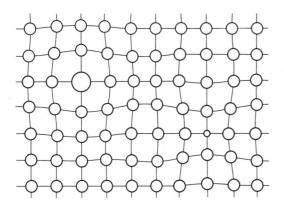

Fig. 1.86. Local deformation of the lattice around impurity atoms of larger or smaller radius

dimensional periodicity with averaged-out periods. Thus, the periodicity is also statistical, but on the average it is accurately maintained. Naturally, deviations of atoms from the ideal positions must be proportional to the difference Δr between the radii of the atoms substituting for each other and must depend on their concentration. When the radii differ by $5\% - 10\%$, Δr are equal to $0.1 - 0.2$ Å. The x-ray determinations of the r.m.s. displacement of atoms from averaged-out ideal positions $\sqrt{\overline{u^2}}$ are of the order of 0.1 Å. Larger values of Δr and $\sqrt{\overline{u^2}}$ evidently themselves lead to lattice instability and prevent the formation of a homogeneous solid solution; a separation of components occurs, which results in new phase formation.

It should be mentioned that the similarity of atomic sizes (with the same type of bond) does not always ensure isomorphism by itself. Moreover, a pair of atoms which substitute for each other isomorphously in one structural type, may not do so in another. This is not surprising because isomorphism is a property of structures as a whole, and not of the individual atoms themselves. If the structure has a complicated chemical formula and a large cell, the requirements regarding the differences in sizes of the atoms substituting for each other are slightly milder, since there are greater opportunities for preserving the same equilibrium of interatomic forces by small shifts of other atoms of the cell. Atoms replacing each other isomorphously must also be similar in the type of the bonds they form, which can be characterized by the ionicity fraction ε (1.20) [1.58, 59]. The stability of an isomorphous mixture with respect to decomposition is determined by the value of the interchange energy u_{int}; the higher this energy, the higher is the decomposition temperature, and the smaller the substitution limits at a given temperature. In turn, u_{int} depends on the squares of the differences in interatomic distances, i.e., the differences in the radii of the ions replacing each other, Δr, and the differences in the degrees of ionicity of the bond of the components, $\Delta \varepsilon$, $u_{\text{int}} \sim a(\Delta r)^2 + b(\Delta \varepsilon)^2$. Because of the large difference $\Delta \varepsilon$, pairs of elements such as Na and Cu^I, and Ca and Hg do not replace one another in practice, despite the similarity of their ionic radii.

The phenomenon of substitution of some atoms for others in the structure of crystals is not necessarily associated with the fact that both components, when mixed, are isostructural and produce a continuous series of homogeneous solid solutions. Limited solubility, when the components are nonisomorphous or nonisostructural, is still more common. The relevant phase diagram is presented in Fig. 1.82b. Here, too, the statistical replacement of atoms of one component by those of another is called isomorphous substitution (although, as indicated above, the components mixed are themselves no longer necessarily truly isomorphous); it proceeds according to the same scheme as for true isomorphism (Fig. 1.9), except that the variable-composition compound, for instance, $A_{n-x}A'_x B_m$, exists at limited, rather than at any arbitrary x, so that $V_{\text{max}} < n$. Naturally, x_{max} depends on the thermodynamic parameters, i.e., the temperature and pressure. If the components are mutually insoluble, we obtain a phase diagram of the type shown in Fig. 1.82 c. The requirements concerning the possibility of the isomorphous replacement of atoms which are not structure isomorphous are less stringent, although the limitation $\Delta r < 15\%$ remains.

Different versions of isomorphism and isomorphous substitutions are possibile according to the number of components and the complexity of the chemical formulae of the phases forming a given phase diagram (or diagram section). Isomorphous substitutions or complete isomorphism occur for one, two, or several sorts of similar atoms, i. e., occupancy of one RPS by similar atoms of A and A', of another by B and B', of a third by C and C', etc. On the other hand, atoms not of two, but of three or more elements A, A', A'' ... may be arranged in a single RPS.

Atoms replacing each other often have an identical valency; this is isovalent isomorphism. Heterovalent isomorphism is also possible, when the solute atoms replacing the solvent ones have a different valence. It is then necessary that the lattice as a whole be neutral, i. e., that the valence (charge) be compensated. For instance, divalent ions of A can be replaced by a set of a univalent A' and a trivalent A'' ion, provided, of course, that the condition of allowance for the radii and a similar nature of the bond is preserved. Vacancies may also play the role of charge compensators; then, for instance, a univalent ion can be replaced by a divalent or a trivalent one with charge compensation by the corresponding number of vacancies. In covalent semiconductor structures of Ge and Si, which belong to group IV of Mendeleyev's system, host atoms can be replaced by corresponding numbers of atoms of groups III and V; on the other hand, if we introduce atoms only of group III or only of group V, electrons or holes will become compensators, which will yield n or p types of semiconductor crystals.

Compensation may occur not only with respect to atoms occupying the same regular point system, but also with respect to different systems. For instance, $Fe^{2+}(CO_3)^{2-}$ and $Sc^{3+}(BO_3)^{3-}$ are isomorphous.

The required approximate equality of the radii restricts the possibilities of isovalent substitution of atoms along the columns (groups) of Mendeleyev's system. This requirement is better fulfilled by diagonal shift, which explains the geochemically and mineralogically important heterovalent isomorphism along Fersman's diagonal rows: $Be - Al - Ti - Nb$, $Li - Mg - Sc$, $Na - Ca - Y$ (P 3), and Th (Zr).

Isomorphous substitutions are very significant in natural and synthetic crystalline substances. Many elements are contained in mineral ores precisely as isomorphous impurities − for instance, rare-earth elements − in silicates, where they replace Ca; Co or Ni often replace Fe in iron-containing minerals. There are many such examples.

The principle concerning the introduction of certain atoms in a lattice (matrix) and thus changing its properties is essential to the production of many technically important crystals and materials, beginning with high-strength alloys and ending with crystals for quantum electronics and semiconductor technology. Thus, in laser crystals a small part of host cations is replaced by an active ion; for instance, ruby consists of Al_2O_3, in which 0.05% of Al atoms are replaced by Cr; in yttrium-aluminium garnet, $Y_3Al_5O_{12}$, up to 1.5% of Y is replaced by the rare-erath element Nd. In fact, practically all crystals contain some dissolved atoms (although in small amounts).

As a specific variant of substitutional solid solutions one can consider *subtractional solid solutions*, where an atom of one of the components does not fill the possible positions completely. An example is $A_{1-x}A_y'B$, where $y < x$ or $y = 0$. Here the solution has a formula $A_{1-x}B$. The unoccupied positions can be regarded as "vacancy-populated" $A*$, and the formula of the compound can be written as $A_{1-x}A_y'A_{x-y}^*B$.

Elements are mixtures of isotopes. Taking this feature of atoms into consideration, the crystals of almost all substances can be regarded as isotopic substitutional solid solutions. Whenever it is possible to single out monoisotopic isomorphous crystals, one can detect slight differences between them. The lattice parameters of hydrides and deuterides (and also of ordinary and heavy ice) differ but slightly, by hundredths of an angstrom. When hydrogen bonds are present, their lengths, thermal vibration parameters, etc., change slightly.

Substitutional solid solutions are also observed in organic crystals, in which the molecule, rather than the atom, must be regarded as the structural unit of the crystal. The requirement regarding the approximate equality of the size and shape of the substituted molecules should be fulfilled in this case, too.

For example, the structure of anthracene [anthracene structure] exhibits limited

solubility of some of its chloroderivatives or of phenanthrene [phenanthrene structure]

The molecules of naphthalene [naphthalene structure] also replace those of anthracene

in its lattice, and the part of space which accommodated the third benzene ring remains unoccupied. One large impurity molecule sometimes can replace two matrix molecules under suitable geometric conditions.

1.6.4 Interstitial Solid Solutions

Isomorphous substitution within the structure of a solid solution is due to the possibility of complete or partial statistical replacement of atoms A by atoms A' within a given regular point system (Fig. 1.83). Another basic type of solid solution is an interstitial solid solution. Its phase diagram is the same as for substitutional solutions with limited solubility (Fig. 1.82b), but the "dissolving" atoms enter the interstices between the host atoms, statistically populating a new, previously unoccupied regular point system of the space

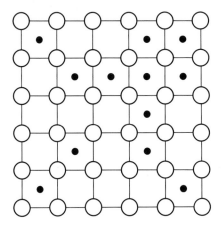

Fig. 1.87. Idealized scheme of the structure of interstitial solid solutions

group describing the matrix crystals (Fig. 1.87). The set of RPS occupied by atoms of the host lattice is often called crystal lattice points; it is then said that the atoms entering the space between them enter the interstices. Sometimes transition to another space group, a subgroup of the first, is also possible. Then we can no longer speak of a new compound as being isostructural with the initial one, although the point symmetry group and the simple forms of the single crystal (if it is obtained) may be preserved. Interstitial-type solubility is usually low, being normally just a few per cent and only rarely reaching 10%. Classical examples of interstitial solid solutions are austenite (a solid solution of carbon in γ-iron), where C atoms settle statistically in the octahedral voids of the cubic face-centered structure of the γ-Fe (Fig. 1.88), and the structures of many carbides, nitrides, borides, hydrides of metal, etc., which are called interstitial phases. Since the atoms of a solute enter the spaces between the matrix atoms (often octahedral or tetrahedral holes of closest packings), their sizes should obviously be close to those of the interstices, and not of the host atoms, as in the case isomorphous substitutions, i.e., they must be smaller than the host atoms. It is therefore easy to understand why interstitial solid solutions are often formed by such atoms as H, B, N, and C, which have small radii. If the size of the interstitial atom does not correspond to that of the interstices, the region of the matrix lattice around them may be greatly distorted. Indications have been found that H molecules enter lattice interstitials in metals, such as Zr and W.

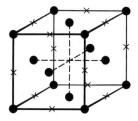

Fig. 1.88. Austenite structure. Crosses represent the positions of the C atoms being incorporated into the γ-Fe lattice

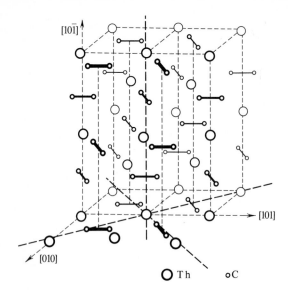

Fig. 1.89. Structure of ThC$_2$

[10$\bar{1}$]

[101]

[010]

○ Th   ○ C

It should be noted, that owing to the metallic nature of the bond in alloys and many interstitial phases, charge redistribution occurs cooperatively throughout the volume of the crystal. The interstices may be filled, not only in accordance with a phase diagram with limited solubility (Fig. 1.82b), but also through attaining the stoichiometric composition, when all the holes of this type are filled. Then a new structure with a definite formula is formed, for instance AB_x, at $x = n$, where n is an integer (Fig. 1.89). Thus the structure AB_x, at $x < n$, can be interpreted in two ways: either as an interstitial solution of the atoms of B in structure A, or as a defect structure AB_n, in which some of the possible positions of the atoms of $B(n-x)$ are vacant (a subtraction solution).

In is also known that, within significant concentration ranges, series of homogeneous solid solutions exist between substances with different formulae (with charge compensation, if the atoms contained in the lattice have different valence). A classic example is the system $CaF_2 - YF_3$ (Fig. 1.90), with YF_3 solubility up to 40%. This system is also of interest as one of the matrices,

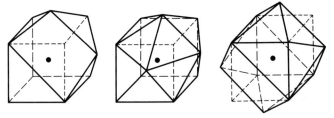

Fig. 1.90. Possible coordination polyhedra in a solid solution of $CaF_2 - Y^{3+}F_3$; the dashed line shows the initial polyhedron (cube) [1.60]

i.e., crystalline media for optical quantum generators, where Ca^{2+} ($r = 0.97$ Å) or Y^{3+} ($r = 1.04$ Å) are replaced by one another or by other ions of rare-earth elements TR^{3+}, for instance Ce, which are approximately identical in radius. Excessive positive valences are compensated by an increase in the number of F^- ions. One anion F^- of the initial structure, which is situated at a vertex of the cube (Fig. 1.75b), is supplemented by another F^- per each TR^{3+} ion; the indicated pair of F^- ions settles in the vacant cubes of the fluorite structure, being shifted in them by 1/6 of the body diagonal. The coordination polyhedron around the metallic atom transforms from a cube into a more complicated figure (Fig. 1.90).

Thus, there are many different ways in which impurity atoms can enter the crystal lattice, beginning with isomorphous substitutions in a series of homogeneous solid solutions and ending with incorporation of atoms into new positions with a change in the structure as a whole. Accordingly, the space symmetry of the crystal structure may be retained or changed (reduced in most cases); the decrease in symmetry may also extend to the point group, and transitions to the lower syngonies are also possible.

It is noteworthy that the process of the formation of substitutional or interstitial solid solutions requires a description of not only an ideal crystal structure but also of a real one, with an indication of the distortions in the ideal lattice, the way the atoms enter it, etc. Different types of incorporation of impurity atoms into the host lattice can be observed. Indeed, the crystallochemical similarity of atoms replacing one another and the thermal motion are the factors permitting the formation of substitutional solid solutions. But the crystallochemical similarity is merely a similarity, not an identity.

The atomic interaction in solid solutions excludes a completely random distribution of atoms of sorts A and A' in the substitutional "sublattice" [1.61]. If atoms of sorts A and A' attract each other, then there is an "atmosphere" around each atom of sort A due to the preferential distribution of atoms of sort A'. This phenomenon is called the *short-range order*. In the opposite case, when the atoms of sorts A and A' repel each other, an atmosphere of the same kind of atoms is formed around the atoms of sort A. This phenomenon is usually called *short-range decomposition*. But in both cases we deal with a correlation in the mutual arrangement of atoms A and A'. The formation of such linear, two- or three-dimensional associations of related atoms ("block isomorphism") is also possible; they can interact with other lattice defects, such as vacancies and dislocations.

1.6.5 Modulated and Incommensurate Structures

In two-phase systems arising with the decomposition of metallic or other solid solutions, the segregations of the new phase are sometimes arranged regularly, rather than randomly with a certain periodicity, as is evidenced by electron-microscopy, x-ray, and electron diffraction data. The shape, orientation, and

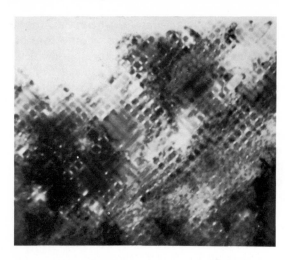

Fig. 1.91. Electron-microscopic image of the two-dimensional modulated structure of a ticonal-type alloy, (001) plane, × 32,000 [1.62]

periodicity of these segregations depend on the initial structure and its orientation relationships with new phase. Such periodic distributions of inclusions are called modulated structures. The periodicity can be observed along one, two, or all three directions. Modulated structures (Fig. 1.91) are observed, for instance, in Au – Pt, Al – Ni, Cu – Ni – Fe, and many other alloys.

The consistent periodic arrangement of the precipitations of the new phase is ascribed to the fact that a substantial energy contribution to such a system is the strain-induced energy due to the mismatch of the crystal lattice of the two phases. The total energy minimum just corresponds to the modulated structure, whose formation results in the dissapearance of the long-range fields of elastic stresses.

The most general case is the so-called incommensurate structures in which the distribution of some structural parameter, say, charge, magnetic moment, or deformation, is periodic, (in one, two, or three dimensions), but the period A_i is not multiple to the integral number of crystal lattice periods a_i, i.e., $A_i \neq k a_i$. In particular, such structures are observed in ferroelectrics and ferroelastics. The periods A_i depend on temperature and other external conditions. At a certain temperature a transition to the commensurate structure ($A_i = k a_i$) can take place.

1.6.6 Composite Ultrastructures

In recent years it has been found, due to the use of high-resolution electron microscopy (see Sect. 4.10 in [1.6]), that many structures with a macroscopically nonstoichiometric formula are actually a combination of parts of related structures having stoichiometric formula. A classic example are the oxides of transition metals $M_n O_{3n-m}$ (tungsten, molybdenum, niobium, etc.)

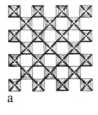

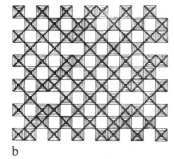

a

b

Fig. 1.92a, b. Ideal structure of the WO$_3$ type made up of octahedra (a) and a structure with *CS* along the {103} plane (b)

and the intermediate structures resulting from their oxidation or reduction [1.63 – 65]. The stable elements of these structures are columns of MO$_6$ octahedra which share the vertices in the ideal case of MO$_3$. Contact by the edges produces the so-called crystallographic shear (CS) (Fig. 1.92). CS along {102} yields the structures M$_n$O$_{3n-1}$, and along {103} M$_n$O$_{3n-2}$, etc. Thus, the coexistence of these structures in the same crystal results in a macroscopically nonstoichiometric formula of the type M$_n$O$_{3n-x}$ (Fig. 1.93. Figure 4.106 in [1.6] may serve as another example.) Coexistence of blocks of different composition in a crystal is known for silicates (Fig. 1.94) and for many other compounds. In all these cases one observes a continuous crystal structure, but one

Fig. 1.93. Electron micrograph of a crystal of overall composition Ti$_{0.03}$WO$_3$ containing mostly [103] CS planes. Short segment of [102] GS plane is arrowed [1.65]

Fig. 1.94. Electron micrograph of three-row Na – Co silicate [theoretical formula $Co_4Na_2Si_6O_{16}(OH)_2$]. Besides the regular structure consisting of 3 silicon – oxygen chains, one can see insertions consisting of 4 or 5 chains. On the schematic representation of the structure the SiO_4^{-2} tetrahedra are denoted by triangles [1.66]

can no longer speak of the same lattice throughout the entire volume of the crystal; there are different, although related, lattices alternating at the micro-level and forming a macroscopically unified crystalline aggregate. If this alternation is uniform and regular throughout the whole volume, one can speak of a superlattice. If it is irregular, there is no long-range order in the whole volume; it exists only in individual blocks. Ultrastructures may natural-ly show various structural defects (Chap. 4).

2. Principal Types of Crystal Structures

At present several tens of thousands of crystal structures are known. They can be classified according to definite features of their enormous diversity. First of all, we shall consider the structures of elements in which different types of bonds are encountered, since these structures most prominently display the crystallochemical properties of the atoms of a given element; these properties are very often inherited by the structures formed by compounds of these atoms as well.

There are many principles for classifying the structures of compounds: by the ratios of components in the formula (AB, AB_2 etc.), by the structural types, by the iso- or heterodesmicity of the bonds, by the dimensionality of the structural groupings, etc. We shall adhere to the most widespread classification, namely that according to the types of chemical bonds. We shall consider structures with a metallic, ionic, and covalent type of bond and, finally, molecular structures with van der Waals binding forces between molecules. Owing to the enormous diversity and specificity of definite classes of molecules, we shall single out, in the molecular structures, common organic compounds, then high-molecular compounds – polymers, liquid crystals, and, finally, biological structures.

2.1 Crystal Structures of Elements

2.1.1 Principal Types of Structures of Elements

The basic features of the crystal structures of elements depend on their position in Mendeleyev's table and are of periodic character as all their basic properties. The structures are schematically represented in Fig. 2.1 in accordance with their position in the table.

They can be divided into two large groups: metal and nonmetal structures. The structure of typical metals is determined by the metallic nature of the bond and its nondirectional character, therefore, these structures are based on close atomic packings. As we move right- and downwards along the periodic table, the fraction of covalent effects in the interatomic bonds increases

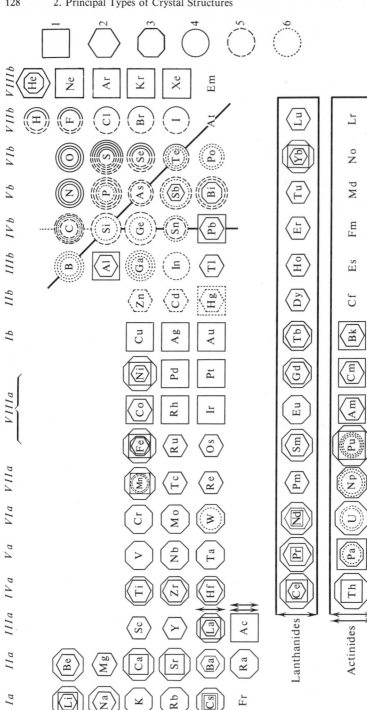

Fig. 2.1. Structures of the elements arranged according to the periodic system. The structures are denoted by the following symbols: (*1*) cubic close packing; (*2*) hexagonal close packing; (*3*) body-centered-cubic packing; (*4*) molecular structure; (*5*) structure with coordination number 8 − *N*; (*6*) other structures. If a substance has several modifications, the transition from the external to the internal symbol corresponds to a transition from a high- to a low-temperature modification and, then, to the modification existing at high pressures. Some modifications which are less common or have not been investigated enough are not presented

gradually, the structures of the metals becoming more complicated and their bonds acquiring a directional character.

On the right side of Mendeleyev's table, beginning with B, Si, Ge, and Sn, we have typical covalent structures. Then there are structures in which covalently bonded atoms form the groupings linked by van der Waals bonds. Finally, the atoms of inert elements have no strong bonds between them and are packed owing to the van der Waals forces. Thus, in addition to the typical metals, it is possible to isolate in the periodic system covalent-metallic structures, typically covalent structures, the molecular heterodesmic structures, and structures of inert elements.

Many elements have several polymorphous modifications. The phase diagrams of some of them are given in Fig. 2.2 [2.1]. When pressure increases, the main trend is a transition to more densely packed structures, and at very high pressure metallization occurs. An increase in temperature, which loosens the orientation of the bonds, usually increases the symmetry.

Let us now consider some typical structures of elements. Hydrogen forms H_2 molecules. In the crystal structure, which exists at a normal pressure below 3 K, the centers of the molecules are arranged according to the close cubic packing, and on heating, to close hexagonal packing. Here we evidently have free or restricted thermal rotation of molecules, respectively; their proper cylindrical symmetry may affect the "choice" of hexagonal packing.

Helium, which forms crystals only under pressure, has a hexagonal close-packed structure, but body-centered cubic and cubic close-packed modifications of ^4He and ^3He were also observed at definite $p - T$ conditions (Fig. 2.2a). In helium crystals quantum phenomena are observed; defects of the lattice in these quantum crystals must be interpreded as "smeared out" over a volume. The crystal structure of other solidified noble gases is the closest cubic packing, with van der Waals interactions between atoms.

At very high pressures (about 1 Mbar), solid Xe becomes metallized and shows superconductivity. Crystals of the elements of subgroup Ia, alkali metals, open up the series of typically metallic structures. The structure of these, the alkaline earth group IIa, and many other metals belong to one of the three types of packing: closest cubic, closest hexagonal (packing coefficient 74%; 13% of all the structures of elements belong to each of them), or body-centered cubic packing, which also exhibits a high density (packing coefficient 68%) (Figs. 1.44a, b, 63a). The common cause for the appearance of these types of structures lies in the nondirectional character or weak orientation of the metallic bond. It is, however, very difficult to explain which of these three types of structure will be realized for an element under given thermodynamic conditions because their energies are very close and depend on fine details of their electron and phonon spectra.

A classical example of polymorphism is iron (Fig. 2.2c); $\alpha -$Fe, the bcc lattice with a ferromagnetic spin ordering transforms to $\beta -$Fe at 770°C, retaining the same bcc structure, but losing ferromagnetism; then, at 920°C, transition to a close-packed fcc structure of $\gamma -$Fe occurs, but at 1400°C the

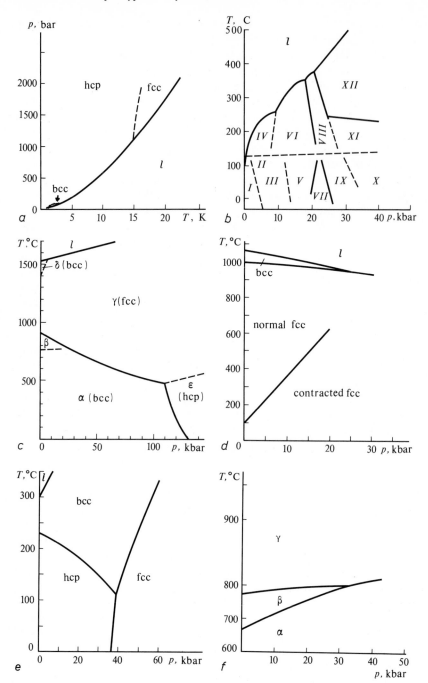

Fig. 2.2a – f. $p - T$ phase diagrams for some elements. (a) helium, 4, (b) sulphur, (c) iron, (d) cerium, (e) thallium, (f) uranium

bcc structure of δ – Fe appears again. Two modifications are observed in Na, Be, Sc, Co, etc., and three, in Ta (Fig. 2.2e), Li, and La. In the hexagonal close-packed structures of metals the c/a ratio is usually slightly lower (down to 1.57) than "ideal" (1.633); for β – Ca and α – Ni this ratio is about 1.65. A change in pressure also results in polymorphous transformations in many metals. Thus, an interesting transition takes place in Ce at 12.3 kbar (Fig. 2.2d). The close-packed cubic structure (fcc) is preserved, but the cell period decreases from 5.14 to 4.84 Å. This is attributed to the transition of one $4f$ electron to the $5d$ level.

As we move right- and downwards in Mendeleyev's table, the covalent interaction plays an ever-increasing role. The whole variety of metal structures which manifest covalent effects to some degree can be divided into several groups. Cd and Zn are examples of deviations from hexagonal close packing with an increase in c/a ratio to 1.86 and 1.89, respectively; this points to non-equivalence of the bonds between the atoms in a layer and between the atoms of different layers. In the metal structures with a still larger fraction of covalent bonding, conduction electrons characteristic of metals are preserved, but the trend towards the formation of directional bonds results in different, sometimes extremely complex structures, whose "background" is close packing, as before. These are "metal – covalent" structures. For instance, β – W (Fig. 2.3a) has a peculiar structure with coordination numbers 12 and 14. Magnesium has three modifications: α with 58, β with 20, and γ with 4 atoms in the unit cell. The phase α – Mn has a distorted fcc structure (Fig. 2.3b). The complexity of this structure is evidently due to the different valency states of magnesium atoms. The phase β – Mn (Fig. 2.3c) can be allocated to the "electron compounds" (see Sect. 2.2.2), and γ – Mn has the structure of γ phases (see Fig. 2.12). Actinides (Pa, U, and Np) have a great variety of polymorphous modifications (examples are presented in Fig. 2.5d, e). Plutonium has six modifications, including bcc and fcc structures, structures based on their modifications, and so on.

A conventional boundary between metals and nonmetals can be drawn between subgroups IIIB and IVB. Where the atomic numbers are low (B), however, the boundary is shifted to the left, and where they are high (Sn, Pb, and Bi), to the right.

A remarkable example of the diversity of polymorphous modifications is offered by B. Its atoms form electron-deficient, directional bonds with a c.n. of 5 or more; c.n. 5 promotes the formation of space groupings with a fivefold icosahedral symmetry. The B – B distance in icosahedra varies between 1.72 and 1.92 Å. Figures 2.50, 66 and 72b in [1.6] exhibit complex strcutures of interlaced icosahedra, or more complex groupings of boron atoms. The following modifications of the boron structure are known (the figure indicates the number of atoms in the unit cell): B – 12, B – 50, B – 78, B – 84, B – 90, B – 100, B – 105, B – 108, B – 134, B – 192, B – 288, B – 700, and B – 1708, which have extremely diverse forms of space symmetry.

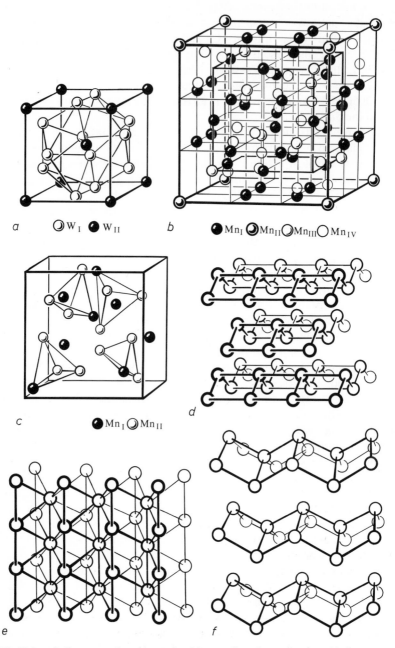

a   ○ W$_I$ ● W$_{II}$ b   ● Mn$_I$ ◖Mn$_{II}$○Mn$_{III}$○ Mn$_{IV}$

c   ● Mn$_I$ ○Mn$_{II}$ d

e   f

Fig. 2.3a–f. Structure of some metals with complicated coordination. (a) β-tungsten, (b) α-manganese, (c) β-manganese, (d) α-uranium, (e) protactinium (modification stable under ordinary conditions), (f) β-neptunium

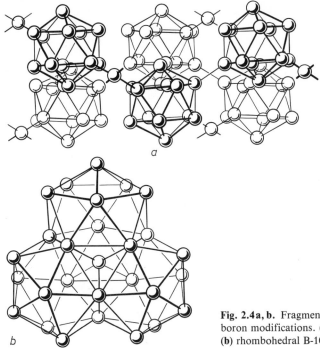

Fig. 2.4a, b. Fragments of the structures of two boron modifications. **(a)** tetragonal B-50 and **(b)** rhombohedral B-105

In covalent structures element's valency determines the number of nearest neighbors and, hence, the extent of structural groupings, i.e., the ability of the atoms of some element or other to form a three-dimensional pattern, two-dimensional layers, one-dimensional chains, or zero-dimensional (molecular) groupings. Groupings of specific dimensionality are mutually bonded owing to van der Waals or partially metallic bonds. The electron spectrum of these structures is often of a semiconductor nature.

In conformity with the principles of the formation of a two-electron bond, a simple rule is obeyed in covalent structures. The coordination number $K = V$, where V is the valence, or $K = 8 - N$, where N is the number of the group. Thus, in the fourth group $K = 4$, C, Si, Ge, and $\alpha - $Sn (grey tin) have a tetrahedral diamond structure (Fig. 2.5a) and $\beta - $Sn (white tin) has a distorted diamond structure (Fig. 2.5e).

Carbon has two principal modifications, diamond and graphite, and two varieties of these (Fig. 2.5a – d). In graphite, which is stable at normal temperature (Fig. 2.5c), the strong covalent bonds between the atoms in the plane net are hybridized; they are one-third double, the C – C distance being equal to 1.420 Å. The bond between the layers is van der Waalsian; the distance between them being 3.40 Å. The graphite structure is double layer. There is also a three-layer structure, namely, the rhombohedral modification of graphite (Fig. 2.5d).

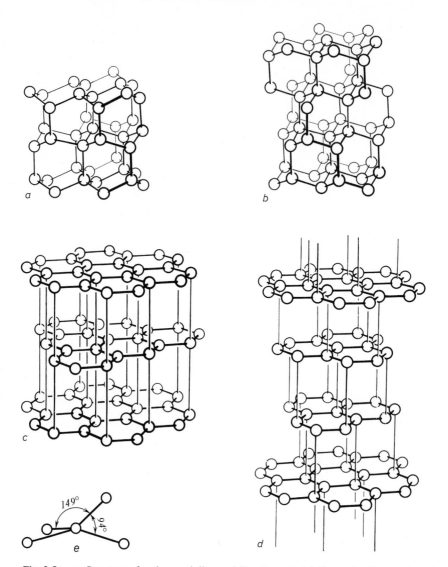

Fig. 2.5a – e. Structure of carbon and diamond-like elements. (**a**) diamond, silicon, germanium; (**b**) hexagonal modification of diamond – lonsdeilite; (**c**) normal hexagonal graphite; (**d**) rhombo-hedral graphite; (**e**) flattened tetrahedron in the structure of white (β) tin

The diamond modification of carbon (Fig. 2.5a), with 1.54 Å distance for a single $C - C$ bond, is metastable; the equilibrium region of its existence lies at high pressures (70 kbar) and temperatures (2000°C). We also know the hexagonal modification of diamond, lonsdalite, in which C retains its tetrahedral coordination, but the packing of the tetrahedra is a three-layer hexagonal, as in wurtzite (Fig. 2.4b).

In the fifth group $K = 3$. White phosphorus forms tetrahedral molecules. Black phosphorus (Fig. 2.6a), which also has c.n. = 3, forms a layer structure. The elements As, Sb, and Bi, with three pyramidal bonds, also crystallize in layer structures (Fig. 2.6b), which can be regarded as distorted hexagonal close packings, in which the distances between the atoms of neighboring layers are 10% – 20% larger than in the intralayer structures.

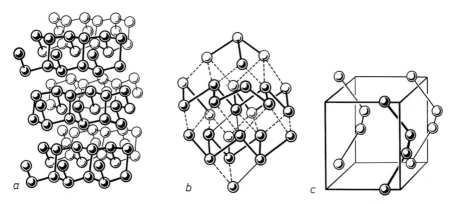

Fig. 2.6a – c. Some covalent structures satisfying the rule $K = 8 - N$. **(a)** black phosphorus; **(b)** α-As (grey), Sb and Bi have the same structure; **(c)** γ-Se and Te, chain structures

In the sixth group, $K = 2$, one of the modifications of S, β – Se, and Te forms chain structures with corner p bonds (Fig. 2.6c). Hexagonal γ – Se and Te may also be regarded as distorted close-packed structures, in which the bonds between the atoms inside the layer are shorter than between the atoms of neighboring layers. In the other sulphur modifications (they are very numerous, see Fig. 2.3b) the chains of atoms are closed into six or eight-membered (Fig. 2.7a, b) rings, and various molecular structures arise.

Oxygen O_2, ozone O_3, and nitrogen N_2 (Fig. 2.7c) do not obey the rule $K = 8 - N$ because of the multiple bonds in their molecules. These elements form molecular structures and have different polymorphous modifications. Finally, in conformity with the rule $K = 8 - N$, the halogens F, Cl, Br, and I form diatomic molecules, which also pack into molecular structures (Fig. 2.7d).

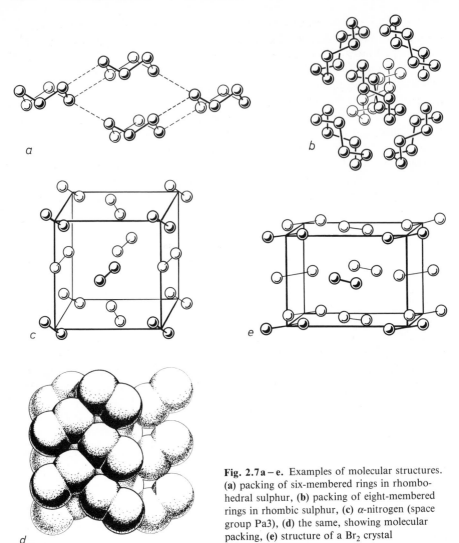

Fig. 2.7a – e. Examples of molecular structures. (a) packing of six-membered rings in rhombohedral sulphur, (b) packing of eight-membered rings in rhombic sulphur, (c) α-nitrogen (space group Pa3), (d) the same, showing molecular packing, (e) structure of a Br_2 crystal

2.1.2 Crystallochemical Properties of Elements

This concept implies those properties of atoms of elements which are retained in compounds. Thus, for intermetallic compounds, the metallic type of bond and the presence of free electrons remain the characteristic properties. The elements located near subgroup IVc form covalent structures and have a predominantly covalent bond in intermetallic compounds as well. This nature of the bond is best preserved in combinations of elements equidistant to the right and to the left from subgroup IVc, for instance, BN and ZnS.

Compounds of metals with elements of the covalent group usually have a metallic bond, although covalent interactions also play a certain part in them. Compounds of elements considerably remoted from each other to the right and to the left of vertical IV, i.e., compounds of metals with oxygen, sulfur, and halogens, already exhibit a predominantly ionic, polar bond. Many of these structures, for instance, oxides and halides of metals, are insulators.

Finally, the combinations of elements forming molecular structures − H, C, N, O, F, Cl, Br, and I − are themselves molecular with covalent intramolecular and van der Waals intermolecular bonds.

The increased complexity of the electron shells of atoms, which promotes the formation of complex hybrid orbitals involving s, p, and d electrons, helps to distinguish another large series of elements of subgroup VIIIa and neighboring subgroups in the periodic system; they are the complex-formation elements. They form predominantly covalent bonds with halogens and some other elements. The complex groupings themselves are bound with each other and with other atoms contained in such compounds by ionic or van der Waals forces.

Thus, the traditional crystallochemical division of the structures into intermetallic, covalent, ionic, complex, and molecular has its foundation in the periodic system, too. Naturally, the allocation of structures to one of the types of chemical bond is often conventional, and compounds with an intermediate-type bond − heterodesmic structures with different types of bonding between certain atoms contained in them − will be formed depending on the complexity of the formula, which may include atoms of elements most varied in their chemical features.

2.2 Intermetallic Structures

Metals may form different types of structures when they combine with each other. Since the metallic bond is due to the free electrons, the individual properties of the constituent atoms do not substantially affect the formation of such structures, as compared with ionic or, all the more so, covalent structures. This means that they are more tolerant to the kind of neighboring atoms, as well as to the stoichiometry of the components. Therefore, intermetallic structures show a considerable trend towards the formation of solid solutions and, during the formation of compounds, towards large deviations from the stoichiometric composition; these structures have many defects. Metastable phases are formed rather readily.

2.2.1 Solid Solutions and Their Ordering

Solid solutions composed of metals are substitutional solid solutions. A continuous series of solid solutions is formed (in accordance with the rules discussed in Sect. 1.6) when two metals are characterized by an identical structure, by similar atomic radii, deviating by not more than 10%, and by similar chemical properties in respect to their proximity in the periodic table. Thus, solid solutions form over the entire concentration range, in the systems Au – Au (both r_m = 1.28 Å), Co – Ni (1.25 Å) (but not in systems Ag – Co and Au – Co), K – Rb (2.36 and 2.48 Å), Ir – Pt (1.35 and 1.38 Å), and so on.

As noted above, a continuous series of solid solutions is characterized by such a distribution of atoms over the positions in the crystal lattice that the atoms of both components (for a binary solid solution) occupy positions with certain probabilities. These probabilities are equal to the atomic fraction of the components. Such a distribution of atoms is called disordered. Since the mutual arrangement of atoms in the structure depends on their interaction, it is natural to ask when a disordered atomic distribution takes place. It follows from physical considerations that disordered distribution should arise when the typical thermal energy kT essentially exceeds the typical value of the energy U of the atomic interaction, which determines the ordering (the interchange energy) ($U_{int}/kT \ll 1$). In the opposite case ($U_{int}/kT \gg 1$), the effects of thermal disordering can be neglected; then, the atomic distribution is determined by the condition of the minimum of the total interaction energy. If the interatomic potentials are such that each atom of sort A tends to be surrounded by atoms of the sort B, an ordered structure arises in which atoms of sorts A and B alternate.

If, however, the atomic interaction potentials are such that each atom tends to be surrounded by atoms of the same sort, decomposition into two phases takes place. One of these phases is enriched in component A, and the other, in B. Formally, these two cases are associated with opposite signs for the interchange energy. The transition from a disordered solid solution, which is realized at high temperatures, to an ordered distribution is a phase transition and occurs at temperatures T_0, at which $kT_0 \approx |U_{int}|$. Decomposition of a disordered solid solution into two phases also takes place at temperature T_0, at which $kT_0 \approx |U_{int}|$.

A classical example of ordering is the Au – Cu system. Both metals have a fcc structure; their radii are close, 1.28 and 1.44 Å, respectively; and they form a continuous series of solid solutions (Fig. 2.8). The disordered structure (Fig. 2.9a) can be stabilized by quenching. At low temperatures, however, it becomes metastable. At temperatures about 400°C the phenomenon of ordering in some concentration ranges is observed; a completely statistical arrangement of Au and Cu atoms over the positions of the fcc lattice, i.e., according to a single regular point system of group $Fm\bar{3}m$, is replaced by an ordered distribution over two RPS, into which the initial RPS is decomposed (with a reduction of symmetry). Ordering is accompanied by a change in the

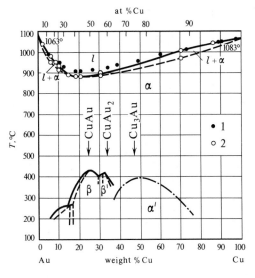

◁ **Fig. 2.8.** Phase diagram of the Cu – Au system

▽ **Fig. 2.9a – c.** Ordering in the Cu – Au system. (a) disordered structure, (b) ordered structure of Cu₃Au, (c) ordered structure of CuAu

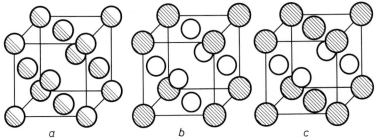

physical properties, for instance, electrical conductivity. In the composition CuAu (β phase) these atoms alternately occupy "floors" perpendicular to axis 4 (Fig. 2.9c), and the structure becomes tetragonal (but with a pseudocubic relation of the lattice parameters $a' \sqrt{2} \approx c$).

With the composition corresponding to Cu₃Au (α' phase), the Cu atoms occupy the centers of the faces of the cubic cell, while Au are situated at its vertices (Fig. 2.9b). But these structures are not realized in the ideal form: each RPS is not completely occupied by the definite sort of atoms; statistically, the atoms of the second component are also present there. For instance, in Cu₃Au the face centers are occupied by about 85% of Cu atoms and, hence, the vertices are occupied by about 55% of the Au atoms. Consequently, gold atoms occur preferentially on alternate planes (Fig. 2.10). The deviation from the stoichiometric composition within certain concentration limits preserves the ordering, and the percentage of each of the positions for each component changes in accordance with the concentration. The degree of ordering also depends on the annealing temperature. The ordered structure consists of domains, whose boundaries are out of phase (Fig. 2.10). The

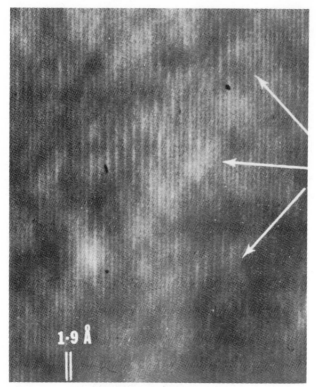

Fig. 2.10. Electron micrograph of a partially ordered structure of Cu_3Au showing planes parallel to the cubic face. Domains are arrowed [2.2]

existence of substitutional solid solutions within the entire concentration range or within only a part of it, as well as order – disorder phenomena, is very common in binary and multicomponent alloys. Some systems form subtractional solid solutions. For instance, in NiAl (structure of the type CsCl) with an Al content of over 50%, some of the Ni positions being vacant.

2.2.2 Electron Compounds

Metals of the subgroup Ib – Cu, Ag, Au, and some others – when forming structures with metals which have more than one valence electron, form an interesting group of alloys, the so-called electron compounds, or Hume-Rothery phases. Their structures are characterized by a definite electron concentration, i.e., by the ratio of the number n_e of valence electrons in the alloy unit cell to the number n_a of atoms in it.

The type of the electron compound mainly depends on the electron energy. The Brillouin zone corresponding to a given structure and determining the

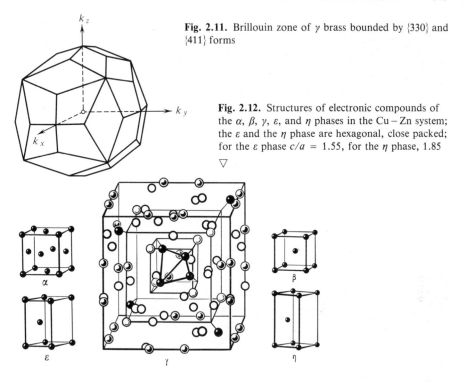

Fig. 2.11. Brillouin zone of γ brass bounded by {330} and {411} forms

Fig. 2.12. Structures of electronic compounds of the α, β, γ, ε, and η phases in the Cu – Zn system; the ε and the η phase are hexagonal, close packed; for the ε phase $c/a = 1.55$, for the η phase, 1.85

maximum electron energy is stable only when the number of electrons is proportional to the volume of the Brillouin polyhedron. The "overcrowding" of the zone results in the structure instability and the formation of a new one with a more capacious Brillouin polyhedron. Thus, the limit to electron concentration for the fcc lattice of the α phase is $7/5 = 1.4$. For the first polyhedron of the γ phases, which is bounded by the simple forms {330} and {411} (Fig. 2.11), $n_e/n_a = 22.5/13$; a more accurate calculation of the polyhedron shape yields precisely the value of $21/13$. This is how the limits to the solubility (with the formation of the fcc structure) of one metal in another are determined, and these limits are lower for higher valence metals. For instance, the solubility of metals Cd, In, Sn, and Sb (with 2, 3, 4, and 5 valence electrons, respectively) in Ag should be 40, 20, 13.3, and 10%, which is in good agreement with experiment. The structures of electron compounds and some of their representatives are given in Fig. 2.12. For β phases (bcc lattice) and β' phases (cubic cell with 20 atoms), $n_e/n_a = 3/2$. In γ phases, $n_e/n_a = 21/13$; their structure is related to the β phases and can be obtained by tripling with respect to all the axes of the latter unit cells; of these 27 cells with 54 atoms, two atoms – at the center and at the corners of the large cell – are absent, while the remaining 52 atoms are slightly displaced with respect to the β phase structure. For hexagonal ζ and η phases, $n_e/n_a = 7/4$. In a number of electron compounds, ordering phenomena are observed.

2.2.3 Intermetallic Compounds

Such compounds with a definite stoichiometric ratio of the components (and a certain range of concentrations near this ratio) may be due to the ordering of alloys in the solid state as well as to direct crystallization from the melt. The factors determining their structures are the ratio of the atomic radii, the electron concentration, and the presence of ionic or covalent components of the metallic bond between atoms in alloys.

The basic structures of intermetallic phases are shown in Fig. 2.13. Many of them are based on the structures of the constituent pure metals. The unit cell of a compound may either correspond to the metal unit cell with reduced symmetry or be based on a multiple of it. There are also compounds, such as CuZn, in which the structure of the intermetallic phase does not correspond to that of any of the constituent metals.

All these compounds have high coordination numbers. Compounds with a high coordination and packing coefficient may also be formed by atoms with considerably different atomic radii, owing to the filling of the interstices in the packing of large atoms by smaller ones. Thus, in the cubic structures of $MgCu_2$ (Fig. 2.13c), which is an example of so-called Laves phases, the Mg atoms occupy the points of the diamond lattice, whereas the Cu atoms form a continuous string of tetrahedral groups along the free octants.

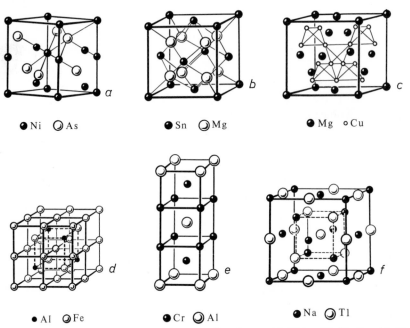

● Ni ○ As   ● Sn ○ Mg   ● Mg ○ Cu

● Al ○ Fe   ● Cr ○ Al   ● Na ○ Tl

Fig. 2.13a–f. Some structures of intermetallic phases. (a) NiAs; (b) Mg_2Sn, (c) $MgCu_2$, (d) Fe_2Al, (e) Cr_2Al, (f) NaTl

The compounds with a low coordination-number usually arise when there is a considerable difference in the atomic radii and the fraction of covalent or ionic bonds increases significantly. Among them are structures of the type NiAs with c.n. 6 (octahedron or trigonal prism); the centers of these polyhedra are occupied by an atom of a more covalent element (Fig. 2.13a). The structures of $PtAl_2$ and $AuIn_2$ are of the type CaF_2 (Fig. 1.75b); here too, the more covalent atom has a tetrahedral coordination. The compounds of electropositive with electronegative metals (for instance $MgPb_2$, type CaF) can be regarded as being fractionally ionic or covalent.

A number of intermetallic compounds are characterized by complicated coordination polyhedra (Fig. 2.14).

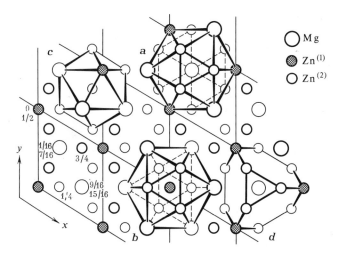

Fig. 2.14a−d. Structure of $MgZn_2$ (projection onto the xy plane). Coordination polyhedra of Mg (a), $Zn^{(1)}$ (b), and $Zn^{(2)}$ (c), and the Laves 12-vertex polyhedron (d) are shown

2.3 Structures with Bonds of Ionic Nature

2.3.1 Structures of Halides, Oxides, and Salts

Halides, oxides, silicates, many chalcogenides, and salts of inorganic acids are structures with bonds of an ionic nature. When allocating, as according to the crystallochemical tradition, these structures to the ionic type it must be borne in mind that they also always show covalent interaction, to some extent. This is manifested very weakly in some structures, more strongly in others, and there are structures in which it appears to be predominant, as was demonstrated in Sect. 1.2.

Classical representatives of compounds with an almost completely ionic bond are halides of alkali metals, in which univalent ions are practically fully ionized. The effective charge of di-, tri, and, all the more so, multivalent ions is always below the formal value of the valence and usually does not exceed one or two electrons. For instance, for some cations the population of the valence orbitals and the effective charge are [2.3]

	s	p	d	Effective charge
Si^{4+}	0.9	1.7	0.7	+0.7
S^{6+}	1.0	2.2	0.9	+1.9
Cr^{6+}	0.2	0.8	4.4	+0.6

(According to other data the effective charge of Si^{4+} is $1.0-2.0$ e).

Most of the halide structures are built according to the geometric scheme of close packing; they include structures of the types NaCl (Fig. 1.49), CaF_2 (Fig. 1.75 b), $CdCl_2$ (Fig. 1.52), etc. Another simple type of ionic structure, CsCl (Fig. 1.75 a), is encountered more rarely. Many oxides − MgO (type NaCl), Al_2O_3 (Fig. 1.72) − are also built according to the close-packing principle; the close packing in the structure of rutile TiO_2 is slightly distorted (Fig. 2.15). The structures of complicated oxides, for instance, perovskite $CaTiO_3$ (Fig. 2.16), spinels (Fig. 2.17), which include oxides of iron and other metals, and some garnets (Fig. 2.18), can also be considered on the basis of the geometrical scheme of close packing of anions with the filling of some of the holes according to the chemical formula and the ion size.

Compounds of this type often have special physical properties and are of technical importance. Thus, the perovskite type (Fig. 2.16) includes barium titanate and many other ferroelectric crystals; the crystals of several complicated oxides of trivalent metals of the spinel (Fig. 2.17) and garnet (Fig. 2.18) types are valuable magnetic materials. The yttrium − aluminium garnets $Y_3Al_5O_{12}$ mentioned above and other similar compounds are used as laser materials.

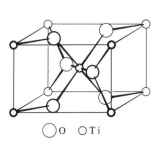

○ O ○ Ti

Fig. 2.15. Structure of rutile TiO_2

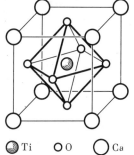

◉ Ti ○ O ○ Ca

Fig. 2.16. Structure of perovskite $CaTiO_3$

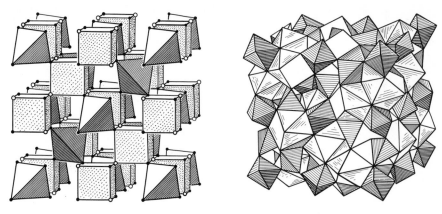

Fig. 2.17   Fig. 2.18

Fig. 2.17. Structure of spinel-type oxides with a general formula $Me_2^{III}M^{II}O_4$. It includes Fe_2MgO_4, Fe_3O_4, Al_2NiO_4, Cr_2ZnO_4 and other compounds. In the closest cubic packing of O atoms (32 per unit cell), metal M^{III} populates (in the ideal case) 16 out of 32 octahedral voids, and metal M^{II}, 8 tetrahedral voids; this arrangement is shown in the figure

Fig. 2.18. Unit cell of the crystal structure of grossular garnet $Ca_3Al_2Si_3O_{12}$ in polyhedra, tetrahedra around Si, octahedra around Al, and twisted cubes around Ca

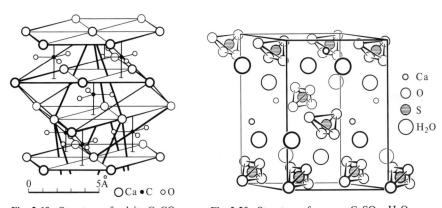

0   5Å
⊙Ca •C ○O

○ Ca
◐ O
⊜ S
◯ H₂O

Fig. 2.19. Structure of calcite $CaCO_3$

Fig. 2.20. Structure of gypsum $CaSO_4 \cdot H_2O$

In the structures of the salts of acids with a complex anion, for example, CO_3^{2-}, NO_3^{1-}, SO_4^{2-}, PO_4^{3-}, the bonds between atoms in the anion are close to purely covalent, the anion entering the crystal structure as a structural unit. In these structures, for instance, those of calcite $CaCO_3$ (Fig. 2.19), and gypsum $CaSO_4 \cdot H_2O$ (Fig. 2.20), the principle of the maximum filling of space is well maintained, but the voids are now filled not only with spheres, but also with more complicated figures.

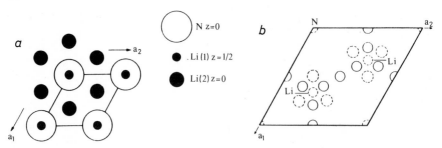

Fig. 2.21. (a) Structure of Li_3N. **(b)** Difference electron density map for Li_3N at 20°C ($z = 0$) calculated under the assumption of the presence of Li^+ and N^{3-} ions. Full and broken lines correspond to positive and negative densities, respectively. The lines are drawn at 0.05 $e\text{Å}^{-3}$ [2.4]

The presence of a cation that freely gives up electrons may lead to the appearance of unusual anions in crystal structures. An interesting example is the Li_3N structure (Fig. 2.21 a), in which the plane of a layer of N atoms and two Li atoms alternates with a layer consisting only of Li atoms (this structure is an ionic conductor, see Sect. 2.3.3). Experimental data [2.4] — the calculation of deformation difference syntheses (Fig. 2.21 b) — showed that both the cation and aninon are almost fully ionized, i. e., the Li^+ cation leads to the appearance of an N^{3-} anion.

In complicated ionic compounds, especially those with large cations, close packing is more rarely the geometric basis of the structure; the c.n. of the cations increases.

When analyzing the structure of ionic compounds L. Pauling introduced the concept of "bond valence strength", i. e., the cation valence divided by its coordination, and formulated the "electrostatic-valence rule". This rule says that the sum of the valence strengths s converging on an anion is (approximately) equal to its valence

$$\Sigma s \simeq V. \tag{2.1}$$

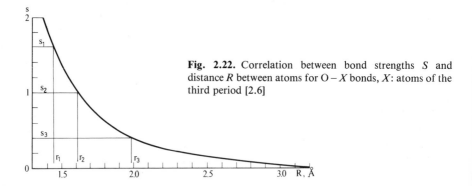

Fig. 2.22. Correlation between bond strengths S and distance R between atoms for $O-X$ bonds, X: atoms of the third period [2.6]

This rule is reasonably well obeyed by many simple and complex ionic structures, and deviations from it do not exceed 1/6.

On the basis of the concept of bond valence strength s between ions of opposite signs in "ionic" structures, some authors proposed empirical quantitative relationships between the values for a given bond and its length R. Thus, for cation – oxygen bonds [2.5]

$$s = s_0 (R/R_0)^{-N} \tag{2.2}$$

where s_0, R_0, and N are the parameters of a given cation for a given c.n.

Thus, for bonds between oxygen and second-row atoms (from Na to S) the relationship $s = s_0 (R/1.622)^{-4.29}$ (Fig. 2.22) holds true. For bonds with other cations, individual parameters were chosen; s_0 fluctuates between 0.25 and 1.5, R_0 between 1.2 and 2.8, and N between 2.2 and 6.0. For example, for Al^{3+}, $s_0 = 0.5$, $R_0 = 1.909$, and $N = 5.0$. The Pauling rule holds true here both for cations and anions, and also for any coordination of the cation. It should be stressed that the use of expressions of the type (2.2) is by no means based on the assumption that the structure is actually built up of "genuine" ions.

Another Pauling rule states that in ionic structures the presence of common edges and, especially, faces in coordination polyhedra is unlikely. The essence of this rule is simple. Electrostatic repulsion between cations in polyhedron centers tends to move them away from each other, and this can be achieved when the polyhedra are linked together at their vertices and, to a lesser extent, at their edges. Linking at faces signifies the maximum mutual approach of cations, and therefore is rarely observed. Octahedra are linked by their edges more frequently than tetrahedra.

2.3.2 Silicates

The most important class of compounds which are traditionally called ionic are silicates, although, in fact, the bond in the principal building block of these structures – the tetrahedron SiO_4 – is basically covalent, as indicated above.

A compound built up exclusively of tetrahedra SiO_4 is silica SiO_2. It has a number of modifications (Fig. 2.23). The effective charge of the anion O^{2-} is about $1e$. The principal modification – hexagonal quartz (density 2.65 g/cm^3) – transforms at 870°C to tridymite (also hexagonal, density 2.30 g/cm^3), and at 1470°C to cubic cristobalite (density 2.22 g/cm^3). Each modification exists in two forms, low and high temperature (α and β, respectively). The $\beta - \alpha$ transition in quartz is represented in Fig. 2.76b (showing only the positions of Si atoms). Research in meteorite craters revealed a denser modification, coesite (Fig. 2.23d), and the densest modification, stishovite, of the rutile type structure; they had formed under high pressures and tempera-

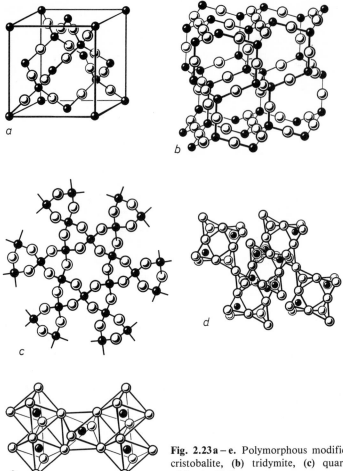

Fig. 2.23a – e. Polymorphous modifications of silica. (**a**) cristobalite, (**b**) tridymite, (**c**) quartz, (**d**) coesite, (**e**) stishovite [2.7]

tures. Stishovite (Fig. 2.23 e) had been obtained under laboratory conditions at 1200 – 1400°C and 160 kbar, the Si coordination in it is nearly octahedral.

In the structure of $AlPO_4$ which is similar to that of quartz, half of the Si atoms are replaced by Al, and the other half by P. The deformation difference syntheses of electron density for aluminium phosphate and quartz (Fig. 2.24a, b) showed that the charges in aluminium phosphate are as follows: Al $+1.4$, P $+1.0$, and O $-0.6e$; in quartz Si $+1.22$, and O -0.61. In both compounds an electron density corresponding to the covalent fraction of the bond is observed. In $AlPO_4$, the electron density of the covalent bridge in the P – O bond is somewhat stronger than in Al – O. Lone-pair densities of O atoms are displaced from the bonding plane, and oxygen appears to be sp^2 hybridized [2.8, 9].

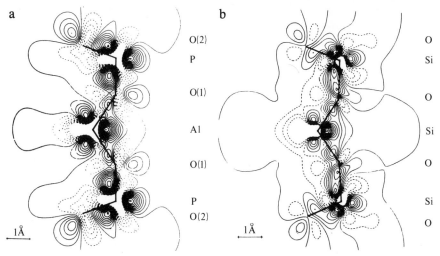

Fig. 2.24a, b. Deformation difference electron density maps for $AlPO_4$ **(a)** and quartz SiO_2 **(b)**; sections through bonded atoms. Interval 0.1 $e\mathring{A}^{-3}$; (*broken lines*) negative densities [2.8, 9]

The grouping SiO_4^{2-}, in chemical combination with oxides of various other metals, forms silicates, the basic minerals of the earth's crust. The silicon – oxygen tetrahedra in them may be isolated or connected at the vertices. The possibility of connection at one, two, three, or all four vertices yields a large variety of spatial motifs. The formation of one of these motifs generally depends on the other cations, to whose coordination polyhedra the silicon – oxygen tetrahedra adapt. Therefore, silicates are the most complex inorganic structures. The foundations of the crystal chemistry of silicates were laid by the classic x-ray structure investigations of *Bragg*'s school in the 1920s and 1930s [2.10]; the crystal chemistry of silicates with large cations was developed in the postwar years by *Belov* and co-workers [2.11 – 13].

As regards the nature of the silicon – oxygen tetrahedra linking, the silicates can be divided into four groups. They are zero-dimensional, $m = 0$, silicon – oxygen groupings, one-dimensional, $m = 1$, (chain-type) structures, two-dimensional, $m = 2$, (layer-type) structures, and three-dimensional groupings with a spatial net of SiO_4 tetrahedra (see Sect. 1.5).

Examples of structures with insular groupings are olivine $MgFe[SiO_4]$[1] (Fig. 2.25), topaz $Al_2[SiO_4](OH, F_2)$, which possesses an isolated orthogroup $[SiO_4]$. Examples of silicates in which the double tetrahedron – diorthogroup $[Si_2O_7]$ – is the final grouping are tilleyite $Ca_5[Si_2O_7](CO_3)_2$ (Fig. 2.26) and seidozerite $Na_4Zr_2TiMnO_2[Si_2O_7]_2F_2$ (Fig. 2.27). In beryl $Be_3Al_2[Si_6O_{18}]$ (Fig. 2.28) the basic structural unit is a classical six-membered ring. Many of the

---

[1] The brackets indicate the silicon – oxygen radical characteristic of the structure.

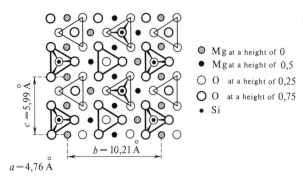

Fig. 2.25. Structure of olivine

⊘ Mg at a height of 0
● Mg at a height of $0,5$
○ O at a height of $0,25$
◯ O at a height of $0,75$
• Si

$c = 5,99$ Å

$b = 10,21$ Å

$a = 4,76$ Å

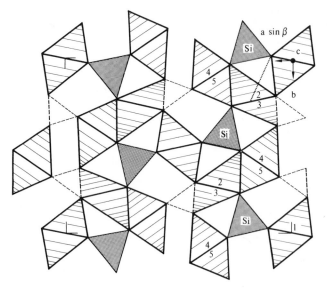

Fig. 2.26. Structure of tilleyite. Double tetrahedra [Si_2O_7] are projected into triangles denoted by "Si". They are surrounded by Ca octahedra (*hatched*)

Fig. 2.27. Structure of seido-zerite; [Si_2O_7] groups are linked with large-cation octahedra

prototype structure of zeolites — hydrosodalite — is cubooctahedral (Fig. 2.36a), each cubooctahedron having eight hexagonal faces ("windows") made up of six-membered rings. In other molecular sieves the windows are built of eight- and twelve-membered rings, and their size may be as large as 13 Å (Fig. 2.36b, c).

Various silicates, which crystallize in the earth's crust from the melt or during hydrothermal processes, contain atoms of a variety of elements which enter the structure stoichiometrically or as isomorphous substituents. Thus, aluminum, which is mostly arranged in octahedra, in aluminosilicates also occupies tetrahedral positions, imitating silicon; tetrahedra are also formed by Be and B. Among the anions, in addition to O, one encounters hydroxyls (OH and F), some complicated anions (such as CO_3 and SO_4), water, etc. The isomorphism in silicates (see Sect. 1.6) vividly confirms the diagonal row rule.

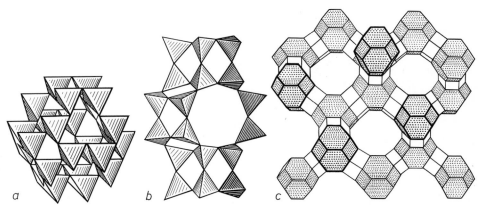

Fig. 2.36 a – c. Structure of skeleton silicates. (a) Cubooctahedral "lantern" built up of SiO_4 tetrahedra in hydrosodalite $Na_8[(OH)_2(AlSiO_4)_6]$; (b) "windows" in the structure of phillipsite Cu_5FeS_4, (c) faujasite — Linde's molecular sieve passing molecules not exceeding ~8 Å in diameter

An important determining factor for silicate structures is the size of the cations. The typical coordination polyhedron for Al, Mn, Ti, Fe, and Mg, i.e., an octahedron with an edge of 2.7 – 2.8 Å, combines directly with orthogroups SiO_4, whose edges are 2.55 – 2.70 Å (Fig. 2.37a). For large cations — Ca, Na, K, Ba, Nb, Zr, and rare-earth elements — the size of the edge, which is 3.8 Å, is, on the contrary, incommensurate with the edges of the orthogroups, but matches well those of the diorthogroups Si_2O_7 (Fig. 2.37b), i.e., with the distance between the oxygen atoms of the coupled tetrahedra, 3.7 – 4 Å. Thus, according to [2.11] the main building blocks of silicates with large cations are the diorthogroups Si_2O_7. This is how complicated chains with alternation of ortho- and diorthogroups (Fig. 2.37), eight-membered rings, etc., appear. If polyhedra with a coordination number exceeding that of the octahedron arise, some of their edges may be foreshortened again to become

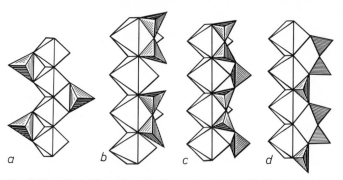

Fig. 2.37a – d. Joining of octahedra and tetrahedra [2.11]. (a) Joining of octahedra formed around small cations with orthogroup $[SiO_4]$, (b) joining of octahedra formed around large cations with diorthogroup $[Si_2O_7]$, (c, d) combined joining

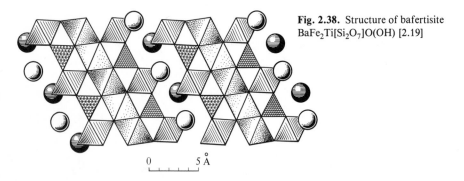

Fig. 2.38. Structure of bafertisite $BaFe_2Ti[Si_2O_7]O(OH)$ [2.19]

0   5 Å

commensurate with the tetrahedron of SiO_4. An example of the structure of such silicates is bafertisite (Fig. 2.38).

2.3.3 Superionic Conductors

In recent years attention has been drawn to a special variety of the crystals of ionic compounds, termed ionic or superionic conductors. They are also called solid electrolytes. As a rule, these are defect structures. Some cations in them are very weakly bound with the lattice of the other atoms; their thermal-vibration amplitudes are so large that they are comparable with the distances between the possible crystallographic positions which these ions may occupy. As a result, some cations can migrate through the crystal, this property being reflected in the term "solid electrolyte." Their ionic conductivity σ is comparable with, and for some compounds even higher than that of liquid electrolytes. Such crystals are promising for various technical applications.

Examples of superionic conductors are AgI, AgBr, CuCl, $RbAg_4I_5$, Ag_2HgI_4, and β-alumina, which is a nonstoichiometric compound based on aluminum oxide containing Na, with an ideal composition $Na_2O \cdot 11 Al_2O_3$.

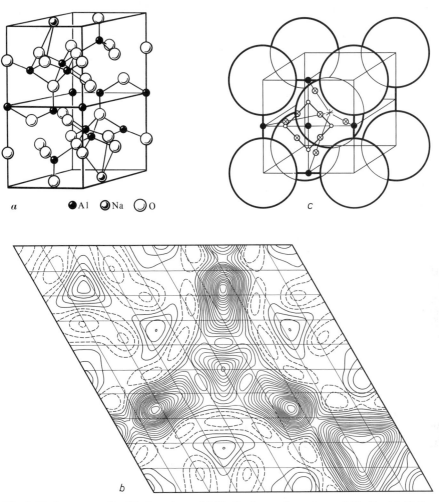

Fig. 2.39. Structures of solid electrolytes. (a) idealized crystal structure of $Na_2O \cdot 11\ Al_2O_3$; (b) distribution of Na atoms in the conducting $xy0$ plane of this structure [2.20]; (c) structure of α-AgI. Spheres: I-ions. Positions of Ag: ($\circ$): coordination number 4, ($\otimes$): c.n. 3, ($\bullet$): c.n. 2

As indicated above, in a crystal of a solid electrolyte the carrier ions migrate through the lattice of the "matrix." In other words, the sublattice of the carrier ions is disordered, but, in distinction to the ordinary case of disordered structures where the atoms are "tied" to the positions which they occupy (on the average, statistically with respect to the volume), in a superionic conductor the disorder is dynamic.

Let us consider one of the structures of this type, β-alumina (Fig. 2.39a). This hexagonal structure is built up of rigid three-layer spinel blocks of Al_2O_3 with Na_2O layers in between. Sodium occupies one crystallographic position in the idealized structure. It has been recently established, however, that these cations can also occupy two other positions in the indicated plane, but with a lower probability of occupancy. The distribution of Na cations over these positions, obtained by neutron diffraction, is given in Fig. 2.39b. In a real structure the number of these ions exceeds the idealized composition by 15% − 20%, which is compensated by the corresponding excess of oxygen. Thus, the indicated plane between the spinel blocks is a conducting plane, in which Na cations "flow" along the "channels" shown in Fig. 2.39b. Here, the ionic conductivity is two dimensional and anisotropic in the conducting plane.

In other ionic superconductors, ion migration is possible along different spatial directions as well. Thus, in the cubic structure of α-AgI large anions I$^-$ form a cubic body-centered packing, while the relatively small cations Ag$^+$ statistically occupy three types of positions with coordination 4, 3, or 2 (Fig. 2.39c). The barriers of the transitions between these positions are small, which enables the cations to "flow" through the anion lattice.

The thermal motion of ions − charge carriers − can be characterized by the probability density function with due regard for the anharmonic temperature factor described by tensors of high ranks. Thermal motion and potential along the diffusion path can be determined with sufficient accuracy from diffraction experiments. Such experiments have been done for the ionic conductor Li$_3$N [2.4, 20a] (cf. Fig. 2.21).

Because the bond between the carrier ions and the matrix lattice is weak, the ionic conductivity strongly depends on the temperature. For the same reason, phase transitions are often observed in such structures. They often permit isomorphous replacements, both of the carrier cations themselves and of the "fixed" cations of the matrix lattice. The properties of solid electrolytes can be changed and improved by introducing isomorphous atoms.

Ionic conductors in which anions are the charge carriers are also known. Examples are CaF$_2$ − YF$_3$ solid solutions, in which some positions of F$^-$ are disordered, and the ZrO$_2$ − CaO system.

2.4 Covalent Structures

Compounds having a covalent bond are mostly formed by the elements of the group IVb and groups close to it. Since the covalent bonds are strong, crystals of these compounds usually show high hardness and elasticity; they are semiconductors or dielectrics, according to their electron spectrum. In a number of crystals the covalent bond should be regarded as somewhat ionic or metallic. The properties of such crystals change accordingly.

A classical example of a covalent crystal is diamond (Figs. 1.30, 2.5) with its uniquely strong lattice. Let us now replace the C atoms of diamond, with their four hybridized sp^3 orbitals in each (Fig. 1.18), by B (3s and p electrons) and N (5s and p electrons) atoms. This is the cubic diamond-like structure of borazon BN, which is obtained at high pressures and whose mechanical properties are similar to those of diamond; it refers to the structural type of zinc

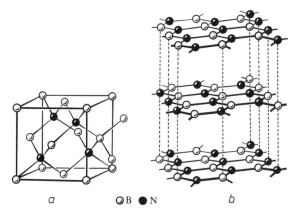

a ○ B   ● N b

blende (Fig. 2.40a). The electron shells of B and N in this structure show the same tetrahedral hybridization as those in C atoms; they form four, strong covalent bonds from spin-antiparallel electron pairs. It is immaterial that the nitrogen atom made a greater contribution to the electron pair of each bond, and the boron atom, a smaller contribution (whereas in diamond each carbon atom contributed equally, one electron). A hexagonal (wurtzite) modification of tetrahedral BN, similar to hexagonal diamond, is also known (see Fig. 2.5 b).

The graphite structure (see Fig. 2.5 a), which corresponds to trigonal sp^2 hybridization of the electron cloud of the C atom (see Fig. 1.17 d) with the participation of π orbitals (Fig. 1.23 c), also has its analog in hexagonal boron nitride, another polymorphous modification of BN (Fig. 2.40 b).

We shall now consider another tetrahedral covalent structures. Similar to the transition from tetrahedral C to BN, it is possible to form compounds with a covalent bond from elements equidistant to the right and to the left of the IVb group of the periodic table – compounds $A^{III}B^V$, for instance GaP, GaAs, CaSb, InAs, and AlP; compounds $A^{II}B^{VI}$, for instance BeO, ZnO, ZnS, ZnSe, CdS, CdSe, CdTe, and HgSe; and compounds $A^{I}B^{VII}$, for instance CuCl, CuBr, and AgI. For all these compounds the ratio of the number of valence electrons to that of the atoms appearing in the formula (or per cell) is $n_e/n_a = 4$. With the distance from the IVb group along the horizontal of the periodic table, and also with the descent along the verticals, i.e., with an increase of the number of electrons in the inner shells, the strength of the bonds decreases, and these become more ionic or metallic. For instance, the bond in a BN crystal is typically covalent; yet estimated by the difference in electronegativities, it is to a small degree – about 15% – ionic. In beryllium oxide BeO the ionic component is already considerable, about 40%, but the coordination remains tetrahedral. On the other hand, the transition to the cubic structure of LiF (the fraction of the ionic component is about 90%) is accompanied by the appearance of octahedral coordination. Figure 2.41 shows the radial distribution functions $D(r)$ (1.7) of the valence orbitals in

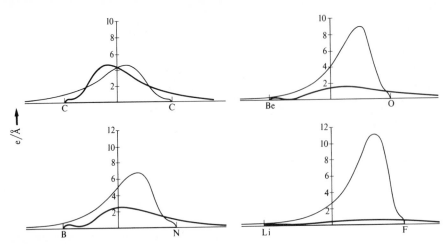

Fig. 2.41. Radial electron density distribution functions $D(r)$ for neutral atoms in crystals of C, BN, BeO, and LiF. (————): left-hand atoms; (———): right-hand atoms [2.21]

these structures. The fraction of ionicity increases with the expansion of the metal orbital and with the coincidence of its maximum with the nonmetal atom orbital. The presence of a fraction of ionicity in $A^{III}B^V$, $A^{II}B^{VI}$ compounds is also indicated by quantum-mechanical calculations (see Fig. 1.27 d).

The covalent-ionic bonding is also observed in semiconducting compounds of the gallium arsenide type (see Sect. 1.2.8). Experimental x-ray data on the electron density distribution reveal the presence of covalent bridges with a density of up to $0.2 - 0.3 \ e/\text{Å}^3$ [2.22].

At the same time, according to various data, the charge of Ga in GaAs is from $+0.21$ to $+0.5 \ e$ and that of As, from $-0.21 \ e$ to $-0.5 \ e$. In tetrahedral structures the gradual change in the nature of the bond as we go from the IVb column is also manifested in an increase of the width of the forbidden band ΔE from fractions of 1 eV to $2-3$ eV.

The tetrahedral coordination for binary compounds is realized in two principal structural types: cubic diamondlike sphalerite ZnS (Fig. 2.42a) and a hexagonal structure with the same formula (ZnS) wurtzite (Fig. 1.73b, c). Many ternary, quaternary, and multicomponent compounds also exhibit tetrahedral coordination of atoms and possess structures similar to those indicated above. Their formulae can be obtained from those of binary compounds by horizontal and diagonal substitution in the periodic system, such that the ratio $n_e/n_a = 4$ is satisfied. For instance, replacing In in indium arsenide by Cd and Sn we obtain $CdSnAs_2$ and replacing Cd in cadmium selenide by Ag and In, $AgInSe_2$. This is how families of the type $A^{II}B^{IV}C_2^V$, $A^IB^{III}C^{VI}$, etc., arise. They have structures of the type of chalcopyrite $CuFeS_2$ (Fig. 2.42b, c), which possesses characteristic tetrahedral bonds of sulfur or related atoms. Various polytypes of SiC (Fig. 1.73d) have been built on the

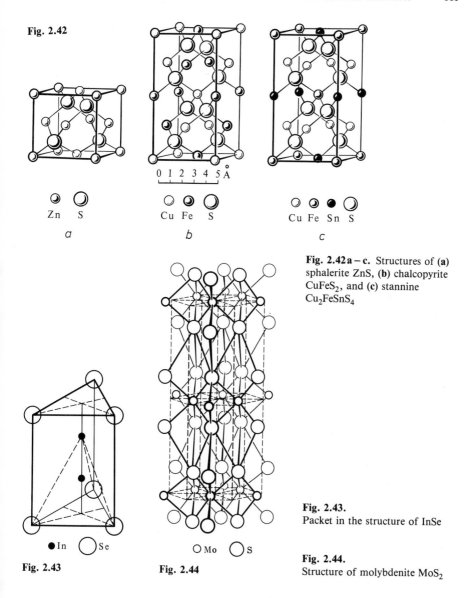

Fig. 2.42

Zn S
a

Cu Fe S
b

Cu Fe Sn S
c

0 1 2 3 4 5 Å

Fig. 2.42 a – c. Structures of (a) sphalerite ZnS, (b) chalcopyrite CuFeS$_2$, and (c) stannine Cu$_2$FeSnS$_4$

● In   ◯ Se

Fig. 2.43

◯ Mo   ◯ S

Fig. 2.44

Fig. 2.43. Packet in the structure of InSe

Fig. 2.44. Structure of molybdenite MoS$_2$

basis of hexagonal interlinking of tetrahedra. A peculiar tetrahedral structure with the preservation of In – In pairs is inherent in InSe (Fig. 2.43); on the other hand it is related to the structure of molybdenite MoS$_2$ (Fig. 2.44) with a coordination in the form of a trigonal prism.

The interatomic distances in tetrahedral structures are well described by a special system of tetrahedral covalent radii, suggested by *Pauling* and *Huggins* [2.23] (Table 2.1).

Table 2.1. Covalent radii [Å]

Tetrahedral

	Be 1.07	B 0.89	C 0.77	N 0.70	O 0.66
	Mg 1.46	Al 1.26	Si 1.17	P 1.10	S 1.04
Cu 1.35	Zn 1.31	Ca 1.26	Ge 1.22	As 1.18	Se 1.14
Ag 1.53	Cd 1.48	In 1.44	Sn 1.40	Sb 1.36	Te 1.32
Au 1.50	Hg 1.48	Tl 1.47	Pb 1.46	Bi 1.46	
	Mn 1.38				

Octahedral

			C 0.97	N 0.95	O 0.90
	Mg 1.42	Al 1.41	Si 1.37	P 1.35	S 1.30
Cu 1.25	Zn 1.27	Ga 1.35	Ge 1.43	As 1.43	Se 1.40
Ag 1.43	Cd 1.45	In 1.53	Sn 1.60	Sb 1.60	Te 1.56
Au 1.40	Hg 1.45	Tl 1.73	Pb 1.67	Bi 1.65	
	Mn 1.31				

Representatives of a large series of structures with octahedral coordination, which have predominantly covalent bonds, are compounds of the types PbS and Bi_2Te_3. The ionic bond character is slightly stronger here than in tetrahedral structures. In octahedral hybridization, not only s and p orbitals of electrons may be included in the bonds, but also the d orbitals which are close to them in level. Because of the increased ionic (and sometimes also metallic) nature of the bonds it is possible to use the terms of the theory of close packing of atoms in describing and interpreting these structures. Since the atomic radii are similar for many compounds, they can here be regarded as joint packings of approximately identical spheres. The structure of lead selenide PbSe and its analogs – PbS, SnTe and SnAs – belongs to the NaCl type (Fig. 2.45). A large family of octahedral structures belongs to the Bi_2Te_3 type (Fig. 2.46a). It includes Bi_2Se_3, Bi_2Te_2S, etc. According to the periodic table, horizontal, vertical, and diagonal replacements can also be made in these compounds, and the resulting structures often have much in common with their precursors. Thus, interesting multilayer packings are formed, for instance, in $GeBi_4Te_7$, $Pb_2Bi_2Te_2$, and $AgBiTe_2$ (Fig. 2.46, see also Fig. 1.67).

The interatomic distances in octahedral structures can be approximately described by the sums of the effective ionic radii and also by the sums of the atomic or atomic – ionic radii (Tables 1.8, 9). But sums of r_i are, as a rule, larger than the observed distances, and more accurate results are obtained by using the system of octahedral radii of *Semiletov* [2.25] (see Table 2.1).

The rule $n_e/n_a = 4$ and the octet rule for semiconductor compounds with a nonquarternary coordination may be supplemented by the Mooser – Pearson rule [2.26]

$$n_e/n_{an} = 8 - b, \tag{2.3}$$

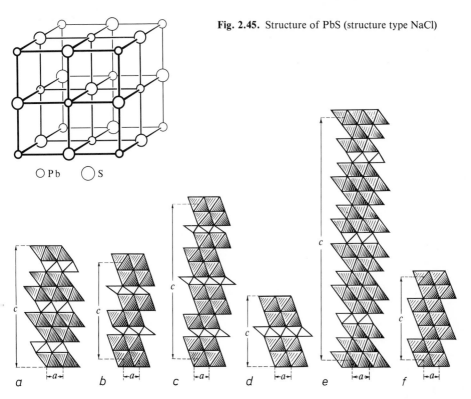

Fig. 2.45. Structure of PbS (structure type NaCl)

○ Pb   ○ S

a   *b*   *c*   *d*   *e*   *f*

Fig. 2.46a – f. Structural motifs of Bi_2Te_3 **(a)** and of its complicated analogs, $GeBi_4Te_7$ **(b)**, $GeBi_2Te_4$ **(c)**, $Pb_2Bi_2Te_5$ **(d)**, $Ge_3Bi_2Te_6$ **(e)**, and $AgBiTe_2$ **(f)** [2.24]

where b is the number of bonds formed by the atoms of the same kind. This rule holds for many structures, for instance, $Mg_2Si (n_e = 8, n_{an} = 1, b = 0)$; $Li_3Bi(8, 1, 0)$; $Mg_3Sb(16, 2, 0)$; $AgInTe_2(16, 2, 0)$; $TiO_2(16, 2, 0)$; $BaTiO_3(24, 3, 0)$; $In_2Te_3(24, 3, 0)$; and $PbS(8, 1, 0)$. It is, however, sometimes violated. The generalized Pearson rule for complicated compounds with a covalent bond has the form

$$\frac{\sum V + b_{an} - b_{cat}}{n_{an}} = 8 . \tag{2.4}$$

Here, $\sum V$ is the sum of the valencies of all the atoms in the chemical formula of the compound, b_{an} is the number of electrons involved in the anion – anion bond, b_{cat} is the number of electrons of the cations which are not involved in the bond (or which form cation – cation bonds), and n_{an} is the number of anions in the chemical formula of the compound.

It should be borne in mind that the atoms which enter into a covalent bond with neighboring atoms do not necessarily use up all their valence electrons. Thus, in compound PbS (Fig. 2.45) lead has a formal valence of 2 and, hence, only two of the four valence electrons are involved in the bond. For instance, for PbSe $(10 + 0 - 2)/1 = 8$, and for ZnP_2 $(12 + 4 - 0)/2 = 8$.

As the formulae become more complicated various structures arise in the course of formation of compounds which do not satisfy the above rules of substitution according to the periodic table. Their common feature is the relatively low coordination of atoms, not exceeding 6. An example of this type of structure is the chain structure of SbSI (see Fig. 1.78), a compound exhibiting valuable ferroelectric and semiconductor properties.

In conclusion we note that compounds of this type tend to form various defective lattices, for instance subtraction structures. The introduction of tiny fractions of atomic impurities having either an excess or a deficiency of electrons (for instance, As or Ga in Ge and binary structures of the type GaAs) into semiconducting tetrahedral structures results in their n (electron) or p (hole) conduction, which is extremely important for manufacturing various semiconductor instruments and devices, including diodes, transistors, and integral circuits.

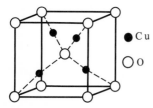

●Cu

○O

Fig. 2.47. Structure of Cu_2O

The covalent nature of the bond is also observed in some metal oxides. An interesting example is cuprite Cu_2O (Fig. 2.47) (Ag_2O has the same structure). The bonds here are directional; the O atom is surrounded by four Cu atoms according to a tetrahedron, while the Cu atoms have a linear coordination of 2, forming bridges between the oxygen atoms. This compound often has vacancies in O positions; some of its properties indicate a partly metallic nature of the bond. This structure can be described as one formed by the entry of O atoms into the tetrahedral voids in the packing of metallic copper having an fcc structure, with an increase in Cu – Cu distances, of course.

2.5 Structure of Complex and Related Compounds

2.5.1 Complex Compounds

Complex compounds always contain stable groupings made up of a central atom, and of nonmetal atoms or molecules – ligands – surrounding it; the bonds of the central atom usually have high symmetry and a stable geometric configuration. Classical examples of complex compounds are offered by platinum, which forms octahedral (for instance, in K_2PtCl_6, see Fig. 1.77) or square (for instance, in K_2PtCl_4, Fig. 2.48a) complexes. Octahedral complexes are also produced by Co in ammoniates, nitrates, halides, and hydrates (Fig. 2.49), and by many other elements. Tetrahedral complexes are produced by Zn (Fig. 2.50). The atoms forming these complexes include Pt, Pd, Rh, Co, Cr, Mn, Fe, Ni, Zn, W, and Mo, i.e., elements belonging mainly to the transitional group VIIIa and those close to it. Complexes observed in the crystal structure are often stable in solution as well, but not necessarily. The ligands of the complex may either not be covalently bonded with the other atoms, or be a part of an inorganic or organic molecule. With the aid of the common atoms, ligand complexes may join into chains, $PdCl_2$ (Fig. 1.79) and $CoCl_2 \cdot 2H_2O$ (Fig. 2.51), and form a space network PtS (Fig. 2.48b).

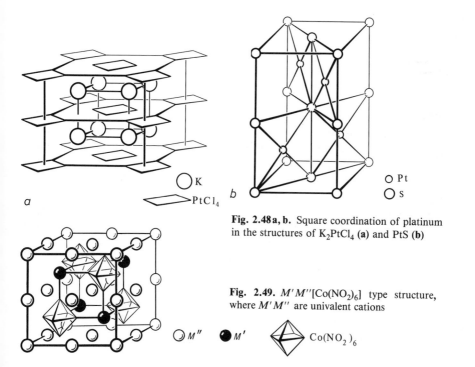

O K

PtCl₄

a

b

O Pt
O S

Fig. 2.48a,b. Square coordination of platinum in the structures of K_2PtCl_4 (a) and PtS (b)

Fig. 2.49. $M'M''[Co(NO_2)_6]$ type structure, where $M'M''$ are univalent cations

◯ M''   ● M'   ◇ Co(NO₂)₆

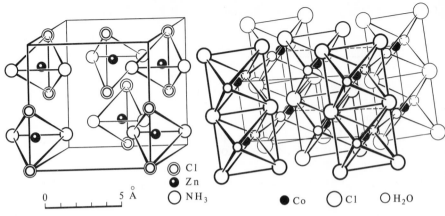

Cl
Zn
NH₃

0 _____ 5 Å

● Co   ○ Cl   ○ H₂O

Fig. 2.50. Structure of Zn(NH₃)₂Cl₂   **Fig. 2.51.** Structure of CoCl₂ · 2H₂O

The ligands contained in a complex may be of two or more sorts; therefore, chemically and structurally different compounds are formed because of their different spatial position. Thus, the complex of dichlorodiamino-platinum may have two forms, *trans* and *cis*

$$
\begin{array}{cc}
\text{NH}_3 & \text{Cl} \\
| & | \\
\text{Cl–Pt–Cl} & \text{Cl–Pt–NH}_3 \\
| & | \\
\text{NH}_3 & \text{NH}_3
\end{array}
$$

Naturally, the higher the coordination and the greater the number of sorts of ligands, the more stereoisomers can appear. Such complexes can be distinguished spectroscopically and also by optical rotation according to sign and magnitude.

The coordination number in a complex is always higher than the formal valence of the complexing metal. For instance, tetravalent platinum gives octahedral (c.n. 6) complexes, divalent platinum, square (c.n. 4), and divalent cobalt, octahedral (c.n. 6) complexes. Note that, in forming a coordination sphere from ligands of some charge, one and the same metal may form complexes of unlike charges. For instance, the complex $[Pt^{IV}Cl_6]^{2-}$ in the crystalline structure of K_2PtCl_6 is bound by ionic forces with the K^+ cations, while a similar octahedral complex $[Pt^{IV}(NH_3)_4Cl_2]^{2+}$, but already positively charged, in the structure of $[Pt(NH_3)_4Cl_2]Cl_2$ is already packed with Cl^- anions owing to the ionic forces. The Cl atoms of the inner coordination sphere play quite a different role here than the Cl^- anions of the outer sphere, and it would be senseless to write the chemical formula of this structure as $Pt(NH_3)_4Cl_4$. Complexes also exist with an internal charge compensation − molecular complexes − for instance, the octahedral grouping $(C_6H_5)_2SbCl_3 \cdot H_2O$.

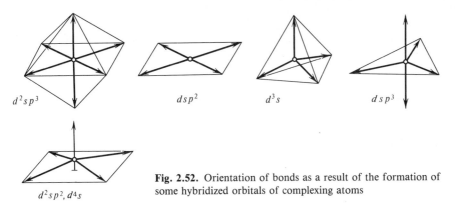

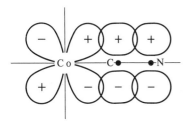

Fig. 2.52. Orientation of bonds as a result of the formation of some hybridized orbitals of complexing atoms

Fig. 2.53. Scheme of the superposition of AO in one of the six octahedral π bonds of $Co - CN$

The complexes are characterized by a strong mutual influence of ligands in chemical reactions; in particular, the so-called trans-influence effect is known, i.e., when a strongly bonded ligand affects, through the complexing metal atom (located in the center), the centrosymmetric (trans) ligand, thereby weakening its chemical bond.

The relatively high strength of the bonds in complex compounds points to their specific nature. Such bonds are called coordination bonds because of a number of their specific features and in accordance with their main characteristic, i.e., the formation of stable coordinations. The bonds may be two electron (covalent) or one electron; in all cases the central metal atom is a donor of bond electrons. It can be seen from the scheme of the levels of atomic orbitals (Fig. 1.8) that in elements with a sufficiently large number of electrons, including transition elements, the d levels are similar energywise to the next s and p levels. This similarity permits the formation of directional hybridized orbitals of the central atom. Thus, octahedral hybridized orbitals are formed from the electrons d^2sp^3, and square orbitals, from dsp^2, etc. (Figs. 2.52, 53).

Each of the AO electrons of the central atom is coupled with an electron of a ligand atom to form characteristic dielectronic covalent bonds. Hybridization of 9 orbitals d^5sp^3 of transition metals may result in a ninefold coordination with angles of about 74° and 120 – 140° between the bonds [2.27].

This approach, however, cannot explain all the characteristics of coordination bonds. The magnetic properties of some of them, for instance, point to the presence of lone electrons; high-spin compounds of this type exist along with low-spin ones.

A more detailed consideration is based on taking into account the effect of the ligands on the energy levels of the complexing atom. These compounds are characterized by a substantial participation of the d and f states in the chemical bond, the energy sublevels of these states being close. At first, the splitting of these degenerate (in a free atom) levels caused by the electrostatic field of the surrounding atoms is taken into account. Level splitting depends on the field symmetry and can be considered on the basis of point group theory [Ref. 1.6, Sect. 2.6]. A more general theory – the ligand field theory – treats, with due regard for symmetry, any possible interactions of the central atom with the ligands, proceeding from the molecular-orbital concepts. This theory makes it possible to calculate the bonds and energy levels of a system and compare them with the data from optical- and radio-spectroscopy.

The color of many complex compounds depends on the light absorption by electrons during the transitions between levels resulted from their splitting in the ligand field. The direction and strength of the resulting bonds thus also depend on the structure of the shell of the complexing atom and on the strength of the ligand field. The stronger fields are produced by small compact anions, such as CN^- and NH_3^+, while large anions, say, I^- and Cl^-, yield weak fields.

The strength parameter of the ligand field is the quantity Δ, whose value fluctuates between 1 and 4 eV. With weak-field ligands the bonds are usually one electron and mostly ionic; the splitting of the levels of the central atoms is insignificant. In the case of strong-field ligands the electrons are paired, using the split levels, and the bonds become covalent. A characteristic feature of

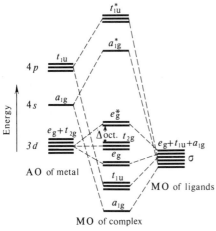

Fig. 2.54. Splitting of energy levels on the formation of MO in octahedral complexes

octahedral and tetrahedral complexes is the formation of molecular σ orbitals. The scheme of level splitting for such octahedral bonds is given in Fig. 2.54.

In crystals and molecules with coordination bonds, the adiabatic approximation, in which the electron states are considered within the framework of a stable configuration of nuclei, is sometimes inapplicable, and the nuclear motion must also be considered. This is the essence of the so-called Jan-Teller effect; the symmetry of the complex diminishes, which manifests itself in the fine structure of optical and EPR spectra.

2.5.2 Compounds with Metal Atom Clusters

Experimental investigations of organometallic and complex compounds have led to the discovery of one more interesting class, in which metal – metal bonds are observed within the complex. Clusters of two, three, or more atoms are formed. The bonds between metal atoms do not resemble the ordinary metallic bonds in the structures of these metals themselves; they are shorter, stronger, and directional, i.e., they are of a covalent nature. For instance, the length of the Hg – Hg bond in dimercurochloride ClHgHgCl is equal to 2.50 Å, whereas the minimum distance in the metal is 3.00 Å; in the complex structure of $K_3[W_2Cl_3]$ the W – W distance is 2.40 Å, which corresponds to a double bond, whereas in the metal the W – W distance is 2.80 Å. In the $PtCl_4$ columns in the structure of K_2PtCl_4 (Fig. 2.48a), the Pt – Pt distance is shortened. Interesting examples are rhenium compounds, in which the Re – Re bonds (Fig. 2.55) are considerably shortened. Thus, in the dinuclear complex $[Re_2Cl_8]^{4-}$ (Fig. 2.55a) the Re – Re distance is equal to 2.22 Å, which is less than those in the metal, while the Re – Cl distances are also shortened. A three-nuclei Re complex is shown in Fig. 2.55b. Me – Me bonds in clusters are attributed to the interaction of orbitals which are not realized in pure

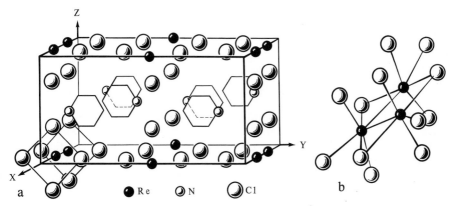

a   ● Re   ◑ N   ◐ Cl   b

Fig. 2.55a, b. Structure of (PyH)HReCl₄ with the groups $[Re_2Cl_8]^{4-}$ in it (a), and of trinuclear complex ion Re_3Cl_{12} (b) [2.28]

metals. More intricate complexes with metal – metal bonds are also known. As an example, Fig. 2.56a shows the octahedral complex of molybdenum atoms with a cube of Cl atoms found in the structure of $[Mo_6Cl_8]Cl_4 \cdot 8H_2O$; the Mo – Mo distance of 2.63 Å corresponds to single bonds (in the metal, this distance is 2.78 Å). The electron pairs responsible for the Mo – Mo bond are localized on the edges of the Mo octahedron. Considerations based on the MO method shows that $d_{3z^2-1}-$, $d_{xy}-$, $d_{xz}-$, $d_{yz}-$, and p_z-AO of each atom take part in their formation. Similar octahedral complexes are formed by Ta and Nb.

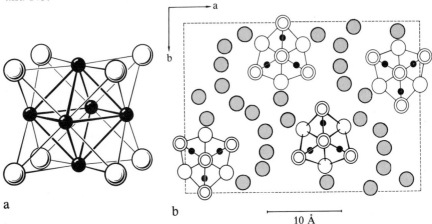

Fig. 2.56. (a) Structure of complex ion $[Mo_6Cl_8]^{4+}$; (b) the projection of $Cs_{11}O_3Rb_7$-structure. (*Open circles*) Cs; (*black circles*) O; (*shaded circles*) Rb [2.28a]

Clusters of metal atoms of the V – VII groups are usually surrounded by weak-field ligands, while clusters of metal atoms of the VII and VIII groups, by strong-field ligands.

Other compounds with clusters of metal atoms have different types of bonding – ionic and metallic. These are Rb and Cs suboxides with unusually high amounts of metal atoms, for example, Rb_9O_2, $(Cs_{11}O_3)Cs_{10}$, and $(Cs_{11}O_3)Rb_7$ (Fig. 2.56b). Metal atoms form groupings with O atoms or may be situated between these groupings [2.28a].

2.5.3 Metal-Molecular Bonds (π Complexes of Transition Metals)

The consideration of the bonds in complexes from the standpoint of molecular-orbital theory has explained the structure of complexes which were absolutely incomprehensible to classical chemistry. In these structures the bonds of the central atoms are directed, not towards the center of the ligand atom, but, say, towards the "middle" of the bond between a pair of atoms, towards the "middle" of the five-membered ring, etc. For instance, in the square complex $[PtCl_3(C_2H_4)]$ the fourth bond of Pt is directed between two carbon atoms of ethylene (Fig. 2.57). This bond can be described in terms of

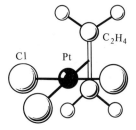

Fig. 2.57. Structure of complex $PtCl_3(C_2H_4)$

the deformation of the binding π orbital of ethylene (Fig. 1.25 a) and its over-lapping with the AO of the metal. In other cases, the antibonding π orbitals of the ligand and the d orbitals of the metal overlap, etc.

A remarkable example is the structure of "sandwich" molecules consisting of cyclic hydrocarbons with metals Fe, Ni, Co, Cr, Ti, Ru, Os, and some others, in which the central atom is squeezed between two parallel-oriented rings of an aromatic hydrocarbon. A classic representative of this class of compounds is iron dicyclopentadienyl $Fe(C_5H_5)_2$, or ferrocene (Fig. 2.58). This is a case of five-membered rings, but the rings may also be six membered, as, for instance, in $Cr(C_6H_6)_2$ and $Mn(C_5H_5)(C_6H_6)$; besides, sandwich compounds with four-, seven-, and eight-membered rings are known. In ferrocene, the symmetry of the molecule is $\bar{5}m$, all the carbon atoms are equivalent, and the Fe atom and both rings are nearly neutral (but, for instance, in nickelocene it was found that Ni is charged by about $+1.4$ e and each C atom, by -0.14 e). The bond in sandwich compounds is ascribed to the overlapping of their π orbitals, lying "under" and "over" the rings (cf. Fig. 1.25 b), with the $4s$ and $3d$ orbitals of the metal. As a result, nine bonding orbitals are formed, in which 18 electrons are involved.

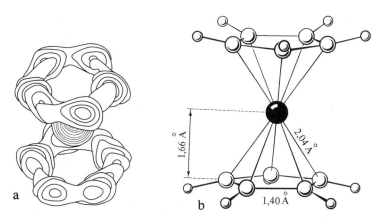

Fig. 2.58 a, b. Three-dimensional Fourier synthesis map of a molecule of ferrocene $Fe(C_5H_5)_2$ (a) and its structure (b) [2.29]

2.5.4 Compounds of Inert Elements

Early in the 1960s, compounds of inert elements XeF_2, XeF_4, $XeCl_2$, oxides of these elements, etc., were discovered; they had formerly seemed impossible from the standpoint of the classical theory of valence. Figure 2.59 pictures the crystal structure of XeF_4. The XeF_4 molecules are flat, roughly square, the $Xe-F$ distances being 1.92 Å. The existence of such molecules is attributed to the interaction of $2p_\sigma$ orbitals of fluorine atoms and $5p_\sigma$ orbitals of xenon atoms; as a result, MO is formed, on which single electrons of the fluorine and an unshared pair of xenon electrons are located.

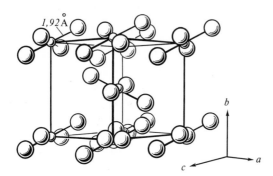

Fig. 2.59. Structure of XeF_4 [2.30]

2.6 Principles of Organic Crystal Chemistry

The crystal chemistry of organic compounds, or organic crystal chemistry, is unquestionably the richest area of crystal chemistry as regards the number of crystals and substances embraced by it. The number of investigated crystal structures runs into many thousands (Fig. 2.60) (30 000 by 1981)[2], but this is a drop in the ocean of the entire diversity of organic molecules, of which about 3 million are now known.

The structural unit of crystals of organic compounds is the molecule. The bonds within the molecule are covalent; their strength greatly exceeds that of the weak van der Waals intermolecular bonds. Therefore most of the organic structures are of the finite-groupings, insular type. In some compounds, not only van der Waals interactions between the molecules exist, but also hydrogen (or, more rarely, ionic) bonds, which are usually stronger than the van der Waals's. This often promotes the formation of chain or layer structures (see Sect. 1.5).

---

[2] Currently more than 3000 organic structures are being determined each year

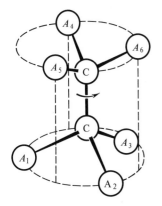

2.6.1 The Structure of Organic Molecules

The structure of organic molecules depends on the covalent bonds between the atoms of the molecule. The lengths of these bonds are well described by the sums of the covalent radii (Table 1.7). The multiplicity or intermediate order of the bonds is then taken into account. For instance, in aromatic compounds the order of the bonds in the rings is between $1\frac{1}{2}$ and $1\frac{1}{3}$. In other cyclic or chain groupings with formal alternation of single and double bonds conjugation phenomena arise, which is accompanied by some equalization of the bond lengths. The bond lengths in different molecules and the order of the bond are given in Fig. 1.46.

Knowing the chemical formula of a molecule and using the tables of covalent radii and graphs of the type of Fig. 1.46, it is possible to predict the interatomic distances of bonds of an intermediate nature within the molecule with an accuracy of about $0.02 - 0.05$ Å. Taking into consideration the orientation of the bonds − their tetrahedral, trigonal, or linear nature − one can often also predict, more or less unambiguously, the stereochemistry of the molecule, its space structure. Several versions are sometimes geometrically possible; then one can find the structure of the molecule from energy considerations. A fairly good approximation is in this case possible according to the "mechanical" model of a molecule, a system of atoms bound together by rigid, slightly extensible elastic hinges with spring fixing the angles between them.

The energy of the molecule in this representation [it should not be confused with the energy of the packing of molecules in the structure, i.e., with the lattice energy (1.52)] can be written as follows:

$$U_{\text{mol}} = U_{\text{nv}} + U_{\text{b}} + U_{\text{ang}} + U_{\text{tors}} \tag{2.5}$$

$$U_{\text{nv}} = \sum_{i,j} f_{ij}(r_{ij}), \tag{2.6}$$

$$U_b = \frac{1}{2} \sum_i K_i (l - l_0)^2, \tag{2.7}$$

$$U_{ang} = \frac{1}{2} \sum_i c_i (\alpha - \alpha_0)^2, \tag{2.8}$$

$$U_{tors} = \frac{U_0}{2} (1 + \cos n\varphi). \tag{2.9}$$

The interaction of non-valence-bound atoms of the molecule is described by the term U_{nv}, the expressions (1.45) or (1.16) being taken for the potential. Deviation of the bond lengths from the ideal l_0, or of the valency angles from the ideal α_0 results in energies U_b and U_{ang}. The term U_{tors} covers the energy of rotation of the molecules about the single bonds, and U_0 is the so-called internal rotation barrier.

The quantity U_{mol} can have one minimum (which defines the conformation of the molecule) or several minima similar in depth (then several conformations are possible, i.e., different conformers of the molecule). Note that the mechanical model of the molecule also helps to calculate its elastic properties, vibrations, etc. If the angles and distances are near to the ideal and there is no forced proximity of non-valence bound atoms, the molecule is unstrained, and its structure (provided no single bonds are present) can be predicted from geometric considerations. The situation is different in the presence of single bonds, about which rotations are possible. The mutual orientation of the parts of the molecule linked by such bonds depends on the nonvalence interactions U_{nv} between these parts, on the term U_{tors}, on contacts with neighboring molecules, and on the hydrogen bonds, if any.

The rotation about a single bond is restricted according to the requirement that the atoms of neighboring parts of the molecule linked by this bond should not approach each other by distances less than the sum of the intermolecular radii or, in any case, should be spaced as far as possible. For instance, atoms of two tetrahedral groupings linked by a C–C bond (Fig. 2.60) are stabilized in positions corresponding to inversion-rotation symmetry $\bar{3}$. But since atoms A_1, A_2, A_3, A_4, A_5, and A_6 may be different,[3] three rotational isomers are generally possible here; A_1 and A_6 are opposite one another (trans-configuration) or A_6 is turned to the right or to the left through 120° (two "gauche" configurations). In cyclical structures the closure condition limits the number of versions. Thus for six-membered rings of the type $(CH_2)_6$, the "chair" and "tub" conformations are possible (Fig. 2.61 a, b).

In such molecules with single bonds, and also in some polymers, the particular structure of the molecules in a crystal often depends on their mutual packing. To put it differently, the energy of a molecule (neglecting its constant

---

[3] These may be atoms of hydrogen, carbon, nitrogen, oxygen, etc. Other atoms may, in turn, be added to them.

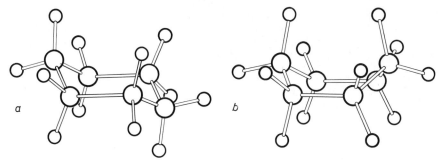

Fig. 2.61a, b. Idealized schemes of "chair" (a) and "tub" (b) formations in cyclohexanes

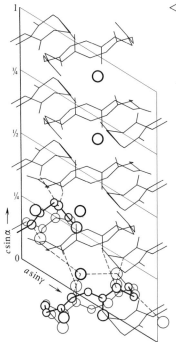

◁   **Fig. 2.62.** Structure of cyclohexaglycyl, in which three molecules occupying symmetrically nonequivalent positions differ in conformation [2.31]

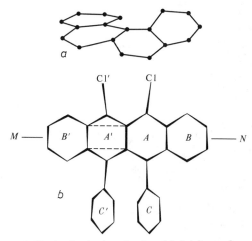

Fig. 2.63a, b. Strained molecules. (a) 3,4 benzophenanthrene $C_{18}H_{12}$ [2.32], (b) 5,6-dichloro-11,12-diphenylnaphtacene $C_{30}H_{18}Cl_2$ [2.33]

contribution U_{nv} and U_{ang}) and the energy of transition between conformers, and also the packing energy, are comparable in magnitude, hence, the energy minimum characterizing the structure as a whole determines both the packing and the conformation of the molecules. A part of the thermal vibration energy also may exert an influence. Sometimes it is found that chemically identical molecules in a given structure have several different conformations (Fig. 2.62); they form "contact isomers."

A number of organic molecules have a so-called sterically strained or "overcrowded" structure. The angles and bond lengths in them are distorted as compared with equilibrium. This is due to the steric hindrances, namely, the mutual approach of non-valence-bound atoms caused by the configuration of the valence bonds (Fig. 2.63).

It should be mentioned that the energy of the distortion of valence angles U_{ang} is much lower than the energy of extension of the bonds U_v; the angles readily deform by $10-20°$, while the bond lengths change by $0.02-0.03$ Å. In complicated strained aromatic molecules, the atoms which have moved towards each other find it more convenient to "twist out" in different directions from the plane of the molecule in order to increase their spacing, thereby also distorting the flatness of the molecule. Thus, all the terms in (2.5) play a part here (except U_{tors}, which is not used for aromatic molecules, having no single bonds).

Another factor stabilizing the definite configuration or causing some distortions is the intramolecular hydrogen bonds, say, in s-methyl-dithizone (Fig. 2.64, see also Fig. 2.75). Naturally, the mechanical model of the molecule is only a convenient approximation, and a more rigorous picture can be obtained by the $SCF-CAO-MO$ method with the aid of quantum-mechanical calculations (Sect. 1.2). But these calculations are extremely complicated.

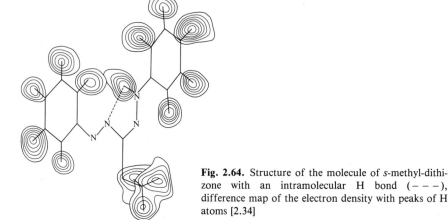

Fig. 2.64. Structure of the molecule of s-methyl-dithizone with an intramolecular H bond $(---)$, difference map of the electron density with peaks of H atoms [2.34]

The structure of molecules is only one of the aspects of organic crystal chemistry which is of special interest to the chemists. Although conformation analysis enables one in most cases to predict the structure of a molecule on the basis of the chemical formula, whenever the presence of different conformers is possible, the final and accurate data on the stereochemistry of the molecule, the order of the bonds, and the nature of thermal vibrations are obtained by

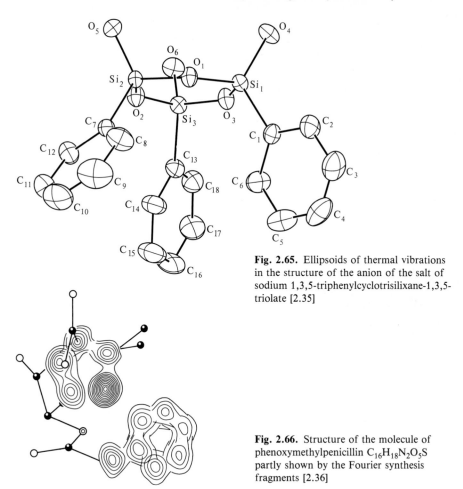

Fig. 2.65. Ellipsoids of thermal vibrations in the structure of the anion of the salt of sodium 1,3,5-triphenylcyclotrisilixane-1,3,5-triolate [2.35]

Fig. 2.66. Structure of the molecule of phenoxymethylpenicillin $C_{16}H_{18}N_2O_5S$ partly shown by the Fourier synthesis fragments [2.36]

x-ray structure analysis (Figs. 2.65, 66, see also Figs. 2.62, 64, 75). Moreover, x-ray structure analysis, which does not, in principle, require the knowledge of the chemical formula (which only facilitates structure determination), is sometimes a method for refining or establishing the chemical formula. Thus, the ultimate formulae (together with the structure of the molecule) of penicillin (Fig. 2.66), and vitamin B_{12} (Fig. 2.109) have been established by an x-ray diffraction technique.

2.6.2 Symmetry of Molecules

Molecules with a simple chemical formula are often symmetric, and their point symmetry is not associated with the crystallographic prohibition of 5th-, 7th-, or higher- order axes, although in most cases the order of the symmetry

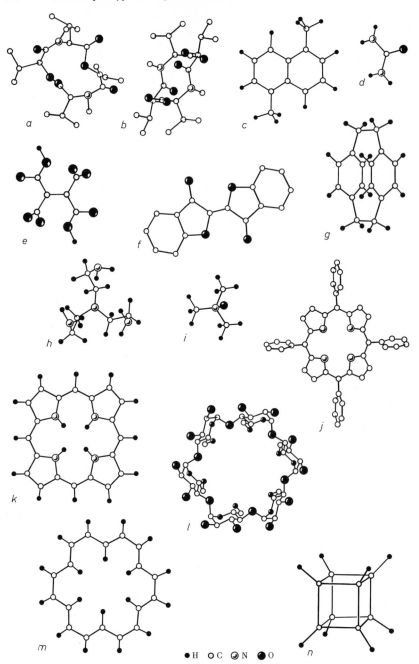

● H   ○ C   ◑ N   ● O

Fig. 2.67 a – n. Examples of organic molecules with different symmetry: **(a)** LDDD-form of cyclic tetradepsipeptide [− (MeVal − HyIv)₂ −], 1; **(b)** LLDD-form of cyclic tetradepsipeptide [− (MeVal − HyIv)₂ −], 1̄; **(c)** 1,5-dimethylnaphthalene 2/m; **(d)** urea, mm2; **(e)** dipotassium ethylentetracarboxylate, 2; **(f)** oxindigo, 2/m; **(g)** di-π-xylylene, mmm; **(h)** ion of 2,2′,2″-tri-

chirality. Therefore, for systematics and statistics of organic crystal structures the concept of the structural class is introduced [2.40]. The structural class is a set of crystals with a given group Φ, in which the molecules occupy identical systems of equivalent positions (i.e., their centers are identical RPS). For instance, a widely spread structural class is that of naphthalene (Fig. 2.69) $P2_1/C$, $Z = 2(\bar{1})$; the molecules occupy one of the systems of centers of symmetry. In another class of the same space group, $Z = 4(\bar{1},\bar{1})$ two systems of centers of symmetry are occupied, a representative being tolan (Fig. 2.70).

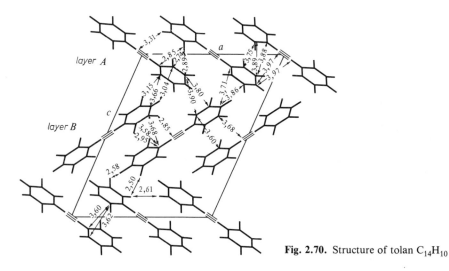

Fig. 2.70. Structure of tolan $C_{14}H_{10}$

The structural classes can be divided into four types according to the chirality or nonchirality of the positions occupied by the molecules forming the crystal. Type A contains groups Φ^I of the first kind, i.e., the structure is built up of chiral molecules (only right-hand or only left-hand). Types *B, C,* and *D* contain groups Φ^{II} of the second kind. In type *B* the positions of the molecules are chiral, but right- and left-hand in equal numbers. In type *C* the positions occupied by the molecules are nonchiral. In type *D* the molecules occupy both chiral and nonchiral positions.

Type *A* crystals include most of the molecular structures of natural compounds. As indicated above, their molecules exist in one of the two possible enantiomorphous forms and, hence, are optically active. Therefore such crystals show no symmetry elements of the second kind, and thus only two groups, $P2_12_12_1$ and $P2_1$, are most probable for them. The predominance of these groups also follows from the usual presence of H bonds in natural compounds. As confirmed by statistics, the most common class in type *A* is $P2_12_12_1$, $Z = 4(1)$; it contains 56% crystals of type *A* (16% of all the homomolecular crystals), and class $P2_1$, $Z = 2(1)$, 25% (and 7%), respectively. Thus, almost all the amino acids crystallize in one of the two indicated groups.

Type B crystals are the most numerous. The most common class among them is $P2_1/c$, $Z = 4(1)$; 52% (28% of all the crystals), it is followed by $P1$, $Z = 2(1)$; 10% (6% of all the crystals); and then comes $Pbca$, $Z = 8(1)$; 7% (4% of all the crystals). These crystals include families of natural and synthetic compounds, i.e., a mixture of equal numbers of l and d enantiomers. In C type crystals, the same group, $2_1/c$, predominates, and the most common class is naphthalene, $Z = 2(\bar{1})$; 39% (6% of all the crystals), which is followed by $P\bar{1}$, $Z = 1(\bar{1})$; 12% (1.9% of all crystals). The six most common structural classes include about 70% of the homomolecular crystals.

There are also certain high-symmetry organic crystals; they form when the molecules with the corresponding high symmetry have a shape suitable for close packing. For instance, the molecule of hexamethylenetetramine has the symmetry $\bar{4}3m$, the space group of the structure being $I\bar{4}3m$ (Fig. 2.71); the molecule of triazidecyanuric acid has the symmetry $\bar{6}$, the space group of the structure being $P6_3/m$ (Fig. 2.72).

In some organic crystals the arrangement of the molecules is such that in addition to space symmetry operations there are some noncrystallographic operations of local symmetry or supersymmetry [Ref. 1.6, Sect. 2.5.5], which do not belong to the set of operations of group Φ. Thus, local noncrystallographic axes 2, by which pairs of molecules are related, and supersymmetry screw axes 2_1, 3_1, and 4_1 were observed [2.41]. It is obvious that noncrystallographic symmetry arises when it is advantageous from the standpoint of close packing of molecules (or linking molecules by hydrogen bonds), while the

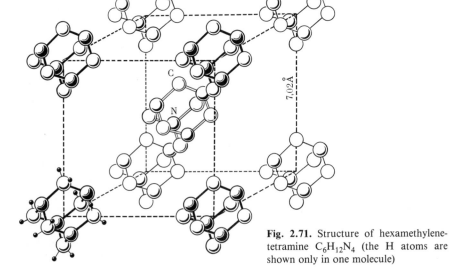

Fig. 2.71. Structure of hexamethylenetetramine $C_6H_{12}N_4$ (the H atoms are shown only in one molecule)

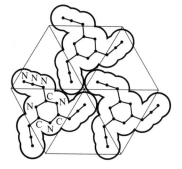

Fig. 2.72. Packing of the molecules of triazidecyanuric acid in a crystal

further uniting of such groupings with proper symmetry occurs within the framework of group Φ, which has already different symmetry operations of its own.

The packing coefficient [see (1.75)] for organic structures does not decrease lower than $0.65 - 0.68$; for most structures it equals $0.70 - 0.75$ and sometimes reaches 0.80. The close-packing principle is only the first, rough approximation, which realizes geometrically the condition for the minimum of packing energy [Sect. 1.5 and (1.52)]. More accurate quantitative results are produced according to the method of energy calculations using atom – atom potential functions [see (1.16, 45)]. In analyzing the mutual arrangement of molecules it is sometimes sufficient to take into account the symmetry of the potential functions, considering energy $U_{\varphi\theta\psi}$ (1.57) when a pair of molecules are spaced at a distance r and have different orientations assigned by angles θ, φ, and ψ.

The most comprehensive physical interpretation of the arrangement of molecules in organic crystals requires thermodynamic considerations with due regard for the lattice dynamics, which helps to calculate, not only the structure of the crystal, but also its main characteristics, such as the heat capacity, compressibility, thermal expansion coefficient, etc. [2.42].

It is noteworthy that, because of weak van der Waals bonds, the thermal vibrations of molecules as a whole and of their constituent atoms in organic crystals are very large. The coefficients B of the temperature factor usually equal $3 - 4$ Å^2, i.e., the mean square displacements are of about 0.2 Å. The displacements of atoms are largely due to the vibrations of a molecule as a whole, and therefore they are greater near the periphery of the molecule. But individual components of the thermal motion of the atoms can also be determined,depending on the structure of the molecule and character of the bonds between its atoms (Fig. 2.65).

Molecular crystals are insulators, as a rule, because the electrons are localized on bonds inside the molecules. However, organic conductors of the quasi-one-dimensional type are known, in which conductivity is achieved due to the packing in stacks of flat organic molecules, which are electron donors.

An example of such a compound is tetrathiotetracene (TTT) and its selenium analog TSeT

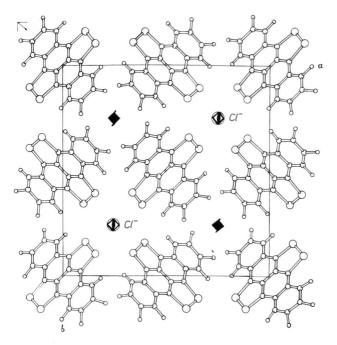

The structure of $(TSeT)_2^+ Cl^-$ is shown in Fig. 2.73. The molecules are packed into infinite stacks extending along the c axis, the spacing between the molecular planes being 3.37 Å. The conductivity in the c direction is 2.10 × 10^3 ohm^{-1} cm^{-1}, i.e., is essentially metallic; it is achieved owing to the maximum overlapping of the π orbitals of the molecules; the conductivity in the other directions is several orders of magnitude lower [2.43]. Because of this high conductivity these compounds are called "organic metals."

Another class of organic compounds − stable nitroxyl radicals − exhibits paramagnetism and nonlinear optical properties. Figure 2.74a shows a map of the difference deformation electron density of a molecule of the organic paramagnetic $C_{13}H_{17}N_2O_4$. Atom O(1) is charged negatively and forms an intramolecular hydrogen bond with H(22). The bond N(4) − O(2) reveals an extended cloud of the π orbital perpendicular to the plane of the molecule (Fig. 2.74b), which evidently carries the nonpaired electron responsible for paramagnetism.

Fig. 2.73.
Structure of "organic metal" $(TSeT)_2^+ Cl^-$. Projection along [001]

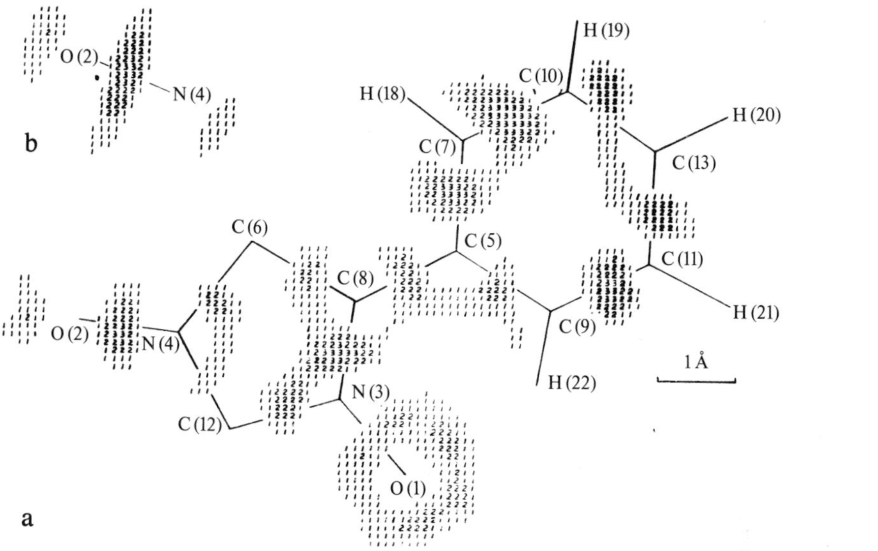

Fig. 2.74. (a) Difference deformation electron density map of paramagnetic compound $C_{13}H_{17}N_2O_2$, **(b)** cross section perpendicular to $N(4)-O(2)$ bond (the numbers indicate the electron density, in 0.1 $e\text{Å}^{-3}$) [2.44]

2.6.4 Crystals with Hydrogen Bonds

The principle of close packing of molecules has to be extended when we refer to structures with hydrogen bonds. These bonds are also formed, in particular, by most molecules of biological origin. Van der Waals bonds determine the close packing in the presence of hydrogen bonds, despite their higher energy, so that a different principle of joining the molecules is fulfilled. The hydrogen bond is directional, and the "projections" of one molecule, namely H atoms of OH or NH groups, draw closer together with the "projections" of the other molecule, i.e., O and N atoms. Therefore, if a molecule of arbitrary shape has these "projections", active in the sense of the hydrogen bond, a convenient mutual arrangement of a pair of such molecules may be realized with the aid of a center of symmetry or a twofold symmetry axis. Often, especially in the case of asymmetric molecules, the disposition on their surface of atoms linked by H bonds is inconvenient for closing into pairs. Then, for instance, NH groups of one molecule combine with O atoms of the other, and the NH groups of this next molecule combine precisely in the same way with the O atoms of a third molecule, and so on. This results in chains which wind around the screw axes of symmetry, mostly around axes 2_1 (Figs. 1.37, 2.75a, b). Sometimes such an axis may be noncrystallographic. If the molecule has more than one hydrogen atom forming H bonds, the structures with two- or three-dimensional networks of hydrogen bonds can arise (Fig. 2.76).

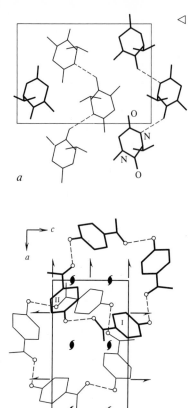

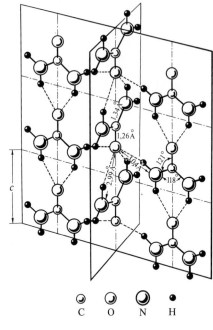

⊲ **Fig. 2.75 a, b.** Chains of molecules united by hydrogen bonds (----): **(a)** in the structure of cyclo-L-valylsarkozyl [2.45], **(b)** in the structure of oxyacetophenone; the chains have the form of helices with local symmetry 4_1. The helix axes are parallel to the *b* axis [2.46]

Fig. 2.76. Structure of urea $CO(NH_2)_2$

C   O   N   H

Thus, in organic crystals with hydrogen bonds the appearance of twofold axes or a center of symmetry could be expected for the packing of molecules possessing symmetry elements of the second kind, and the appearance of simple or screw twofold axes for asymmetric molecules and those with symmetry elements of the first kind, which is actually the case.

The formation of a space network of hydrogen bonds may cause the appearance of stable symmetric configurations. The urea molecule

$$\begin{array}{c} O \\ \| \\ H_2N^{\diagup C}{}^{\diagdown}NH_2 \end{array},$$

for instance, has symmetry $mm2$, while the tetragonal space group $P\bar{4}m$ of its structure (Fig. 2.76) is unusual from the standpoint of close packing. It is the consequence of the arrangement of H bonds. Upon saturation of the hydrogen bonds between molecules, the close-packing principle and van der Waals forces come into play, which results in structures where the hydrogen bonds are saturated and the molecules are packed quite densely.

2.6.5 Clathrate and Molecular Compounds

The clathrate, or inclusion, compounds make up an interesting group among the organic crystals. These are structures composed of two sorts of structural units: organic molecules and some small organic (for instance HCOOH) or inorganic (H_2S, HCl, SO_2, etc.) molecules fitting into the voids in their packing. The bond with these molecules may be not only van der Waals ones, but also hydrogen or ionic. An example of a clathrate structure is presented in Fig. 2.77. An interesting kind of clathrate structure is offered by the filling of voids in icosahedral packings of H_2O molecules by the atoms of inert elements or by small molecules (see Fig. 2.78).

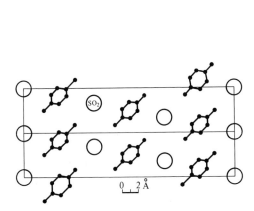

Fig. 2.77. Clathrate compound $3C_6H_4(OH)_2 \cdot SO_2$

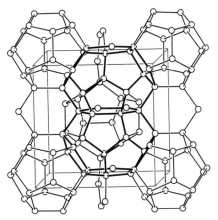

Fig. 2.78. Network of icosahedra and more complex polyhedra formed by water molecules in a clathrate compound

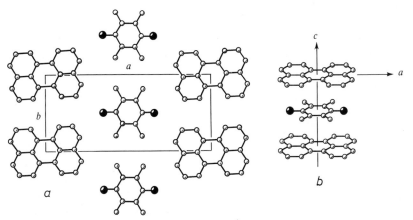

Fig. 2.79 a, b. Molecular complex of perylene $C_{20}H_{12}$ with fluoroanil $C_6F_4O_2$. (a) One layer of the structure located near the plane $z = 0$, (b) superposition of molecules along the c axis

In recent years it has been found that organic molecules differing in nature, but not greatly differing in size, may cocrystallize, forming a crystal with a stoichiometric ratio of the components (mostly 1 : 1). The structure and formation of such molecular compounds (Fig. 2.79) can be explained proceeding from the maximum-filling principle; in this case the shapes of the molecules complement each other to achieve the highest packing coefficient. Such "compounds" exist only in solid phases. Molecular compounds and solid solutions of polymer and ordinary organic molecules are known.

Solid solutions are known to form in organic crystals; however, their solubility limits are low, up to several per cent. For instance, technically important organic crystals − scintillators − are solutions of certain molecules in a matrix of aromatic compounds (say, 10^{-4} wt.% anthracene in naphthalene, 0.5 wt.% phenanthrene in naphthalene, etc.). In a number of organic crystals complete or limited rotation of molecules or their parts has been observed.

2.7 Structure of High-Polymer Substances

2.7.1 Noncrystallographic Ordering

Proceeding to a description of the basic features of the structures of high polymers and liquid crystals, as well as those of biological macromolecules and their associations, we somewhat expand the limits of classical crystallography, which deals (or, more precisely, dealt) exclusively with true crystal structures exhibiting three-dimensional periodicity. At the same time crystallography (if we regard it as a science of the ordered atomic structure of condensed matter) has gradually taken on new objects, which did not necessarily show three-dimensional periodicity, but were characterized by some ordering and a definite symmetry, which were sometimes very close to, and sometimes very far from, those of crystals. Of course, this enormous class required proper theoretical and structural treatment because of the widespread occurrence of these diverse structures, the importance of their functions and properties in organic and inorganic nature, and their increasing application in industry and technology. This task was entrusted to crystallography, which had at its disposal the methods and tools best suited to cope with it. The extension of its field of application to a wider class of objects is sometimes emphasized by using the term "generalized crystallography", implying that attention is focused on structures of live organic matter.

2.7.2 Structure of Chain Molecules of High Polymers

The high-polymer compounds are built of long, chain-type molecules. It is just this feature that explains most of the physico-chemical properties of polymers. However, not only the structure of the molecule as such is important, but also the character and manner of mutual packing, i.e., the aggregation of the chain molecules into a polymer substance.

We shall first consider the structural principles of the chain molecule itself. Such a molecule is a one-dimensional sequence of atoms or radicals linked by covalent bonds, to which side groupings can be attached at certain sites. Here, the term "one dimensional" indicates, of course, only the type of sequence of atoms or radicals in the three-dimensional chain molecule. As a rule, it is possible to distinguish, in such a molecule, a consecutive single chain of atoms, the backbone of the molecule (Fig. 2.80a). The backbone will be more complex if the chain contains cyclic or more complicated groups (Fig. 2.80b).

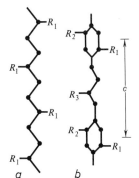

Fig. 2.80a, b. Structure of chain molecules formed by a simple (**a**) or complex (**b**) chain

The chain is uniformly constructed if it consists of one sort of radical and if all its units' links are identical (homopolymers). A chain consisting of various radicals (heteropolymers) is more complicated. Two or more kinds of radicals may be arranged in a random order along the chain. Such chain molecules do exist, although chains of regular structure are for more common. The simplest requirement for ordering − the repetition of some structural unit along the molecule axis − is maintained to some extent in them. The repetition is an essential feature of a chain molecule, although often its occurrence is statistical rather than strict.

Symmetry groups of chain molecules are groups G_1^3, which describe three-dimensional structures, periodic in one direction. The width of chain molecules in the direction perpendicular to translation (i.e., their "thickness") for most of the synthetic polymers does not exceed 10 Å, but in complex biological polymers it may reach several tens of angstrom.

When the number of links in a molecule exceeds 10 or 20, the peculiarities of the structures of the molecules and their packings, typical of high-polymer substances, begin to manifest themselves. In true polymers the number of links runs into the thousands and millions. The simplest examples are paraffins C_nH_{2n+2} and the "infinite" paraffin chain of polyethylene $(CH_2)_n$ (Fig. 2.81), its backbone consists of a chain of carbon atoms ⌢C⌣C⌢C⌣C , the repeating unit being group CH_2.

The structure of a chain molecule is governed by the general rules concerning the direction and length of covalent bonds, which determine the backbone structure and the attachment of side radicals, and by the van der Waals interactions of the side radicals with themselves and with the backbone atoms.

These interactions, as well as those of the chain molecules with themselves in their aggregates, stabilize definite conformations of the backbones of chain molecules, which are highly mobile owing to the single bonds, which are almost always present and permit rotation (see Fig. 2.60). Thus, taking into consideration the repulsion of C atoms in a polyethylene chain it is easy to understand that the trans-configuration is the most stable for it (Fig. 2.81a). The repulsion of adjacent links of the chain practically always tends to "straighten" it out. But the shape of the molecule as a whole is governed by the shape of the individual links at their contact, and such repulsion may lead to a conformation of a more general type, namely to a helical structure (a particular case is a zigzag conformation). Let us consider tetrafluoroethylene $(CF_2)_n$, in which all the hydrogen atoms in the polyethylene chain are replaced by fluorine. Here, the chain configuration changes from zigzag to helical (Fig. 2.81b). The reason is that H atoms with an intermolecular radius of ~1.2 Å

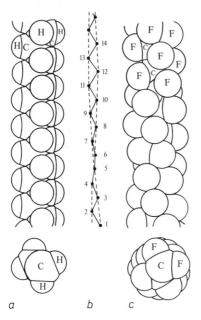

a   b   c

Fig. 2.81a–c. Chain molecules of polyethylene (**a**) and tetrafluoroethylene (**b, c**). (**b**) shows the arrangement of the centers of the C atoms

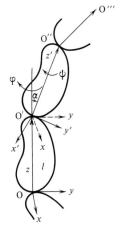

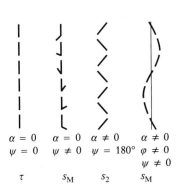

$$\begin{array}{cccc} \alpha = 0 & \alpha = 0 & \alpha \neq 0 & \alpha \neq 0 \\ \psi = 0 & \psi \neq 0 & \psi = 180° & \varphi \neq 0 \\ & & & \psi \neq 0 \\ \tau & s_M & s_2 & s_M \end{array}$$

Fig. 2.82. Mutual arrangement of the links of the chain molecule and the angular parameters characterizing it

Fig. 2.83. Principal types of chain molecules and their symmetry

settle freely one over another, because the periodicity of the backbone of the carbon atoms is 2.54 Å > 2 × 1.2 Å. The fluorine atoms, however, have an intermolecular radius of 1.4 Å, and the indicated periodicity cannot be realized, since 2 × 1.4 > 2.54 Å. These spatial stresses are overcome by a small rotation about the C–C bonds with respect to the trans-position, which ultimately leads to a helical structure of the backbone.

The geometry of the ideal periodic chain molecule can be described as follows. It consists of monomeric units – links. The succession of the links (Figs. 2.80, 82) can be described by the inclination α of the axes of the links, the angle φ of torsion of axis $O'O''$ of a link about axis OO'' of the preceding, and, finally, the angle of rotation ψ of the link about its own axis $O'O''$. The characteristics of the chain are the length of the link l, the value c' of the projection of l on the axis of the molecule, and period $c = pc'$ of the chain molecule (p links per period constitute a unit grouping, an analog of the unit cell of a crystal). If the molecule is helical, the period c does not necessarily correspond to a single turn of the helix; translational repetition may be achieved after q turns, and the general case of the helical symmetry S_M, $M = p/q$ [Ref. 1.6., Fig. 2.58].

The principal types of periodic chain molecules are shown in Fig. 2.83. The simplest case is a simple repetition of monomers: $\alpha = 0$ and $\psi = 0$. At $\alpha = 0$ and $\psi \neq 0$ we obtain a class of screwlike, linear molecules. The case where $\psi = 180°$ and $\alpha \neq 0$ yields a large class of zigzag chain molecules (polyethylene and others). The general case of arbitrary $\alpha \neq 0$, $\varphi \neq 0$ and $\psi \neq 0$ gives different elongated or gently sloping helices, which are thus the most general type of the chain molecule (Figs. 2.84 – 88). Note that the helical structure of a chain molecule is clearly revealed in the diffraction pattern (see Fig. 2.156).

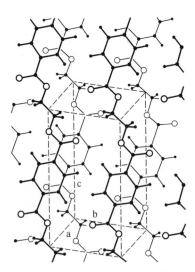

Fig. 2.84. Structure of crystals of polyethyleneterephthalate $(-CH_2-CH_2-CO_2-C_6H_4-CO_2-)_n$. The molecule is formed by translational repetition of the monomer [2.47]

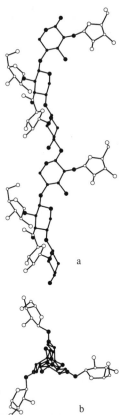

Fig. 2.86a, b. Structure of a polysaccharide: xylane backbone with side groups of arabinose [2.49]. (a) View along the perpendicular to the helix axis, (b) projection along the helix axis

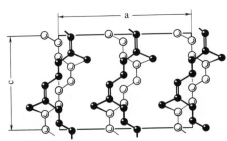

Fig. 2.85. Structure of gutta-percha $[-CH_2C(CH_3)CHCH_2-]$ [2.48]

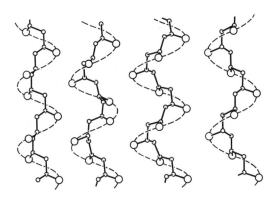

Fig. 2.87. Structure of some regular helical polymers with different helix parameters

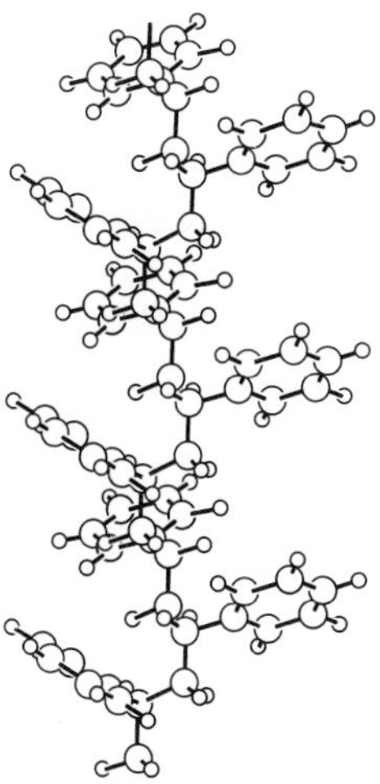

Fig. 2.88. Structure of polystyrene (compare with the first structure in Fig. 2.87)

A number of polymers are built so that two or more sorts of side radicals are attached in some order to the backbone, which has a fixed structure. In this case the description of the chain molecule is more complicated. The possible cases are a) a regular sequence of radicals R_1, R_2 ... with a certain period, b) their statistical distribution described by some distribution function, and c) random distribution of radicals along the chain.

Speaking of the order of attachment of the radicals, or of violations of this order, one must keep in mind not only the individual chemical nature, the "types" of radicals and their relationships with specific atoms of the chain, but also such geometric characteristics as the "right-" or "left-handed" attachment and the possible different orientations of the attached radicals.

Thus, the tetrahedral carbon atom in a chain molecule has two possible positions of attachment (l and d positions)

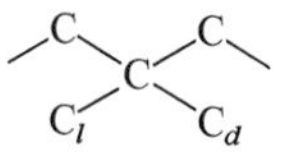

If we consider the possible attachments of radicals to such a chain, we can have a chain of the type ... *lll* ... (or ... *ddd* ...) (isotactic polymer), of the type ... *ldld* ... (syndiotactic polymer), or with random alternation of l and d positions (e.g., atactic polystyrene). These are different cases of stereoisomerism in chain molecules. Finally, we must mention one more kind of possible disordering of polymer chains; this is when they are composed of structurally identical but polar links, i.e., the links have a "head" and a "tail" and, therefore, can enter the chain in two different positions. Complicated chain molecules of biological origin may consist of several chains; their structure will be discussed below.

2.7.3 Structure of a Polymer Substance

Let us now turn to the structure of three-dimensional aggregates of chain molecules. The packing of molecules is described by the familiar principles of organic crystal chemistry. These principles operate, however, within the framework of extremely specific circumstances − the chain structure of molecules and their great length. The former determines the main peculiarity of polymer substances, the parallelism of the packing of molecules. The latter leads to a very strong influence of kinetic factors (conditions of formation) on the chain-molecule aggregate's structure.

Owing to the great length of chain molecules the formation of a three-dimensionally periodic crystalline structure from them is impeded. In the course of crystallization such long and flexible molecules must straighten, either completely or in some of its parts, and occupy strictly definite places in a definite orientation. But interaction with the neighbors, entanglement, twisting, etc. hinder this in every possible way. Therefore, in most of the chain-molecule aggregates an ordered equilibrium state is not achieved, and the degree of ordering depends on the character and time of condensation and further treatment.

Formely, the two concepts "crystalline" and "amorphous" components were used in describing the structure of polymer substances. In some parts of a polymer substance the chain molecules are stacked regularly with respect to each other and form a "crystalline", although maybe not perfect, region several tens or a hundred angstroms in size. The great length of the molecules lead to their entanglement and does not allow the lattice to be built over the entire volume of the object; therefore, crystalline regions interchange with those of amorphous substance, in which no order exists. A single molecule may run through several crystalline and amorphous regions. It is clear at this juncture that the concepts just described are extremely tentative, although they are justified as certain limiting cases, which are sometimes realized. Apart from the crystalline and amorphous structure, various intermediate types of chain-molecule aggregation are possible; they will be treated in Sect. 2.7.5.

2.7.4 Polymer Crystals

Some polymers crystallize under certain conditions forming a crystal lattice throughout their volume or part of it. A necessary (but not a sufficient) condition for the formation of a truly crystal structure is the regular structure of the chain molecule itself, its stereoregularity. Regions with crystalline ordering can, as usual, be characterized by the unit cell, space group Φ, etc. Figure 1.80 presents the unit cell of n paraffins (the same cell as in polyethylene), and Figs. 2.84, 85, that of polyethylene terephthalate and gutta-percha.

Fig. 2.89 a, b. Crystals of polyethylene $(CH_2)_n$. (a) spiral-layer structure, (b) corrugated zone structure [2.50]

Macroscopic-faceted single crystals can be grown from solution of some simple synthetic polymers – polyethylene, polypropylene, polyamides, and others (Fig. 2.89). The crystals have a characteristic lammelar habit and often exhibit clearly defined signs of a spiral (dislocation) growth mechanism. Investigation into the morphology of these crystals with the aid of electron microscopy and x-ray diffraction analysis revealed a remarkable feature of their structure. It was found that, in the course of packing into the crystal structure, the long polymer molecule, with its many thousands of links, bends over many times to form straight parts of the same length of about $100-150$ Å. The result is an elementary layer of a crystal shown in Fig. 2.90a. The thickness of the elementary growth steps on the crystal surface (Fig. 2.89) corresponds to the length (L) of the straight parts of the folded chain molecule. The length L (and hence the thickness) of the layer is constant under given conditions, but is temperature dependent. To a higher crystallization temperature corresponds a larger L. If a formed crystal is heated, L also increases. This indicates that the layers become thicker owing to the passage of molecules through the bending site and longitudinal slipping of the straight parts of the chains relative to each other. Theoretical considerations show that bend-

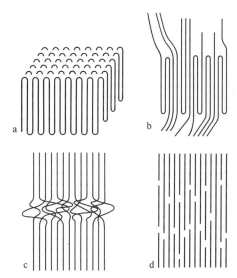

Fig. 2.90 a – d. Structure of the crystalline regions in polymers. (a) Crystals formed by regular folding of chain molecules, (b) polymers crystallizing from the melt or subjected to orienting treatment, (c) crystalline region of a polymer obtained by flux crystallization, (d) crystalline region obtained by polymerization of a monomer single crystal

ing can be ascribed to the contribution from the free surface energy to the total energy of the crystal; the change in L can be explained likewise.

The growth of a single layer of a crystal is achieved by the wrapping of repeatedly folded chains around the lateral face. This is the basis of many interesting features of such crystals, for instance, that their lateral faces and sectors are not equivalent, and that the flat face of the crystal is perpendicular to the molecular axis in each sector. Thus, on transition to the condensed state chain molecules tend to self-aggregation by folding.

Multiple folding is also a basic mechanism in crystallization of polymers from the melt. Since a perfect crystal structure cannot be achieved, much of the volume is occupied by crystalline regions, while in its smaller part the ordering is lower. Crystalline regions are of different shape and size. While they are forming, the polymer molecules can bend over many times or pass through them. The structure of a crystalline region in an aggregate of this type is schematically shown in Fig. 2.90b.

By orienting treatment, i.e., crystallization of a polymer simultaneously with its drawing, stretching of specimens, etc., one can obtain a parallel orientation of molecules over a large distance. This may produce fiber-like (microfibrillar) sections interchanging with "knots" of entangled molecules (Fig. 2.90c). When crystalline polymer specimens composed of crystalline blocks are stretched, individual folded crystals (lamellae) disintegrate and the chains folded into them unbend and stretch into a microfibrillar structure (Fig. 2.91).

One of the possible ways for parallel packing of polymer molecules consists in their crystallization in a melt flux with a velocity gradient. In such cases, and sometimes in crystallization from solutions, fibrillar crystals with lamellar thickenings are produced, which are called "shish kebab" structures (Fig. 2.92a) [2.52]. The appearance of these interesting structures, the packing

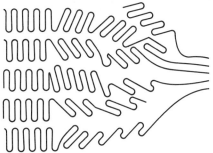

Fig. 2.91. Model of formation of a microfibrillar structure from folded crystals [2.51]

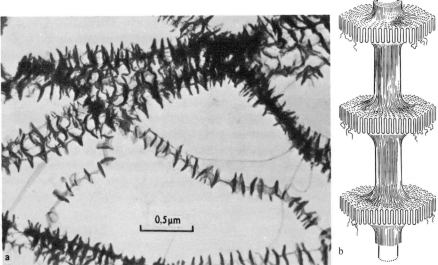

Fig. 2.92a, b. Fibrillar crystals with lamellar overgrowth ("Shish kebabs") of polyethylene. (a) Electron micrograph, (b) molecule packing model

of molecules in which is demonstrated in Fig. 2.92b, is due to the fact that the orienting action is not uniform throughout the system. Presumably, the fibrillar core structure is first formed, which serves as a nucleus for the growth of portions with a folded (lamellar) crystalline structure around it at certain distances from each other.

In natural polymer fibers the orientation and parallel arrangement of the chains are caused by the fact that the crystallization (orientation) process coincides with the synthesis of the chain itself.

A method is known for obtaining crystals of some polymers in which the chain molecules are packed parallel without bends along their entire length. This is so-called directed stereoregular polymerization in the solid phase. The precursor of such a crystal is a single crystal of a monomer. Polymerization – sewing together of monomers – gives rise to an array of chain molecules with the order laid down by the crystalline array of the monomers. For instance, single crystals of trioxane exposed to x radiation give chain molecules of poly-oxymethylene $(-CH_2O)_n$, which are packed according to Fig. 2.90d to form a highly ordered crystal. Such crystals possess a very high strength, close to theoretical.

2.7.5 Disordering in Polymer Structures

Let us now consider aggregates of chain molecules with less-than-crystalline ordering. The determining factor here is the presence of the short-range order in the packing of near-parallel molecules. The following deviations from the ideal three-dimensionally periodic packing may take place (Fig. 2.93): a) parallel shifts of molecules along their axes; b) turns (rotations) of molecules about the axes; c) distortions of two-dimensional periodicity in the projection of molecules on their axes ("net distortions").

These distortions are almost always interrelated. They may be accompanied by nonparallelism or bending of the chains. The distortions are described by functions of a statistical nature. Thus, the statistics of the shifts can be assigned by displacement function $\tau(z)$. This function indicates the probability of displacement of molecules along the axis from some ideal position. If any shifts are equally probable, then $\tau(z)$ has a constant value. Such an arrangement can be characterized by the limiting symmetry operation of an infinitely small translation τ_∞.

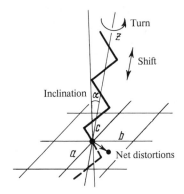

Fig. 2.93. Parameters specifying the position of a chain molecule

"Turns" imply the statistical spread of the azimuthal orientations of the different molecules of a given aggregate about the equilibrium angular position. They are characterized by function $f(\psi)$, which gives the probability of finding a molecule in the azimuthal orientation at an angle ψ.

The angular spread of the molecules, especially if their cross section is nearly circular, readily transforms into a complete set of all possible orientations, and then one can speak of the "rotation" of molecules. In that case $f(\psi) = \text{const}$. The term "rotation", as well as "shift", should be understood

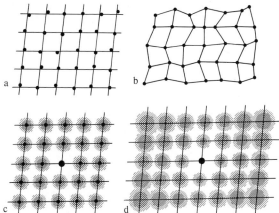

Fig. 2.94a – d. Distortions in the arrangement of the projections of the axes of chain molecules onto the basal plane, (a) first-kind, (b) second-kind, (c, d) corresponding distribution functions

statistically, i.e., as indicating that different molecules are oriented differently with respect to the azimuth. But sometimes (as in paraffins at temperatures near the melting point) torsional thermal vibrations of molecules about the principal axis become so large that one can speak of the real rotation. The statistical symmetry of a "rotating" molecule is characterized by the rotation symmetry of infinite order, ∞.

We shall now consider the distortions in the net of the projections of molecular axes onto a certain plane. The ideal net is characterized by two translations a and b. Its distortions consist in the fact that the system of translations is preserved only statistically. Two cases are possible here, so-called distortions of the first and second kind. Distortions of the first kind can be described as follows. There is an infinite regular system of points of a two-dimensional net, and there is a certain probability (which diminishes with the distance from the point) that the axis of a given molecule may be displaced from this point (Fig. 2.94a). Then, although each of the molecules is not actually at the point, the crystalline ordering throughout the volume is preserved on the average, and it is said that the long-range order is retained.

Distortions of the second kind are of a different geometric nature (Fig. 2.94b). Although the (statistical) average translations $\bar{a}$ and $\bar{b}$ can also be indicated there, these translations give the probable mutual arrangement only of the neighboring molecules and do not extend to the description of the projection net or of the volume as a whole (Fig. 2.94b). Only the short-range order is retained. Such a type of ordering is also called paracrystalline.

Net distortions can be described with the aid of the statistical distribution function $W(r)$, which gives the probability of the arrangement of the neighboring molecules relative to the given one at point r. For first-kind distortions the distribution function $W(r)$ is periodic, but its peaks are smeared out (Fig. 2.94c). With second-kind distortions $W(r)$ has the form shown in Fig. 2.94d.

It clearly shows the distribution of the first nearest neighbors — the short-range order, — because there is a minimum distance to which molecules are allowed to approach each other, and a maximum distance, since no large gaps between molecules can exist, either. It is easy to understand that the distances to the next-nearest neighbors will vary over a wider range, since deviations in translations not only from a given molecule to a neighboring one, but also from the neighboring to the next one are accumulated. The further the neighbor, the more "smeared out" is the distribution function of the second kind, and at a certain distance it becomes practically constant, the probability of finding a molecule being the same everywhere. It can be said that translations $\bar{a}$ and $\bar{b}$ gradually degenerate into continuous ones, τ_∞^a and τ_∞^b.

In many x-ray structure investigations of fibrous materials and polymers, the "unit cell" of the packing is determined. In light of the above-described considerations it is easy to understand that if there is no in an assembly of chain molecules with second-kind distortions the "unit cell" concept has only a statistical meaning, and the "periods" $\bar{a}$ and $\bar{b}$ characterize the short-range order.

Apart from the indicated distortions, polymers show nonparallelism of chains and bends. Bends may be conforming ("curvilinear crystal") or unconforming, when the molecules tangle up, retaining approximate parallelism (Figs. 2.95, 97). Bends can also be characterized by the statistical function $D(\alpha)$, which gives the probability of the deviation of molecular axes by angle α from the principal axis.

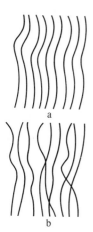

Fig. 2.95a, b. Conforming (a) and unconforming (b) bending of chain molecules

Different types of ordering of polymer substances taking into consideration their mutual effects and indicating the functions characterizing each type are presented in Fig. 2.96. The limiting case of disordering [random orientation $\infty/\infty\ W(r)$] corresponds to the "amorphous" state of polymers. Even in amorphous polymers, chain molecules form regions in some places, where they are approximately parallel to each other (Fig. 2.97).

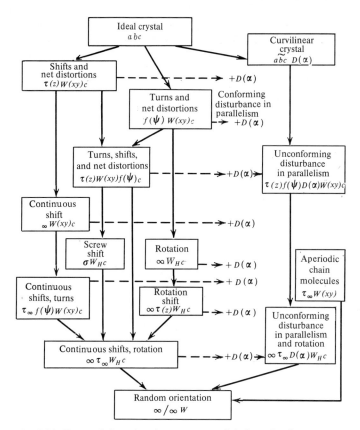

Fig. 2.96. Types of distortions in aggregates of chain molecules

Fig. 2.97. Structure of amorphous polymers

As in polycrystalline solids, macroscopic polymer specimens may consist of regions which are identical as regards the type of ordering, and may be differently oriented relative to each other. In most cases, textures are formed. The ordering regions have no distinct boundaries; there are transition zones between them, whose order is always lower than in the regions themselves.

Thus, high-polymer substances made up of chain molecules have many gradations of ordering both at the level of molecular packing and at the level of packing microregions, which themselves may show different ordering.

2.8 Structure of Liquid Crystals

2.8.1 Molecule Packing in Liquid Crystals

There exists an interesting class of organic substances capable of forming liquid crystals. The ordering of molecules in liquid crystals is neither three-dimensionally periodic, as in a solid crystal, nor random as in a liquid, but is of an intermediate nature. The fact that liquid crystals exhibit a certain ordering of molecules is indicated by the anisotropy of the liquid-crystal substances, although they are fluid. Liquid crystals or, as they are also called, mesomorphous phases arise within a certain temperature range from t_{lc} to t_l, where t_{lc} corresponds to the transition of a solid to a liquid crystal and t_l, to that of a liquid crystal to an isotropic liquid.

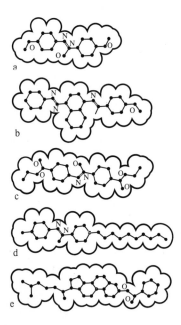

Fig. 2.98 a – e. Shape of some molecules forming the liquid crystal. (a) Para-azoxyanisole, (b) α-benzolazo-α-anizyl-naphthylamine, (c) ethyl ether of para-azoxybenzoic acid, (d) n-n-nonoxybenzolyl-toluidine, (e) cholesterylbenzoate

Molecules of substances forming liquid crystals have an anisometric elongated shape; they have low symmetry or are asymmetric, as a rule (Fig. 2.98). Such molecules are often nonuniform in thickness, but may have some parts approximately equal in cross section. Arrangement of two benzene rings or some other aromatic groupings in the middle of a molecule and the presence of hydrocarbon "tails" on one or both of its ends is characteristic of mesogenic molecules. The main structural feature of the liquid-crystal state is a parallel array of molecules, with a high lability of all or some of the contacts between them. Such mutual packing determines their short-range order and can be characterized by some statistical symmetry.

2.8.2 Types of Liquid-Crystal Ordering

As regards the packing of molecules, the liquid-crystal state is divided, according to *Friedel* [2.52a] into two types, the nematic and the smectic. A variety of the former is the cholesteric type. All the types are characterized by a parallel arrangement of the neighboring molecules within certain microregions of liquid crystals (Fig. 2.99). A region in which a given orientation is preserved is called a domain.

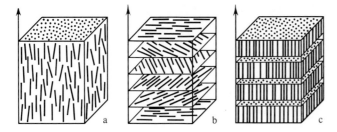

Fig. 2.99 a – c. Arrangement of molecules in liquid crystals. (a) Nematic, (b) cholesteric, (c) smectic

Liquid crystals (their domains) possess strong optical birefringence, which lies at the basis of the methods of their optical investigations (Fig. 2.100, see also [1.7]). Nematic liquid crystals are optically inactive, i.e., they do not rotate the polarization plane. (However, if a layer of the nematic phase is sandwiched between two glass plates, which are then rotated, optical activity arises as a result of gradual twisting of the direction of the molecular axes).

When elongated molecules are packed parallel to each other, their packing is denser than in random orientation, which is characteristic of a liquid. Thus, this geometric factor permits the existence of some energy minima (packing energy plus free energy of thermal motion) intermediate between the true liquid and the true solid state, which actually correspond to the liquid-crystal state. The arrangement of molecules in liquid crystals can be described by statistical functions of the second kind, which were applied above for analyzing the structure of high polymers [2.53].

In the nematic state the centers of gravity of the molecules (with their parallel arrangement preserved) are randomly disposed (Fig. 2.99a). Here, only the factor of the elongated shape of the molecules is important. Statisti-

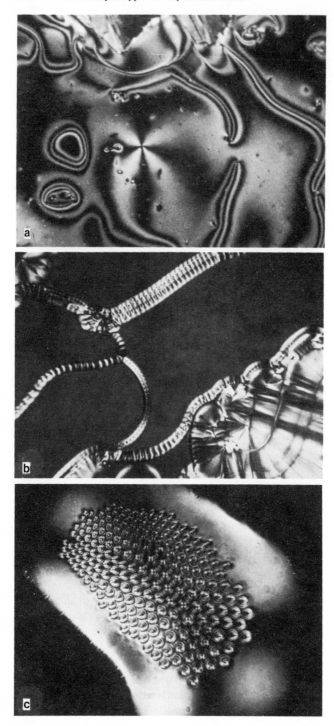

Fig. 2.100a – c.
Textures of liquid
crystals in a polarized
light. **(a)** Nematic,
(b) cholesteric,
(c) smectic

cal rotation of molecules about their long axes occurs. There is no correlation between the projections of the distances between the molecular centers onto the principal axis of the domain. Hence, any shifts of molecules relative to each other are equally probable, which can be described by an infinitely small statistical translation τ_∞, i.e., the axis of continuous translations. If the molecules are polar (and this is almost always the case), their parallel and antiparallel arrays are possible, generally speaking. Here, an additional symmetry operation, which statistically transforms molecules into each other, is axis 2, which is perpendicular to the principal axis of the liquid-crystal domain.

In the projection along the long axes the molecular packing in the nematic phase shows some ordering. This ordering can be described by the distribution function $W(r)$ of the projections of the molecular axes onto the plane perpendicular to the principal axis. The function is of the second kind (see Fig. 2.94b, d).

Because the cross sections of the molecules making up liquid crystals are nonuniform, the distances between the nearest neighbors vary widely. Therefore, function $W(r)$ has, as a rule, less prominent maxima for them than for chain molecules proper.

The mutual azimuthal orientation of molecules is maintained, to some extent, only for the nearest neighbors. Function $f(\psi)$, which describes the orientation (see Fig. 2.93), is strongly smeared out, and any orientations of molecules are encountered throughout the domain as a whole. Therefore, the principal axis of a nematic domain is (statistically) an axis of infinite order ∞, and the whole function $W(r)$ becomes cylindrically symmetric, $W(r)$.

Summing up, the symbol of the structure (symmetry) of nematic liquid crystals can be written as $\infty \tau_\infty W(r)$. This symbol can be detailed by taking into account the proper symmetry of the molecules, the presence or absence of antiparallel packing, etc.

Nematic liquid crystals have the so-called cholesteric variety. The molecules of cholesteric liquid crystals are chiral optically active, and hence. Their aggregate comprises monomolecular layers with nematic ordering, but the natural superposition of layers occurs so that their axes gradually rotate (Fig. 2.99b). The arrangement of the molecules in the layer plane is characterized by operation τ_∞, the symmetry of a set of layers being characterized by a screw axis of infinite order σ_∞. The period of the set is usually equal to several thousand angstroms.

Another basic type of structure of liquid crystals is the smectic one. As in nematic liquid crystals, in smectic phases the molecular axes are parallel to each other, but the molecules are packed in layers (Fig. 2.99c). The higher ordering of the smectic phase follows from the fact that when a given substance forms both a smectic and a nematic phase, the former is the low temperature one. For instance, p,p'-nonyloxybenzaltoluidene (BT) (Fig. 2.98d) is smectic in the temperature range from 70° to 73.5 °C, and nematic from 73.5° to 76 °C.

The formation of smectic layers means that the side interactions between the molecules are strong enough. This may be assisted by the absence of projecting side groupings, which prevent close contacts between the molecules. For instance, p-azoxyanisole and α-benzolaso-(anisole-α'-naphthylamine) (BAN) (Fig. 2.98a, b) form only nematic liquid crystals, as noted above, and p,p'-nonyloxybenzaltoluidene (BT) forms both smectic and nematic liquid crystals. When analyzing the shape of molecules one can conclude that one of the main possible causes for the formation of smectic layers is the convenience of antiparallel packing of molecules (Fig. 2.101a). Thus, smectic liquid crystals show a correlation between shifts of molecules along the axis relative to each other, which is described by function $\tau(z)$, while the layers are stacked at a certain regular distance c, which is close to the length of the molecule. Inside the layers, the arrangement of the molecules is characterized, as in nematic liquid crystals, by function $W(r)$ and, evidently, often by "statistical" axes 2 (which assign antiparallelism) disposed in the layer normal to the principal axis of the molecules. Consequently, the symbol of the smectic liquid crystal structure can generally be written as $\infty c\,\tau(z)\,W(r)/2$. This type of smectic liquid crystals, in whose layers the molecules are parallel to each other and perpendicular (with some spread in orientations) to the plane of the layer, is called smectic A (Fig. 2.101b).

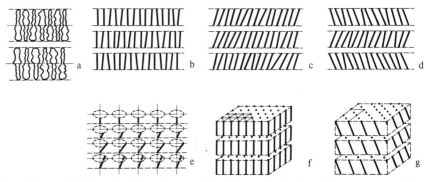

Fig. 2.101a–g. Packing of molecules in smectic liquid crystals. Antiparallel arrangement of the molecules in smectic layers (a). The varieties of smectic phases: smectic A (b), smectics C (c, d, e), smectic B (f), and smectic H (g)

Other types of molecular ordering in smectic modifications are also known. Thus, in smectics C the molecular axes are not perpendicular to the plane of the layer, but are inclined to it. Versions of smectics C are shown in Fig. 2.101; a uniform tilt in the layers (Fig. 2.101c), a tilt that alternates in either direction in each layer (a "herringbone" pattern, Fig. 2.98d), and helical variation of uniform tilt from layer to layer (Fig. 2.101e), which is characteristic of chiral smectics C possessing piezo- and ferroelectric properties. If we imagine that the tilt is 90°, then the chiral smectic C will transform into a cholesteric liquid crystal. Interestingly, similar types of ordering of an entirely different physical nature of the structural units and the interaction between them are observed in magnetics (Sect. 1.2.11).

In smectic B, a higher degree of molecular ordering is observed (Fig. 2.101f). The molecular axes in the layer are arranged in accordance with the law of hexagonal packing, which has distortions described by the function $W(r)$ of the first kind (Fig. 2.94a).

Figure 2.101g depicts the structure of smectic H, which has correlated tilts of molecules in the layers, as distinct from smectic B, and the molecular axes form an orthorhombic net.

Smectic phases exhibit still more ordered structures, which are in fact already particularly plastic three-dimensional crystals, but disordered. Thus, smectic D is optically isotropic and has a cubic structure with a space group $Ia3$. Such a mesophase has been found in only two compounds. One of them − 3'-nitro-4'-n-hexadecyloxybiphenyl-4-carboxylic acid

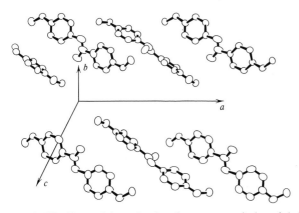

− has a cubic structure with $a = 105$ Å, containing 1150 molecules in the unit cell. Curiously, this phase lies in the temperature range between anisotropic smectic phases of the types C and A. The so-called smectic E, to which the orthorhombic structure is ascribed, is actually also three-dimensionally ordered. Types F and G are known as well. Thus, in liquid crystals it is possible to observe several polymorphous modifications, i.e., different phases with an ordering of the constituent molecules characteristic of each phase.

Fig. 2.102. Shape of the molecules of para-azoxyanisole and their arrangement in the unit cell of a solid crystal [2.54]

Solid crystals which transform into liquid crystals upon melting are called mesogenic. In such crystals approximately parallel arrangement of the molecules already exists. A classic example of a nematic liquid crystal is *p*-azoxyanysol. Figure 2.102 pictures the arrangement of the molecules of this substance in the solid phase. The parallelism of the molecules in a nematogenic crystal, but without a strictly fixed mutual arrangement, is inherited by the liquid-crystal phase as well. Similarly, solid crystals transformed into smectic liquid-crystal phases are called smectogenic.

In those cases where a liquid crystal has some polymorphous modifications, ordering at phase transitions diminishes with rising temperature. When the temperature decreases several other phases may arise. Terephthal-bis(*p*-butylaniline) (TBBA)

$$C_4H_9-\bigcirc-N=CH-\bigcirc-CH=N-\bigcirc-C_4H_9$$

is an example. Its phase transitions' scheme is as follows:

$$\text{solid crystal} \xrightarrow{113\,°C} sm_B \xleftarrow{144\,°C} sm_C \xleftarrow{175\,°C} sm_A \xleftarrow{200\,°C} nem \xleftarrow{236\,°C} \text{liquid}$$

$$\uparrow 63\,°C \qquad \downarrow 88\,°C$$

$$\text{VII} \xleftarrow{74\,°C} \text{VI} (sm_E).$$

When the nematic phase is formed from the smectic on heating, it may, especially at temperatures close to the phase transition point, retain "fragments" of smectic layers, i.e., small molecular associates. An example is the nematic phase of nonyloxybenzoic acid (Fig. 2.103a), which is obtained on melting of the smectic phase *C*.

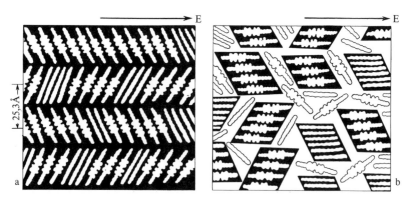

Fig. 2.103a,b. Para-nonyloxybenzoic acid. Molecular arrangement in the smectic (**a**) and nematic (**b**) phase in a constant electric field. The field direction is shown by the arrow

Liquid-crystal phases have been found lately, which are not built of elongated molecules, but, on the contrary, of disklike molecules [2.55; 55a, b]. These disklike molecules, for instance the molecule

$; R = C_4H_9 - C_9H_{19}$

are stacked on top of each other forming columns. These columns play the role of a structural element of liquid crystal (Fig. 2.104). Different kinds of ordering in such columnar phases have been observed; the column axes can be in hexagonal or rectangular order or also can be tilted as in smectic C. Another ordering of disclike molecules is when columns are not formed but the planes of molecules are still parallel (nematic discotics).

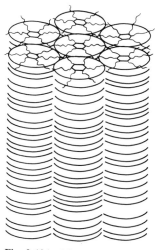

Liquid crystalline polymers form another type of mesophases which exists in two varieties. The first variety is made up of linear polymers displaying the properties of liquid crystalline phases on melting. The other is composed of flexible polymeric molecules with large side radicals of mesogenic nature attached to the backbones. In the liquid crystalline phase the radicals are ordered either according to the nematic or the smectic type.

Various types of order of molecules in liquid-crystal phases are preserved, as noted above, within certain microregions, or domains. Optical observations show that if there are no special orienting effects, liquid-crystal domains are disposed randomly with respect to each other. The special cases of continuous transition from one axes orientation to another and the formation of a special kind of texture are also possible (see [1.7]).

Fig. 2.104. Schematic representation of the mesophase structure built up from disclike molecules [2.55]

Fig. 2.105a – c. X-ray photograph of a smectic liquid crystal (**a**), a model of the smectic structure (**b**), and the optical diffraction pattern from a scaled-down image of the model (**c**)

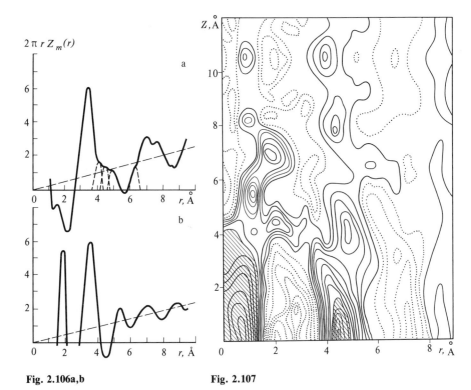

Fig. 2.106a,b   **Fig. 2.107**

Fig. 2.106a, b. Para-azoxyanisole. Function $2\pi r Z_m(r)$ of a nematic liquid crystal oriented by a 16,000 Gs magnetic field (**a**) and by a constant 4,000 V/cm electric field (**b**)

Fig. 2.107. Two-dimensional cylindrically symmetric function of interatomic distances $Q(r, Z)$ constructed for the nematic phase of ethyl alcohol of p-anizalaminocinnamic acid oriented by an electric field [2.56]

Liquid crystals, i.e., their domains, can be oriented, for instance, by stretching, in flux, or by an electric or magnetic field. As a result, an axial texture is formed; its domains which have only slightly defined boundaries merge continuously with each other. The texture has cylindrical symmetry (Fig. 2.105). The same symmetry is exhibited by the intensity distribution in reciprocal space and is revealed by x-ray diffraction. A number of conclusions concerning the structure can be obtained here by constructing a cylindrical Patterson function [Ref. 1.6, Sect. 4.4.3] and its sections at $z = 0$ and $R = 0$. The former defines function $W(r)$, which gives the distribution of distances between the "side" neighbors in an aggregate of parallel molecules. Figure 2.106 shows $W(r)$ for a nematic liquid crystal (p-azoxyanisole) oriented by a magnetic and an electric fields, with an average lateral intermolecular distance of about 4 Å.

Figure 2.107 presents the cylindrical function of interatomic distances $Q(r, z)$ for the nematic phase of ethyl ether of p-anizalaminocinnamic acid oriented by an electric field, and Fig. 2.108, for the smectic phase of the same substance oriented by a magnetic field. This function was obtained by optical diffraction, i.e., by Fourier optical transformation of the observed x-ray diffraction pattern; computation gives the same result. The plot of Fig. 2.108

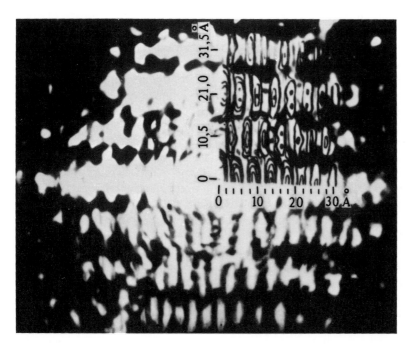

Fig. 2.108. Two-dimensional cylindrical function of interatomic distances $Q(r, Z)$ for the smectic phase of ethyl ether of p-anizalaminocinnamic acid oriented by a magnetic field as obtained by the optical diffraction. Contours of calculated function are also shown (upper right) [2.57]

vividly illustrates the stacking of smectic layers (its maxima are spaced with a corresponding periodicity of 10.5 Å along the c axis).

All the above-described types of liquid crystals were one-component systems formed only by molecules of the given substance. The liquid-crystal state can also be obtained in two-component or more complicated systems, for instance, in aqueous or other solutions of long molecules of fatty acids, lipids, and other organic substances. Such liquid crystals are called lyotropic. Their structures are still more diversified than those of thermotropic liquid crystals. The physical properties of liquid crystals will be treated in [1.7].

2.9 Structures of Substances of Biological Origin

2.9.1 Types of Biological Molecules

The structure of substances of biological origin has been in the focus of attention ever since the advent of x-ray diffraction analysis. In the early twenties the first x-ray photographs of natural fibers – silk and cellulose – were obtained which demonstrated a high ordering of their structures. The extremely complicated structure of most of these substances and the absence of comprehensive chemical data on them hindered their detailed interpretation for a long time. It was not before the postwar period that these investigations gained momentum. The particular complexity of the objects stimulated the improvement of x-ray methods and the techniques of crystallographic structure calculations, and promoted the development of new concepts in the theory of symmetry and in molecular crystal chemistry. The results obtained have brought about a radical turn in the development of biology, which now makes use of data on the atomic structure of a great variety of organic molecules, including proteins, nucleic acids, viruses, vitamins, and so on. A new science – molecular biology – has sprung up, whose most efficient tool is the generalized crystallography with its powerful x-ray technique. Valuable data were also contributed by high-resolution electron microscopy.

The complexity of structures which have become objects of investigation is illustrated by Table 2.3.

Investigations up to the level of atomic resolution are being conducted for almost all the above structures, with the exception of membranes and nucleoproteins. At the same time, electron microscopy is also widely used for studying most of these objects, except for – as they now seem – relatively simple molecules of the first two lines; this technique supplies information on the structure of stable aggregates of atoms and the subunits in various structures, as well as of chain molecules, at a resolution up to several angstroms.

The molecules, and the crystals formed of them, which are listed in the first line of Table 2.3, are typical of contemporary organic crystal chemistry

Table 2.3. Structural characteristics of biological objects

Object	Symmetry of molecules	Type of ordering	Period of cell or chain [Å]	Number of atoms in a molecule or in a unit of a chain molecule	Number of reflections in x-ray diffraction experiment
Amino acids nucleotides, sugars	G_0^3	crystals	5 – 20	up to 10^2	1000
Peptides, steroids, hormones, vitamins, lipids	G_0^3	crystals, liquid crystals	10 – 30	up to $10^2 - 10^3$	3000
Fibrous proteins, polysaccharides	G_1^3	textures, layers	10 – 100	up to 10^2	10 – 100
Globular proteins	G_0^3	crystals, layers	30 – 200	up to $10^3 - 10^5$	up to 100,000
Membranes	G_2^3	layers	50 – 100	$10^2 - 10^4$	10 – 1000
Nucleic acids	G_0^3, G_1^3	textures, crystals	30 – 100	$10^2 - 10^3$	100 – 1000
Nucleoproteids, viruses	G_0^3, G_1^3	textures, crystals	up to 2000	$10^6 - 10^8$	up to 10^6

as regards their complexity. They are, as indicated above, characterized by the presence of hydrogen bonds. These molecules are often represented in nature by only one of the two possible enantiomorphous forms. The enormous importance of these simple molecules lies in the fact that they are building blocks of almost all biological structures. The number of these "blocks", of which all the living beings are made up, is rather small, only 20 amino acids, 4 nucleotides, glucose and some other sugars as well as a number of fatty acid derivatives. The structures of all these molecules and of a great number of their less abundant varieties have been determined by now. Some data have also been obtained for various "small" (from the biological standpoint), but already very complex (for x-ray crystallography) molecules and crystals in the second line, steroids, hormones, vitamins, peptides, etc. These molecules usually play an important role in the control of biological processes in organisms. Figure 2.109 shows the structure of vitamin B_{12}. Water molecules, which fill the empty space between large molecules, play an important part in the packing of biological molecules in crystals.

Lipids make up an interesting class of biological structures. They are elongated molecules with an aliphatic "tail" at one end, which sometimes contains a cyclic group, and a polar group (OH, O, COOH, NH_2, etc.) at the other. The simplest molecules of this type are fatty acids $C_nH_{2n+1}COOH$.

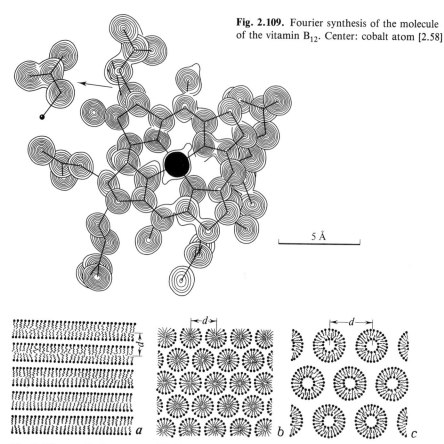

Fig. 2.109. Fourier synthesis of the molecule of the vitamin B_{12}. Center: cobalt atom [2.58]

5 Å

Fig. 2.110a–c. Various types of lipid molecule ordering in solution, **(a)** in layers, **(b)** in mono-molecular and **(c)** bimolecular spheres

They behave peculiarly in aqueous solutions, in which they form layer struc-tures (Fig. 2.110a) similar to smectic liquid crystals. Here, the molecules are extended parallel to each other and perpendicular to the layer, so that the polar groups face the water, while the nonpolar parts come together inside the layer, being excluded from the water (hydrophobic interaction) and attracted to each other by van der Waals forces. Other associations may also arise, de-pending on the shape of the molecules and their concentration in solutions (Fig. 2.110b), mainly due to the hydrophobic and polar interactions.

2.9.2 Principles of Protein Structure

Proteins are large chain molecules of high molecular weight and made up of amino acid residues. The amino acids (they are always L in proteins)

$$H_2N-\underset{\underset{R}{|}}{\overset{\overset{H}{|}}{C}}-COOH \qquad (2.10)$$

differ in their radical R. Figure 2.111 depicts the structure of one of the amino acids, phenylalanine. The vast majority of proteins are built up of 20 "main" amino acids, which are often likened to a "protein alphabet" (Fig. 2.112). The side chains of one group of amino acids, for instance glycine (R = H), alanine (R = CH), phenylalanine (R = CH$_2$⟨◯⟩), are neutral; they are hydrophobic,

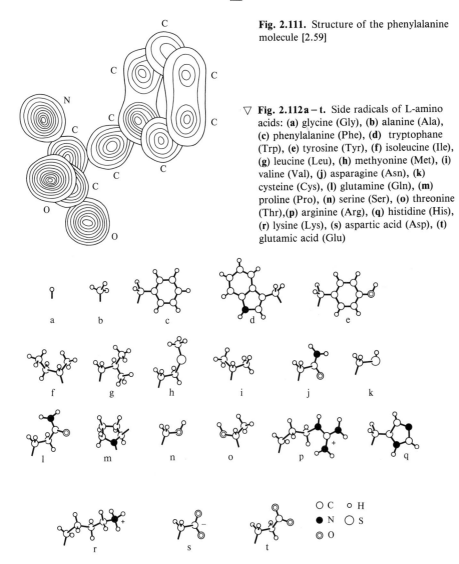

Fig. 2.111. Structure of the phenylalanine molecule [2.59]

▽ **Fig. 2.112a – t.** Side radicals of L-amino acids: (**a**) glycine (Gly), (**b**) alanine (Ala), (**c**) phenylalanine (Phe), (**d**) tryptophane (Trp), (**e**) tyrosine (Tyr), (**f**) isoleucine (Ile), (**g**) leucine (Leu), (**h**) methyonine (Met), (**i**) valine (Val), (**j**) asparagine (Asn), (**k**) cysteine (Cys), (**l**) glutamine (Gln), (**m**) proline (Pro), (**n**) serine (Ser), (**o**) threonine (Thr),(**p**) arginine (Arg), (**q**) histidine (His), (**r**) lysine (Lys), (**s**) aspartic acid (Asp), (**t**) glutamic acid (Glu)

i.e., they are repelled by water molecules. The side chains of other amino acids have polar or charged groups OH, NH_3^+, SH, COO^-, etc. in serine $R = CH_2OH$, asparagine $R = CH_2CONH_2$, in cysteine $R = CH_2SH$, and so on. These groups are capable of forming hydrogen or ionic bonds; water molecules readily combine with them. Amino acids are abbreviated by using the first letters of their names, such as *leu* for leucine and *phe* for phenylalanine.

The detachment of one of the H atoms of the amino group and OH from the carboxyl (with the release of water) enables the residues to join into a poly-peptide chain

$$\tag{2.11}$$

which lies at the base of the protein structure (Fig. 2.113).

Protein molecules may be described by their primary, secondary, tertiary, and quaternary structures. The primary structure is the chemical formula given by the sequence of amino acids in the chain. Such a formula, for instance that of lysozyme (Fig. 2.114), describes the covalent topology of the

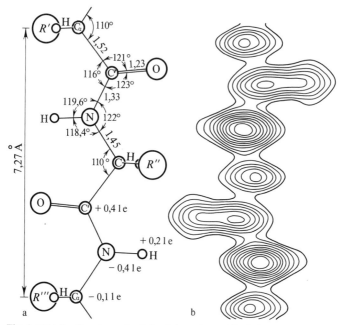

Fig. 2.113a, b. Structure of a link of the polypeptide chain, (a) standard parameters of the chain [2.60], (b) Fourier projection of the potential of synthetic polypeptide poly-γ-methyl-L-glutamate [2.61]

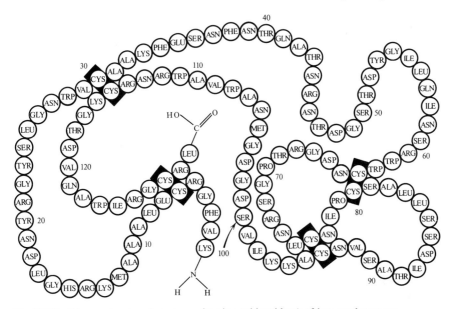

Fig. 2.114. Primary structure (sequence of amino acid residues) of hen-egg lysozyme

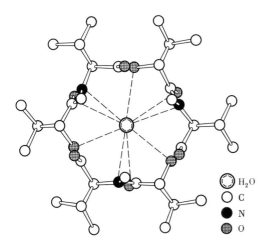

◎	H_2O
○	C
●	N
▦	O

Fig. 2.115. Structure of the molecule of cyclic depsipeptide enniatin *B* with a dumbbell of two H_2O molecules in the center [2.63]

protein molecule. The letters are already composed into a phrase, but this is far from enough for a full description. A chain, each of its units, and each of its side groups are arranged in a distinct way in three-dimensional space.

The secondary structure of protein describes how the polypeptide chain is arranged into a definite conformation stabilized by hydrogen bonds. "Standard" bond lengths and bond angles, characteristic of the unit of the polypeptide chain (Fig. 2.113), have been established on the basis of x-ray structure determinations of amino acids and small peptides [2.60, 62] (Figs. 2.111, 115).

The distribution of valence and deformation electron density in the peptide group is shown in Fig. 1.32b, c. It is seen that the valence bridge density in the $C-N$ bond is higher than in single bonds.

Experimental data, as well as theoretical calculations [2.64] show that atoms of the peptide group are charged, their values being as follows: $N -0.3$ to $-0.4\,e$; H (linked to N) $+0.2\,e$; $C + 0.4$ to $+0.5\,e$; $O -0.4$ to $-0.6\,e$; C_α -0.1 to $+0.1\,e$; H (linked to C_α) is almost neutral.

The amide group $\diagdown N \diagup C \diagdown$ is always plane or almost plane[5], since the

$C-N$ bond in it is not quite single, but is approximately of order $1\frac{1}{2}$, and the diversity of conformations or the chains is ensured by the possibility of rotation about single bonds: through angle φ about $C_\alpha - N$ and about through angle ψ about $C-C_\alpha$ (Fig. 2.116). Yet these rotations are limited owing to the

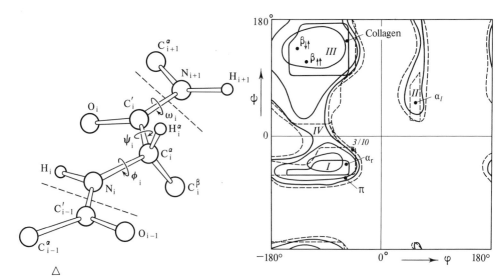

Fig. 2.116. Angles determining the configuration of the polypeptide chain. Angles φ and ψ determine the mutual conformation of neighboring units of the polypeptide chain. Angle ω determines the deviation of the amide group from planarity. Side radical R is represented by carbon atom C^β

Fig. 2.117. Conformation plot [2.65]. The heavy line denotes the parts of fully allowed values of φ and ψ, the dashed line corresponding to the allowed conformation strains. The smooth contours were obtained on the basis of energy calculations with the aid of semiempirical potentials of atomic interactions. Regions I, II: helical conformations. The dots mark the right-handed and the left-handed α helices, π helix with 4.4 residues per turn, 3/10 helix with 3 residues per turn. Region III: extended structures, β layers and collagen, which are marked with dots, and Region IV: intermediate field

---

[5] Small deviations from planarity of the amide group may be described, if they are observed, by the angle of rotation ω about the $N-C$ bond (Fig. 2.116)

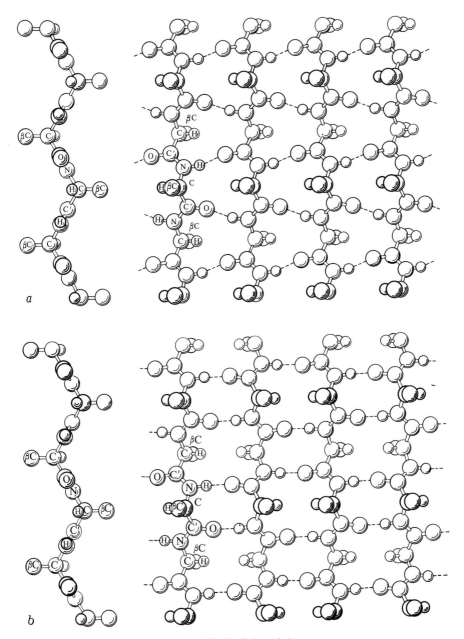

Fig. 2.118 a, b. Parallel (**a**) and antiparallel (**b**) β-pleated sheets

interaction of the atoms of the side groups R adjacent to the chain; these atoms cannot approach each other by less than the sum of their van der Waals radii. Proceeding from this, *Ramachandran* constructed a conformation plot of the permissible ψ and φ values for the polypeptide chain (Fig. 2.117) (the diagram differs for different amino acid residues in some details). The geometric approach can be replaced by the energy treatment, using some interaction potentials (1.35, 2.5). The fixed values of φ and ψ correspond to some stable conformation of the chain or a part of it, i.e., to a definite secondary structure. More complicated folding of the chain may be characterized by the sequence of angles ψ_i and φ_i for each unit.

There are two main types of secondary structures: an extended β and a helical α conformation. In the former the polypeptide chains are parallel to each other and are joined together by hydrogen bonds. The polypeptide chain (2.11) is polar; the sequence of amino acid residues is different when it is read backward. In the layer shown in Fig. 2.118a (or in a section of such a layer) the chains run in the same direction; this is the so-called parallel β structure. The period of a completely stretched chain is equal to 7.34 Å, while in a parallel layer the period decreases to 6.5 Å providing the best H contacts by the rotation about $C_\alpha - N$ and $C - C_\alpha$ bonds. The spacing between the chains is 4.7 Å, this is another period of the layer. In the antiparallel β structure (Fig. 2.118b) the chains are stretched almost completely, $C = 7.0$ Å; the period perpendicular to the chains is now equal to the doubled distance between them, i.e., to about 10 Å, because oppositely directed chains alternate normally to it; region III of the conformation plot of Fig. 2.117 corresponds to the β structures. Figure 2.119 portrays the structure of silk fibroin formed of parallel β layers.

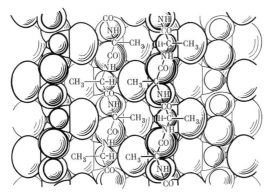

Fig. 2.119. Silk fibroin structure

The other principal type of the secondary structure of polypeptide chains — the α form — was predicted by *L. Pauling* and *R. Corey* in 1951 on the basis of stereochemical data for amino acids and on H bond regularities. The ideas of the helical structure of polymers in general and of polypeptide chains

a   b

Fig. 2.120 a, b. Structure of the right-handed α helix. (a) Scheme, (b) model of the structure of synthetic polypeptide poly-γ-methyl-L-glutamate in the α form

in particular, which haunted the minds of investigators, could not be reconciled with the rules of crystallographic symmetry, according to which only integral screw axes with $N = 2, 3, 4$, and 6 are feasible in crystals. Abandoning this restriction, it became possible to construct a helical structure satisfying all the conformational requirements: near-linear H bonds with satisfactory distances, good van der Waals contacts, and permissible rotation angles [2.66]. For the right-handed α helix $\varphi = -57°$ and $\psi = -47°$. In it, H bonds exist between the NH of each amino acid residue and the O atom of the residue standing fourth from it along the chain, so that these bonds are approximately parallel to the chain axis (Fig. 2.120). The pitch of the helix is 5.4 Å and the projection of the residue onto the axis is 1.52 Å. In an α helix, there are 18 residues per 5 turns ($M = 18/5$), the full period $c = 1.5 × 18 = 5.4 × 5 = 27$ Å. The cross section of the helix is about 10Å. There are conformations close to the α helix: a helix with $M = p/q = 3/10$, and also π helix with 4.4 residues per turn (see Fig. 2.117). The left-handed $α_1$ helix is energetically less advantageous because of the excessively close contacts of the C_β atoms and, therefore, is not realized.

The main regularities of the folding of polypeptide chains and the two principal types of the secondary structure are observed even in the simplest substances of protein type, i.e., synthetic and natural peptides. The peptides are open or cyclic chains, in which the number of amino acid residues ranges from 4−6 to several tens. They include some antibiotics, for instance gramicidin C; ionophores, complexes which handle the transport of metal ions through biological membranes; some hormones; and toxins. Figure 2.115 shows the structure of cyclic hexadepsipeptide enniatin B, whose ring can accommodate a metal cation or a water molecule.

The spatial folding of a polypeptide chain possessing some secondary structure into a fixed complex configuration of the protein molecule is called a tertiary structure. In other words, the tertiary structure is a specific description of the three-dimensional structure of a polypeptide chain in peptides, in globular or fibrous proteins with allocation of some parts of the chain to a definite secondary structure (if any) or to an irregular conformation. The next stage of the hierarchy is the quaternary structure. This is an association of several subunits of biological macromolecules having a tertiary structure; such an assembly usually possesses a definite point symmetry (see Figs. 2.131a, c, 152). Finally, we can consider structures of the fifth order, i.e., associations of large numbers of molecules (they may also have a quaternary structure and be one or several sorts) into assemblies like viruses, membranes, tubular crystals, etc.

The description of the structure of biopolymers with its division into the primary (chemical), secondary, tertiary, and quaternary is applied not only to proteins, but also to nucleic acids, polysaccharides, and some other classes of biological compounds.

2.9.3 Fibrous Proteins

Natural fibers, the building materials for the tissues of higher animals, are mainly fibrous proteins, and, for plants and some organisms, also polysaccharides. In fibrous proteins the number of types of amino acid residues is small, for instance, glycine and alanine in silk; keratin (structural protein of hair, wool, horns, and feathers) is also rich in these amino acids.

X-ray studies of packing of polypeptide chains in fibrous proteins, which were begun in the 1930s, have revealed their two main varieties, which were named the α and β forms. The two principal types of the secondary structure were later named after them. Valuable results were obtained by investigation of model polymers − synthetic polypeptides − which contain only one type of radical R. It was established that in the β form the chains are extended (because the period along the fiber axis was $6.5 - 7$ Å), parallel, and linked with each other by H bonds (Fig. 2.118).

Parallel or antiparallel strands in β proteins form layers. The symmetry of a parallel layer is $p12_11$, and that of an antiparallel one $p2_12_12$, axes 2 being perpendicular to the layer [Ref. 1.6, Fig. 2.63]. The layers in β structures of fibrous proteins are usually packed with small degree of ordering. Thus, it is the two-dimensional layer that is a rather stable structural unit. An example of a fibrous protein with a β structure is silk fibroin (Fig. 2.119).

The α proteins include α keratin, myosin (the muscle protein), epidermin, fibrinogen, etc. They, as well as some synthetic polypeptides, are characterized by a meridional reflection with $d \sim 1.5$ Å on x-ray texture photographs. These had defied interpretation for a long time until the α helical type of secondary structure was found, for which the indicated value was precisely the projection of the screw displacement of one amino-acid residue onto the helix axis. Thus the structure of α proteins was established. It is interesting to note that by extending some of these proteins their structure can be transformed into the β form, in which the chains straighten out.

Chains with a secondary α helical conformation form structures of a higher order in fibrous proteins. Thus, keratin in wool exhibits coiling of the axes of α helices, due to the mutual packing of α helices into a structure of the type of three-stranded "rope" (Fig. 2.121). Three-stranded "subfibrils" are packed in still more complicated fibrils (Fig. 2.121c) in the so-called 9 + 2 structure.

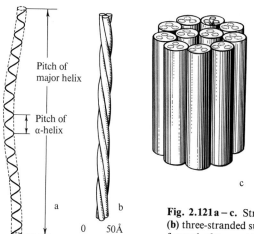

Pitch of major helix

Pitch of α-helix

a

b

0   50Å

c

Fig. 2.121 a – c. Structure of α keratin. (a) Coiled α-helix, (b) three-stranded subfibril, (c) individual fibril of α keratin formed of two subfibrils surrounded by nine others

Another type of a secondary structure of polypeptide chains, different from α and β, is formed in collagen, the protein of the connective tissue, skin, and tendons. The chemical structure of the collagen chain is such that each third unit in it is a glycine residue, while the other, most common residues are five-membered rings of proline or hydroxyproline. The molecule consists of three weakly wound chains ($\varphi \approx -70\,°C$, $\psi \approx 160\,°C$, see Fig. 2.117) joined into a unified system by H bonds. The chains are transformed into each other by a rotation through 108° and a translation by 2.86 Å. Each of the units is twisted as a left-handed screw, and their axes, as a right-handed one. The period of the molecule is 28.6 Å. Its model is depicted in Fig. 2.122. Owing to the complicated primary (chemical) structure of collagen the fibers of this protein have super periods of 85 and 640 Å.

Thus, fibrous proteins have a complicated structural organization, which obeys definite symmetry and conformation laws. This organization also exhibits the hierarchy typical of biological systems in general, namely, a small number of certain kinds of structural units form a unit of the next order of complexity, and so on.

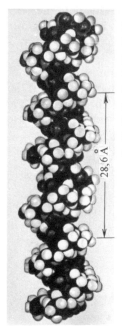

While the fibrous proteins are the basic building materials for most animal tissues, sugar polymers – cellulose and other polysaccharides – play this part for plants, and also for certain animals. For instance, chitin, the substance of the exoskeleton and joints of insects and other Arthropoda, is a polysaccharide. The arrangement of cellulose chains is presented in Fig. 2.123; the fibrils are about 100 Å thick; the packing of the chains in a fibril can be described, to a first approximation, as paracrystalline (see Fig. 2.94).

Fig. 2.122.
Model of the collagen structure

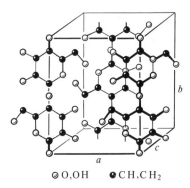

⊘ O,OH   ● C H,C H$_2$

Fig. 2.123.
Cellulose structure

2.9.4 Globular Proteins

The most important biological macromolecules are the molecules of globular proteins, whose polypeptide chain is folded into a compact globule. Many globular proteins are enzymes, i.e., biological catalysts, which carry out an infinite number of metabolic reactions in living nature. Globular proteins in organisms also perform many functions such as the transportation of small molecules or electrons, the reception (for instance, of light), and protection (immunity proteins).

If a globular protein is well purified, it may, as a rule, be crystallized in vitro (Figs. 2.124a – d). Sometimes crystals are formed in organisms (Fig. 2.125). With the aid of electron microscopy it is possible to visualize directly the regular packing of protein molecules in a crystal (Fig. 2.126). Protein crystals ("wet crystals") are extremely peculiar; they contain mother liquor in the empty space between the molecules, and are stable only when in equilibrium with this liquid or its vapors. It has been established by x-ray diffraction techniques that some of the water molecules are bound on the surface of the protein molecule, forming a kind of a "coat" (Fig. 2.127).

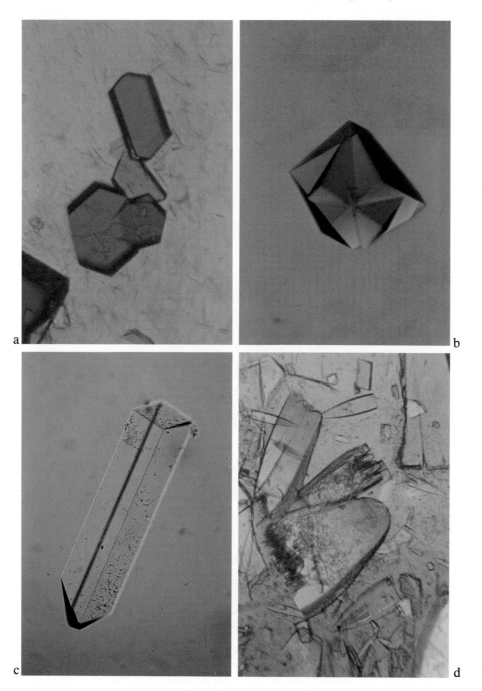

Fig. 2.124 a – d. Protein crystals (× 100). (a) Leghaemoglobin, (b) catalase Penicillium vitale, (c) aspartate-transaminase, (d) pyrophosphatase

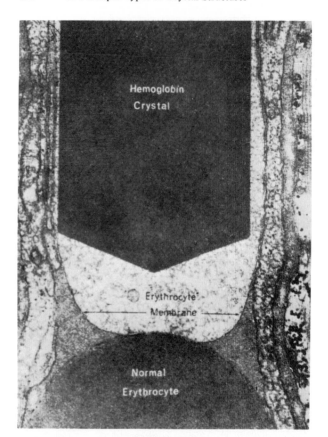

Fig. 2.125. Haemoglobin microcrystal formed in a blood vessel of the cat miocard [2.67]

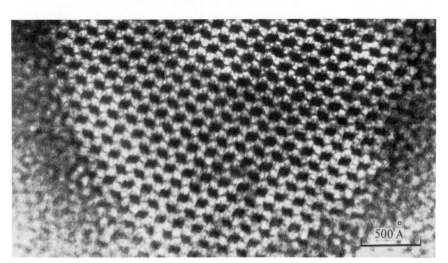

Fig. 2.126. Electron micrograph of orthorhombic ox-liver catalase. (Courtesy of V. Barynin)

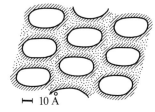

Fig. 2.127. Structure of a crystal of globular protein. The protein molecule is adjoined by a layer of water molecules bound with it (*hatched*); solution molecules are denoted by dots

Crystals can be dried, in that case much of the free crystallization water disappears, the unit cell shrinks, and the crystals become disordered.

The very fact that a formation of protein crystals produces thousands of reflections on x-ray photographs (Fig. 2.128) suggests that all the giant molecules of a given protein are identical and have an ordered internal structure.

The interaction between protein molecules in a crystal is extremely complicated because the protein surface carries various polar residues and the electrostatic interaction is screened by the water molecules. The ions present in the solution and the hydrogen ion concentration exert an influence on the surface of the molecules. Protein crystals are extremely sensitive to these factors and can form various polymorphous modifications.

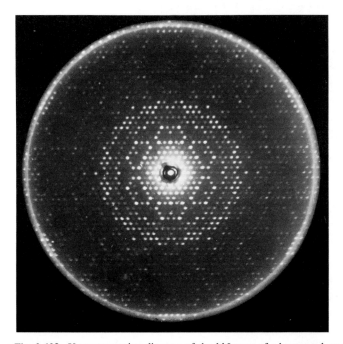

Fig. 2.128. X-ray precession diagram of the $hk0$ zone of a hexagonal crystal of catalase Penicillium vitale (sp. gr. $P3_121$, $a = 145$ Å, $c = 180$ Å), $\mu = 9°$. CuK_α radiation

Because of the weak bonds between protein molecules and the rotational and translational movements in the intracrystalline mother liquor, as well as the lability of the molecules themselves, the displacements of the atoms from the equilibrium position and the values of the temperature factor are very large, $\sqrt{u^2} = 0.5 - 1.0$ Å, $B = 30 - 100$ Å^2. This causes a rapid drop in intensities. The diffraction field boundary and, hence, the resolution is about 1.5 Å at best; in some proteins only a 2.5 Å resolution can be attained. Sometime the diffraction field contains only low-resolution reflections ($5 - 10$ Å), which implies considerable disordering in the molecules or their mutual arrangement.

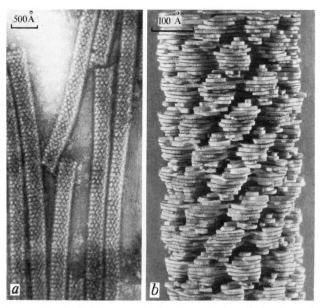

Fig. 2.129 a, b. Tubular crystals of ox-liver catalase. (**a**) Electron micrograph, (**b**) model of the packing of molecules in the tube obtained by the method of three-dimensional reconstruction [2.68]

The packing of molecules of globular proteins in crystals (or of subunits of a given molecule) sometimes shows a noncrystallographic symmetry, i.e., local symmetry elements (for instance, axis 2), which do not belong to the symmetry elements of the group Φ of the crystal. Some proteins were found to form planes of monomolecular layers (see [Ref. 1.6., Fig. 4.112]) and tubes with monomolecular walls (Fig. 2.129, see also [Ref. 1.6, Fig. 2.61]) which exhibit a helical symmetry.

X-ray structure analysis of protein crystals, which have huge unit cells (see Table 2.2), is very complicated and laborious; the main problem consists in determining the phases of the structure amplitudes. This is solved by introducing into these crystals heavy atoms (as part of small organic molecules or inorganic ions), thus attaching them to the protein molecules [Ref. 1.6., Sect. 4.7.5]. It is necessary that the structure of the protein crystal with such additions would not change, i.e., would be isomorphous to the structure of the native protein crystal. The intensities of reflections change owing to the addition of heavy atoms. By constructing the difference Patterson function one can find the positions of the heavy atoms, determine their contribution to the phases, and then define the phases of the reflections of the protein crystal itself. Additional information can be obtained by taking into account the anomalous scattering by heavy atoms. Then a Fourier synthesis is constructed. Fourier maps with a low resolution of ~5 Å yield the shape of the molecule and its subunits as well as the position of the helical segments in it. Synthesis with a resolution of ~2.5 Å allows us to trace the course of the chain and distinguish the amino acid residues, while those with a resolution of 2 Å and higher reveal individual atoms.

The first crystalline proteins whose structure was solved were myoglobin and hemoglobin. Their function consists in the reversible binding of the oxygen molecule O_2. Hemoglobin is present in the erythrocytes of the blood and transports O_2 in the circulation, while myoglobin stores oxygen in the muscles. The molecular weight of myoglobin is about 18,000; it contains 153 amino acid residues, i.e., about 1200 atoms excluding hydrogen. The unit cell is monoclinic: $a = 64.6$, $b = 31.1$, $c = 34.8$ Å, $\beta = 105.5°$, space group $P2_1$. The unit cell contains two protein molecules [2.69, 70]. Fourier synthesis at 6 Å resolution revealed the polypeptide chain (Fig. 2.130a), and that at 1.4 Å resolution (about 25,000 reflections were included) the arrangement of all the atoms (Fig. 2.130b). The molecule of myoglobin, as well as molecules of some other proteins, contains a so-called prosthetic group attached to the polypeptide chain. Here it is a flat porphyrine molecule, the so-called haem, at whose center there is an iron atom with which an O_2 molecule is bonded. A model of the myoglobin structure is given in Fig. 2.130c. Its polypeptide chain is 75% helical. The hemoglobin molecule (molecular weight 64,000) is built up of four subunits (Fig. 2.131), similar to the myoglobin molecule as regards its tertiary structure [2.72, 73]. Investigations into several proteins of this type showed that the "myoglobin" type of folding of the polypeptide chain is maintained in all of them. Figure 2.132a, b pictures a simplified diagram of the structures of myoglobin and plant leghemoglobin. Although these proteins differ drastically in their primary structure and stand wide apart evolutionarily (their common precursor existed about 1.5 billion years ago), their tertiary structure differs only in some details. Figure 2.133 presents the structure of the leghemoglobin molecule established on the basis of x-ray data with a resolution of 2 Å. The leghemoglobin structure differs from other globins in size of the heme pocket; it is larger on the distal histidine side. This permits the attachment of such large ligands as acetic or nicotinic acids to the molecule, and seems to explain the high affinity of oxygen to leghemoglobin.

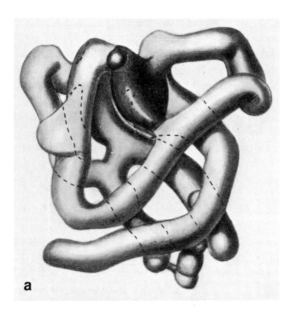

Fig. 2.130a. Myoglobin molecule. **(a)** Model of molecule based on Fourier synthesis at 6 Å resolution. High-electron-density regions (rods) correspond to α helices

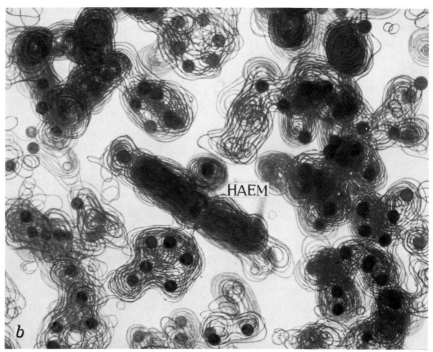

HAEM

Fig. 2.130b. Myoglobin molecule. Part of Fourier synthesis at 1.4 Å resolution in the heme region. Upper left: α helical part projected along the helix axis; right: perpendicular to the axis

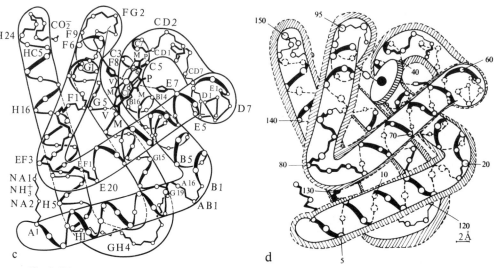

Fig. 2.130 c, d. Myoglobin molecule. (**c**) Tertiary structure. Arrangement of C_α atoms of amino-acid residues in the polypeptide chain is given. A–H: α helices, the numerals indicate the numbers of the residues in the helix, the pairs of letters showing the nonhelical regions. The heme group and the side groups of hystidines $E7$ and $F8$ interacting with it are fully represented [2.69]. (**d**) Intramolecular conformational and thermal displacements in myoglobin (*hatched*) [2.71]

Let us consider the structure of some other proteins and the general principles of their organization. The folding of the polypeptide chain is determined by the successive disposition of its units. The permissible rotations through angles φ about the $C_\alpha - N$ bond and ψ about $C - C_\alpha$ are determined by the conformation plot (Fig. 2.117). A small side group, for instance in glycine ($R = H$), allows a wide range of angles φ, ψ; larger side groups, for instance in tryptophan, reduce the range.

There are several possible ranges for different residues R, and different values of φ and ψ are possible for them. Therefore, an immense number of conformations are possible for the polypeptide chain, whose number of units in globular proteins runs into the hundreds. If we assume that each unit can occupy only two angular positions relative to its neighbors, then each of them is characterized by two angles, and for n units the number of chain conformations is $\sim 2^{2n}$.

However, all the molecules of a given protein with a definite intrinsic primary structure are absolutely identical, i.e., one and only one conformation is actually formed. Under given thermodynamic conditions and in a given medium (i.e., under conditions of a cell) this conformation arises by itself and is equilibrium for this medium. In other words, a biological molecule is constructed according to the self-organization principle. This term does not have any specific biological meaning; indeed, the formation of crystals, too, is essentially self-organization, and in the case of folding of a polypeptide chain the process is much more complicated only quantitatively.

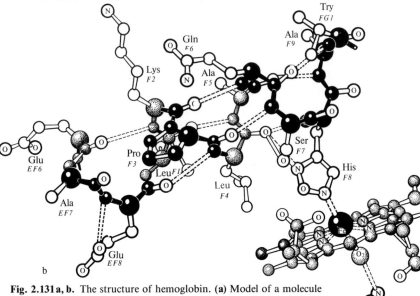

Fig. 2.131 a, b. The structure of hemoglobin. (**a**) Model of a molecule built up of plates corresponding to the sections of a three-dimensional electron density distribution at 5.5 Å resolution; subunits of the molecule have myoglobin folding, so-called α subunits are white, and β subunits are black. The helical segments of the subunits are denoted as in Fig. 2.130 c, the haems are shown as disks, the place of attachment of the O_2 molecule to the β subunit is indicated [2.74]. (**b**) Part of the molecule near the hemogroup

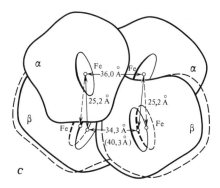

Fig. 2.131c. The structure of hemoglobin. (c) Change in the quaternary structure of the hemoglobin molecule. The β subunits come closer together in the course of oxygenation

The transition in solution filament $\leftrightarrow$ helix $\leftrightarrow$ disordered globule are well known in polymer physical chemistry, but where proteins are concerned it is necessary to explain the origin of the unique and definite globular structure. In other words, if we take a surface of equal energy in a space of about 2^{2n} dimensions (n being the number of units), by analogy with Fig. 2.117 it will show a global minimum which exactly corresponds to the conformation of the protein globule. Denaturation of protein molecules by heating causes their transition to conformations in which they are no longer capable of performing their biological functions.

What are the factors determining the appearance of equilibrium conformation in the allowed ranges of the angles φ and ψ for all the units? One of them is the possibility of forming hydrogen bonds stabilizing α or β structures. Investigations into model polypeptides, and also an analysis of the structure of molecules already determined by x-ray diffraction techniques demonstrated that some amino-acid residues fit well into the α helix and promote its formation (alanine, leucine, lysine, methionine, and tyrosine), while other residues,

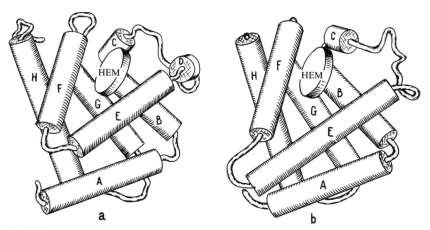

Fig. 2.132a, b. Comparison of the tertiary structure of the molecules of myoglobin (a) and leghemoglobin (b), (α helices are shown as cylinders [2.75]

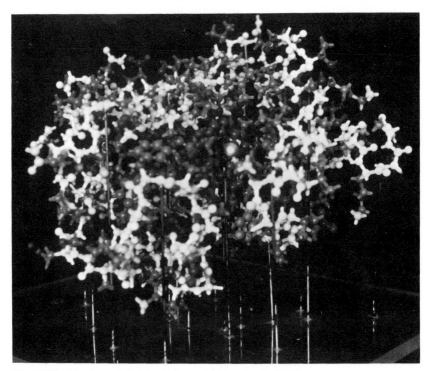

Fig. 2.133. Atomic model of the leghemoglobin molecule [2.76]

which often contain large side groups (valine, isoleucine, cysteine, serine, and threonine) are antihelical, and still others (proline and histidine) are characteristic of kinks of α helices. All these regularities are statistical, however, and cannot serve as a basis for fully reliable conclusions.

An important organizing factor during the formation of a protein globule is the interaction of the chain radicals with the solvent, i.e., water. It has been established that there is a trend towards the collection of nonpolar (hydrophobic) residues inside the molecule, where they are attracted to each other by van der Waals forces, and of polar ones on the globule surfaces, so that they contact the water molecules, as shown in Fig. 2.110b. The formation of a protein globule corresponds to the minimum of free energy and an increase of entropy in the globule + water system, rather than in the molecule as such. Some hypotheses hold that the globule is formed stage by stage: at first, the chain is collected into more stable parts, for instance, α helical segments, and then these segments are assembled into a globule.

Proteins similar in function nevertheless differ greatly in their primary chemical structure. In myoglobin-hemoglobin type proteins the number of residues replacing each other in the corresponding sections of the chain may reach 60% – 90% of their total number, and only three amino-acid residues remain unaltered in the entire set of these proteins. The similarity of the tertiary structure (see, for instance, Fig. 2.132a, b) is amazing. This indicates that the invariant features of R in the chain, primarily their polarity or non-polarity, rather than their uniqueness, are essential for the formation of a definite tertiary structure. Thus, analysis shows that about 30 nonpolar residues, making up the hydrophobic core of the molecule, play the main part in maintaining the basic tertiary structure of the myoglobin type, and the residues replacing each other in the chain at these sites are always nonpolar in various proteins of this family. A similar nonpolar core exists in cytochromes, proteins (Fig. 2.134) which transport electrons in the energy system of the cells.

As we have already mentioned, in a crystal the molecule of a globular protein as a whole experiences translational and rotational vibrations, but the protein molecule itself is also a dynamic system. The motion inside it is due to the thermal vibrations of its atoms and the conformational mobility of the main chain and the side chains of the amino-acid residues. The intramolecular component can be determined by x-ray analysis [2, 71]. For myoglobin, the root-mean-square value of intramolecular displacements $\sqrt{\bar{u}^2}$ is about 0.3 Å, but this value differs for the various parts of the molecule. It is smaller for the atoms of the main chain and larger for those of the side ones; also, it is smaller in the central part of the molecule (in its hydrophobic core) than on its surface (Fig. 2.130d).

By 1981 the number of globular proteins investigated up to the resolution of a tertiary structure is about two hundred.

These investigations have revealed some fundamental features of the structure of globular proteins. They include, first of all, the presence of certain standard building blocks – fragments consisting of α helices and β strands, and, secondly, the presence in many proteins of large isolated parts consisting of one segment of the polypeptide chain; such parts are called domains. A protein globule may contain two or three domains (Fig. 2.135, see also Figs. 2.142, 143, 146). Finally, the structures of many proteins (or their domains) including those not only with similar but also with different functions show a strong similarity in the folding of the polypeptide chain. This is called the homology in the structure of protein molecules.

An analysis of the course of the polypeptide chain in numerous globular proteins revealed a number of definite patterns in the formation and mutual arrangement of α helices and β strands. Three commonly occurring folding units of the polypeptide chain are singled out as $(\alpha\alpha)$, $(\beta\beta)$, $(\beta\alpha\beta)$ (Fig. 2.136a). The $(\alpha\alpha)$ unit is formed by two adjacent α helices, connected by a loop at the common ends. The strand connections in β sheets may be formed either by simple bending (reentering) of the chain – "hairpin" $(\beta\beta)$ connections – or by joining the ends of two parallel strands by irregular chain loops or an α helix – "crossover" $(\beta\alpha\beta)$ connections. Combinations of these types of connections permit one to describe all the topologically different β layers observed in globular proteins. Some of these structures are shown in Fig. 2.136b.

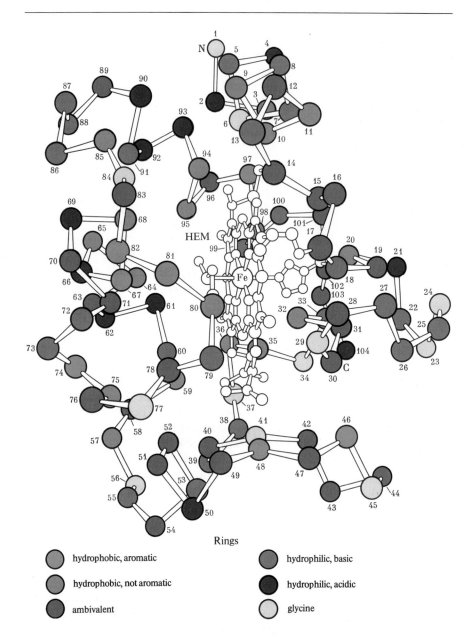

Fig. 2.134. Model of the molecule of cytochrome *C*, a typical representative of the family of the molecules of electron-carrying proteins [2.77]

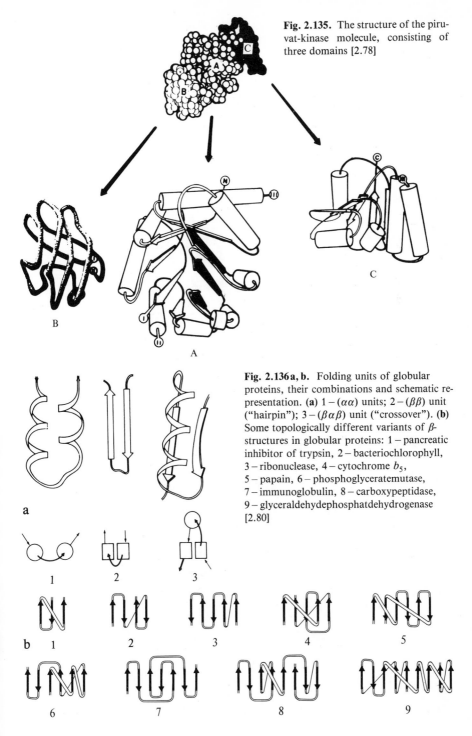

Fig. 2.135. The structure of the piru-vat-kinase molecule, consisting of three domains [2.78]

Fig. 2.136a, b. Folding units of globular proteins, their combinations and schematic representation. **(a)** $1 - (\alpha\alpha)$ units; $2 - (\beta\beta)$ unit ("hairpin"); $3 - (\beta\alpha\beta)$ unit ("crossover"). **(b)** Some topologically different variants of β-structures in globular proteins: 1 – pancreatic inhibitor of trypsin, 2 – bacteriochlorophyll, 3 – ribonuclease, 4 – cytochrome b_5, 5 – papain, 6 – phosphoglyceratemutase, 7 – immunoglobulin, 8 – carboxypeptidase, 9 – glyceraldehydephosphatdehydrogenase [2.80]

If the β structure is formed by several parallel strands, its plane is usually twisted in propellerlike fashion (Fig. 2.137) (the twist is right-handed when viewed along the strands). Remarkable twisted cylindrical barrel configurations were also found (Fig. 2.138a – c). At the same time antiparallel β strands do not exhibit a twist (Fig. 2.139).

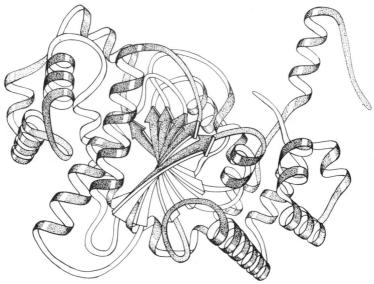

Fig. 2.137. Phosphorilase b, structure of domain 2. Twisted β sheet surrounded by α helices [2.81] (Pictures Figs. 2.137, 2.138a, 2.139 – courtesy of J. Richardson)

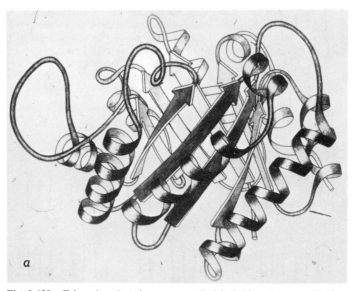

Fig. 2.138a. Triosephosphate isomerase, cylindrical β-barrel surrounded by α helices [2.81a]

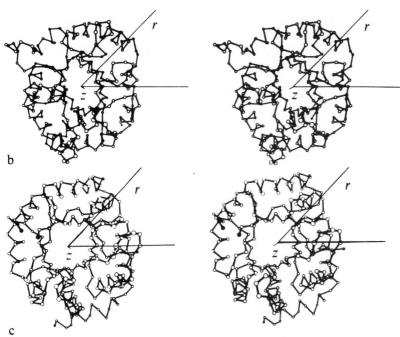

b

c

Fig. 2.138b, c. Stereo diagrams of cylindrical β-barrels (α-carbon backbone). **(b)** Triosephosphate isomerase; **(c)** domain A of piruvat-kinase, cf. Fig. 2.135 [2.78]

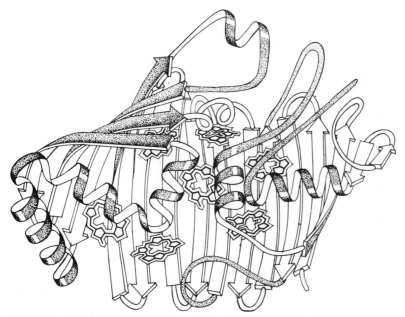

Fig. 2.139. Bacteriochlorophyll protein. Wide antiparallel β sheet, which serves as a support for chlorophyll molecules [2.82]

If we consider the structures of globular proteins from the standpoint of the presence and arrangement of α and β regions, we can single out four principal classes [2.79]. These classes may be described schematically by two-dimensional topology-packing diagrams (Fig. 2.140a – d). Class I consists of α proteins made up of α helices joined by chain segments in irregular conformation (Fig. 2.140a); α helices are not always parallel. Examples of such proteins are globins, hemerythrin TMV subunits; they usually have two layers of α helices. Class II consists of β proteins, which are built of β sheets stacked to form a layer structure; they do not have, or have only a very small part of, α helices. Examples are rubredoxin and chymotrypsin; they usually have two β layers (Fig. 2.140b). The proteins in which both the principal types of secondary structure are observed can be divided into two classes. Class III comprises $\alpha + \beta$ proteins, in which the α and β regions are present, within the same polypeptide chain, as in ribonuclease and thermolysine; they are also "double layered" in most cases (Fig. 2.140c). In class IV, α/β proteins, the α helices and β strands alternate along the chain. These more complicated and larger proteins are usually "three layered"; they generally exhibit the sandwich $\alpha\beta\alpha$ structure. The β layer is flanked by α layers (adenylate kinase and carboxypeptidase) on both sides, but other variants are also possible (for instance, triose-phosphate isomerase, Fig. 2.140d). In these proteins, as in β and in $\alpha + \beta$ proteins, the globule is sometimes subdivided into two distinctly different domains, for instance $(\alpha\beta\alpha)(\alpha\beta)$ in glyceraldehyde-phosphate isomerase.

Some rules that govern the mutual packing of α helices and pleated β sheets in the formation of the tertiary structure of protein globule have been established [2.84]. The helix – helix packing requires that the surface ridges of the helix pack into grooves between the ridges on the second one, which results in some definite angles between the axes of the α helices. Helix – sheet contacts are the best when their axes are nearly parallel. Sheet – sheet contacts depend on the degree of their twist.

Let us consider the homology in protein structures. The homology is observed in the families of proteins which are more or less similar in function; examples are myoglobin-haemoglobin (see Figs. 2.130 – 132), cytochrome (Fig. 2.134), and trypsin-chymotrypsin (Fig. 2.141) families. Proteins showing almost no correlation as regards the primary structure were, however, also found to have domains very similar in structure (see Fig. 2.140a – d). Thus, dehydrogenases (Fig. 2.142), phosphoglyceratekinase, aspartate-transaminase [2.87], catalase *Pen. vitale* (Fig. 2.143), and some other proteins have a structurally similar domain (Fig. 2.144). The structural fragments ($\alpha\alpha$, $\beta\beta$, $\beta\alpha\beta$, etc.) and the various domains, which have lost their similarity as to the primary structure in the course of evolution, but which have proved to be stable as regards their three-dimensional structure, can evidently play a part in the crystallography and biochemistry of proteins, somewhat analogous to that of standard radicals in inorganic (for instance, silicon-oxygen tetrahedra) and organic crystal chemistry.

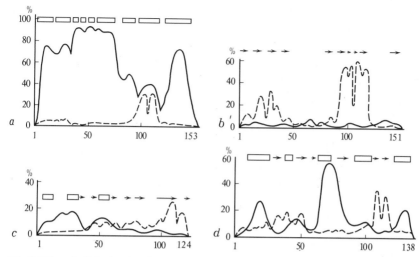

Fig. 2.147a – d. Calculated probabilities of α-helical (———) and β-strand (– – –) states of the polypeptide chain in some proteins. The number of the amino-acid residues is laid off on the x axis. (a) Myoglobin (α protein), (b) superoxide dismutase (β protein), (c) ribonuclease ($\alpha + \beta$ protein), (d) flavodoxin (α/β protein). The real localization of the α and β parts in the structure is shown as rectangles and arrows, respectively [2.91]

polysaccharide which has to be broken there are active radicals of lysozyme: glutamate, aspartate, tryptophane, methionine, tyrosine, and some others. They weaken this bond by changing its electron structure. The lysozyme molecule tears the polysaccharide in two by a slight (1 – 2 Å) shift of the atoms at the active site, ejects both halves, and then returns to its initial configuration and is prepared for activity again. The structure of the $T4$ phage lysozyme, which is homologous in structure (but has an extra loop in the chain) and similar in functional activity to the hen-egg lysozyme, is shown in Fig. 2.149.

In another enzyme, e.g., carboxypeptidase (Fig. 2.145), the displacements of some groups on the molecular surface may exceed 10 Å as a result of this action, i.e., the rupture of the C terminal peptide bonds, but most of the atoms – the foundation of the active site – remain stationary. Thus, the molecular structure of an enzyme has no stable conformation; its functioning results in directional displacements of atomic groups at the active site.

Many proteins possess a quaternary structure. They are built up of not one, but of several globules (subunits) joined together. For instance, the haemoglobin molecule consists of four pairwise identical subunits (Fig. 2.131a, c). Each of them contains a haem binding the O_2 molecule. A remarkable electron-conformational mechanism operates here. The Fe^{2+} atom in the nonoxygenated hemoglobin has a coordination number 5 (Fig. 2.131b). It is in a high-spin state, and its large size prevents it from "squeezing" into the heme plane between the four N atoms. Upon addition

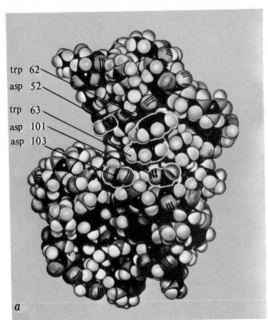

trp 62
asp 52
trp 63
asp 101
asp 103

a

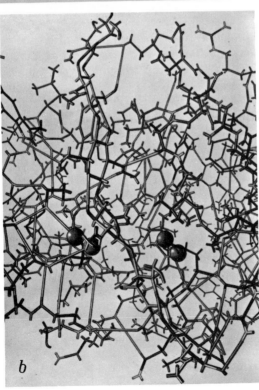

b

Fig. 2.148a, b. Structure of hen-egg white lysozyme. (a) Atomic molecule model illustrating the density of atom packing in the protein molecule and the "crevice" in the region of the active site; (b) skeletal model of the active-center area with a hexasaccharide molecule incorporated into it [2.92]

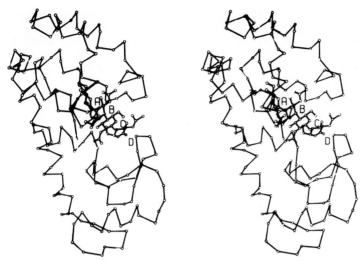

Fig. 2.149. Stereo view of a T4-phage lysozyme molecule with a tetrasaccharide lactone substrate analog. The structure of the binding site of the molecule is very similar to that of hen-egg lysozyme [2.93]

of a sixth ligand – the O_2 molecule – the electron shell of the Fe atom is rearranged (the atom becomes low spin) and its radius decreases, allowing it to enter the haeme plane entraining the residue of His F8. This results in slight mutual shifts of the α helices of the given subunit, which, in turn, changes the quaternary structure (Fig. 2.131c) – the subunits approach each other to a distance of about 6 Å. In this way the addition of oxygen to one subunit increases the oxygen-binding ability of the others.

Let us consider other examples of the quaternary structure of proteins. Figure 2.150 shows the structure of a dimer of aspartate-transaminase (mol. wt. 94,000, the structure of one subunit is given in Fig. 2.141). The subunits of this molecule in the crystal are connected by a twofold noncrystallographic axis. Figure 2.151a, b shows the quaternary structure of the molecule of another protein, leucine aminopeptidase, which was established by electron microscopy. It consists of six subunits and has symmetry 32. The molecule of phosphofructokinase may serve as an example of a complicated quaternary structure (Fig. 2.152). A number of globular proteins form very complicated symmetric structures made up of one or several kinds of subunits with a total molecular weight up to about a million daltons or more (Figs. 2.153a, b). If a molecule contains several sorts of subunits, they may perform different functions, some being regulatory, others substrate-binding, etc.

The slight movements of parts of molecules or their subunits in enzymes promote the process of reactions. But the function of the movement itself is basic to other protein molecule associations. The simplest organs of this type are bacterial flagella which are a helical array of globular molecules (see also Fig. 2.178 – 180).

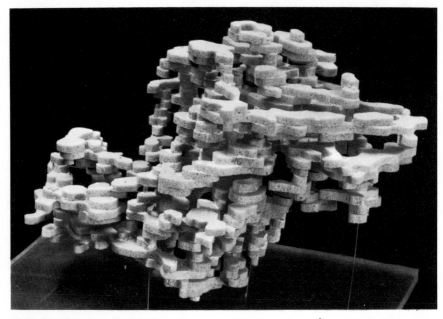

Fig. 2.150. Dimeric molecule of aspartate-transaminase (model at 5 Å resolution) [2.94]

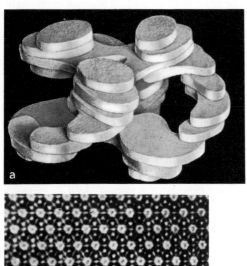

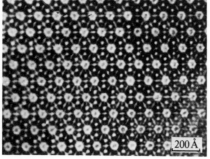

Fig. 2.151. (a) Model of a leucineamino-peptidase molecule consisting of six sub-units; (b) electron micrograph of the packing of molecules in crystal (optically filtered)

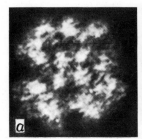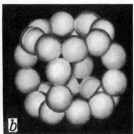

Fig. 2.152. Quaternary structure of phosphofructo-kinase. The molecule consists of 4α and 4β subunits, point symmetry 222 [2.96]

Fig. 2.153. (a) Electron micrograph of lipoyltranssuc-cinylase nucleus of an intricate dehydrogenase complex, (b) model of the arrangement of the protein subunits in the nucleus [2.97]

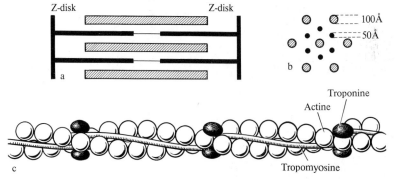

Fig. 2.154a – c. Structure of the sarcomer of a muscle (a, b); (a) 2 disks (membranes) joined by actine protofibrils (about 50 Å thick) with myosine protofibrils (100 Å thick) arranged between them; (b) sarcomer in cross section, two types of protofibrils are disposed hexagonally. (c) Structure of the actine protofibril. The mechanism of muscle contraction consists in the movement of two types of protofibril into the gaps between them, thus drawing the z disks closer together

The universal molecular force apparatus of all living organisms, the muscle, has a more complicated structure. The molecules of the two principal muscle proteins, actin and myosin, have distinctly elongated shapes, and are packed in a strictly hexagonal order (Fig. 2.154). The muscle contraction process consists in the myosin filaments being drawn into the space between the actin molecules.

Another important class of biological structures, membranes, consists of lipids and special proteins. Intricately built membrane layers can let certain ions or molecules through in one direction, either selectively or by active transport, thus organizing their spatially ordered motion in the cell. One of this type of structures, the bacterial purple membrane of rhodopsin, has been studied by electron microscopy (Fig. 2.155). Its framework is built up of parallel α helices.

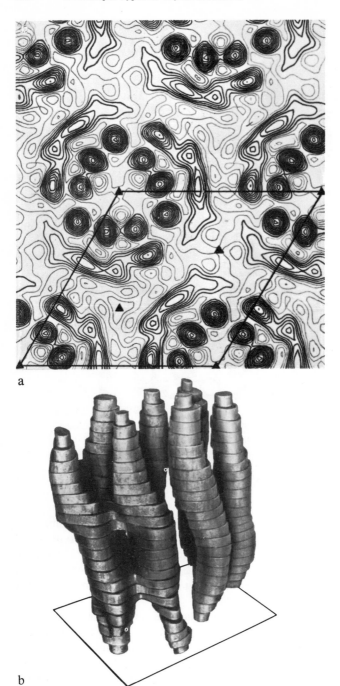

a

b

Fig. 2.155. (a) Fourier projection of the rhodopsin purple membrane, nine clusters represent the projection of α helices; (b) their spatial arrangement [2.98]

Concluding this brief survey of the protein molecule structures, we wish to stress that the primary structure of the polypeptide chain and the levels of the three-dimensional structure − secondary, tertiary, and quaternary − corresponding to it and forming by self-organization, are not accidentally produced chemical and spatial combinations of amino-acid residues. The structure of protein and other biological macromolecules has been improved and perfected by the evolutionary process at the molecular level throughout billions of years of life on Earth. The criterion of this natural selection is the stability of the structure and its ability to perform a specific biological function.

2.9.5 Structure of Nucleic Acids

Another important class of biological macromolecules consists of nucleic acids (Table 2.3). Their function is to store and transfer the genetic information necessary for the reproduction and existence of biological systems. Like proteins, the nucleic acids are chain molecules. The chain has a phosphate − sugar backbone

$$\tag{2.12}$$

The phosphate − sugar group with base R is called nucleotide. In deoxyribonucleic acid (DNA) the sugar is the molecule of deoxyribose, and in ribonucleic acid (RNA), that of ribose. In all, there are four sorts of bases in the polynucleotide chain of DNA. They are the purine (cytosine C and thymine T) and pyrimidine (adenine A and guanine G) bases. In RNA, the cytosine is replaced by structurally similar uracil (U). These four letters A, G, T, C (U) comprise the alphabet of the "nucleic-acid" language. The thickness of a single polynucleotide chain is about 7 Å, on the average.

The structure of DNA was established on the basis of x-ray diffraction investigations carried out early in the fifties by *Franklin, Gosling* [2.99] and *Wilkins* et al. [2.100]. Proceeding from biochemical and x-ray data, *Watson* and *Crick* [2.101] proposed their famous model of this complicated molecule, which explains the biological mechanism of the replication (self-reproduction) of DNA.

Figure 2.156a represents an x-ray photograph of the gel of wet DNA in *B* form, in which the molecules have paracrystalline hexagonal packing (see Fig. 2.99a), and Fig. 2.156b, an x-ray photograph of the crystalline *A* form of DNA. Both patterns, especially the first, reveal the crosslike arrangement of reflections and vacant zones in the meridional region characteristic of helical structures (cf. Fig. 4.41 in [1.6]). The DNA molecule has two chains. The join-

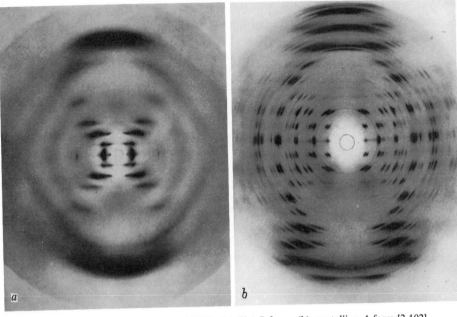

Fig. 2.156 a, b. X-ray photographs of DNA. (a) Wet B form, (b) crystalline A form [2.102]

ing of the chains is due to the hydrogen bonds between the pairs of bases adenine – thymine $A - T$, and guanine – cytosine $G - C$ (Fig. 2.157a, b) because of the suitable disposition of the hydrogen bond donors and acceptors on the indicated molecules. These bonds are formed precisely as shown in Fig. 2.157a, which has been confirmed by investigations of modelling structures (Fig. 2.157b). In other words, pairs $A - T$ and $G - C$, and only these pairs, are complementary. The presence of two (in the $A - T$ pair) or three (in the $G - C$ pair) H bonds ensures the arrangement of each of the pairs of these bases in the same plane. It is due to the contact of such pairs that the double-strand molecule of DNA can be held together.

Joining up with each other selectively, the bases are stacked in the DNA molecule in B form in a plane perpendicular to the helix axis with spacing ~ 3.4 Å (the thickness of a flat organic molecule), and the phosphate – sugar backbones of the two chains remain outside (Fig. 2.158a, b). The backbone chains (which are polar) are antiparallel; they are symmetrically related by axes 2, which are perpendicular to the axis of the helix. The helix in the B form makes a complete revolution after each 10 base pairs, so that the period of the molecule $c = 34$ Å. The symmetry of the DNA molecule is $S_M/2$. The molecule is about 20 Å thick and can be observed directly with an electron microscope (Fig. 2.158c). The period of the DNA molecule in the A form is 28.2 Å; with 11 base pairs for period; and the planes of the bases are inclined to the axis of the molecule at $\sim 19°$. More varieties of DNA structure are known, differing in parameters and structure of the double helix. For instance, in the

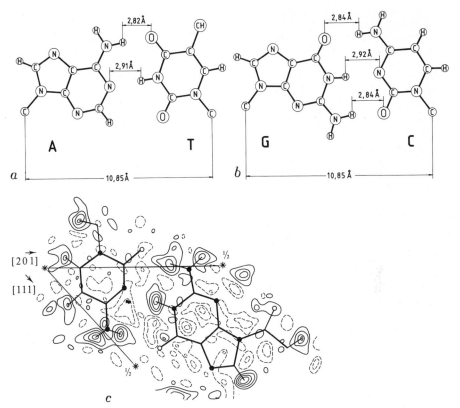

Fig. 2.157a – c. Complementary Watson-Grick base pairs in the structure of nucleic acids. (**a**) Adenine – thymine; (**b**) guanine – cytosine; (**c**) difference Fourier map of the structure of a molecular compound 9-ethylguanine with 1-methylcytosine showing the arrangement of the H atoms of hydrogen bonds in the G – C complex [2.103]

T form (DNA from $T2$ bacteriophage) there are 8 base pairs for the 27.2 Å period [2.104].

In all these forms of DNA the double helix is right-handed. Recently, an unusual left-hand double-helical conformation, Z-DNA, was also found (Fig. 2.158d) [2.105].

The DNA chains in chromosomes (apparently in B form) are actually the storage of genetic information, which is written as a sequence of nucleotides in the chains:

$$... AGCATCCTGATAC ...   (a)$$
$$... TCGTAGGACTATG ...   (a')$$

(2.13)

Because the nucleotides are complementary, both a and a' chains of the DNA molecule are also complementary. It can be said that they are antiequal to one another. This complementarity explains the reproductive mechanism of

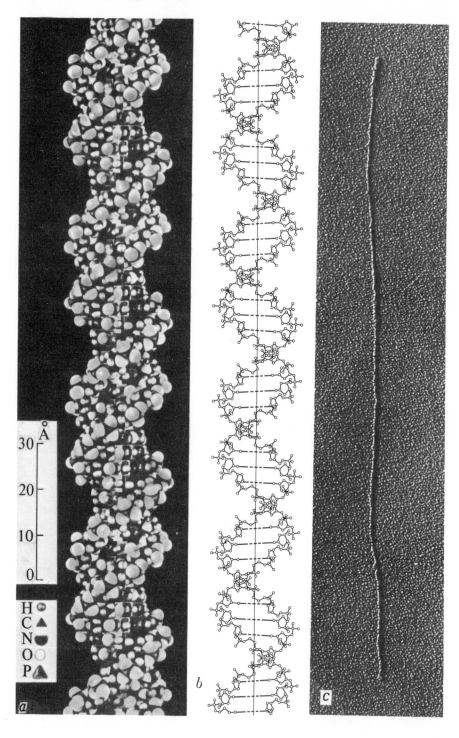

30 Å
20
10
0

H ⊙
C ▲
N ⬯
O ○
P ▲

a

b

c

Fig. 2.158d. Structure of DNA molecule in Z form

d

DNA molecules. The double helix aa' unbraids into single chains a and a' (Fig. 2.159). With the aid of special proteins, the free nucleoside-3-phosphates available in the cell nuclei are added to them one after another, as if crystallizing on matrices a and a'. Owing to the $A-T$, $G-C$ rule, a new sequence a', the same as the former a', crystallizes on matrix a (it is antiequal to that matrix) and a sequence a, the same as the former a, crystallizes on matrix a'.

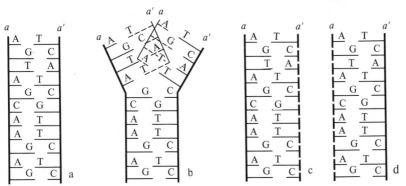

Fig. 2.159a–d. Unbraiding of the double DNA chain and the formation of two new chains identical to the initial one. (a) Initial chains, (b) partial unbraiding of chains, (c, d) two new chains

Fig. 2.158a–c. Structure of DNA molecule. (a) Molecule model, (b) molecule skeleton, (c) electron micrograph of a fragment of a DNA molecule ($\times$ 100,000)

As a result, the two newly formed molecules *aa'* and *a'a* are exactly equal to one another and to their precursor.

In precisely the same way, the RNA molecule formed on DNA "reads out" the sequence of nucleotides with one of the DNA chains (in RNA the thymine T is replaced by uracil U, which is also complementary to A). Now RNA becomes a single chain. The principal function of this, the so-called messenger RNA is the delivery of information stored in DNA into the particles of the cells which perform the synthesis of the protein – the ribosomes. At present, no data are available on the secondary and tertiary structure of m-RNA, which is relatively short lived and evidently exists in combination with special proteins. This m-RNA passes through the ribosome like a punched tape and amino acids are joined in the ribosomes in stepwise fashion, forming a protein.

A code has been established, which determines the correspondence between the nucleotide and protein alphabet; it is called the genetic code. Since the former alphabet has fewer letters, it is clear that only their combination can correspond to the characters of the latter alphabet. It is a triplet code: the nucleotide triplet (codon) determines the amino acid. For instance, UUU, UUC assigns phenylalanine, GGU, GGC, GGA, GGG, glycine, UAU, UAC, thyrosine, and AUU, AUC, isoleucine. The code also has a "fullstop", UAG (UAA, UGA), i.e., a symbol indicating the end of the synthesis of the polypeptide chain.

Amino acids are delivered to the ribosome by comparatively small molecules of still another variety of nucleic acids, transfer RNA (t-RNA, mol. wt.

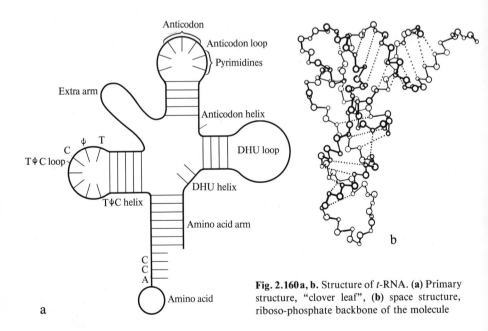

Fig. 2.160a, b. Structure of *t*-RNA. (a) Primary structure, "clover leaf", (b) space structure, riboso-phosphate backbone of the molecule

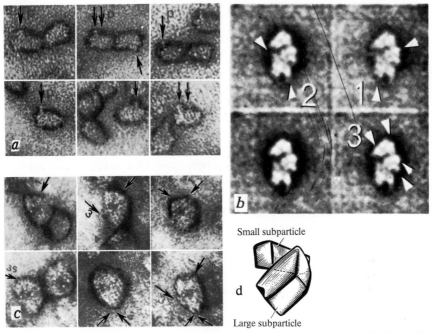

Fig. 2.161 a – d. Rat liver ribosomes (electron micrographs). **(a)** Side projections of a large sub-particle ($\times$ 370,000), **(b)** four filtered images of a small subparticle ($\times$ 450,000); **(c)** ribosomal particles as a whole ($\times$ 420,000), **(d)** scheme of the structure of ribosome [2.108, 109]

30,000). The chemical structure of this molecule, which consists of about 80 nucleotides and whose separate sections are complementary, is described as a "clover leaf" (Fig. 2.160a). X-ray diffraction analysis of crystals of phenylalanine t-RNA, which contain a very large amount of solvent, has established the structure of the molecule (Fig. 2.160b) [2.106, 107]. It is L shaped, its complementary sections are double helical, and the molecule has at its center a stabilizing "lock" composed of three H-bonded nucleotides. Three free nucleotides project at one end of the molecule with their bases turned outwards: they form an anticodon, which is joined complementarily to the codon of the m-RNA in the ribosome, while an amino-acid molecule is attached to the other end.

The ribosome joins the amino acids into a polypeptide chain with the aid of a yet unknown mechanism. Here, the text written in the nucleotide alphabet, consisting of four letters A, C, G, and U, is "translated" into that in the protein alphabet of twenty letters – amino acids – and the genetic information is embodied in the concrete primary structures of the protein, i.e., a polypeptide chain is formed in the ribosome by successive addition of amino acids. The chain, as mentioned above, then self-organizes into a three-dimensional structure of protein molecules. Thus, a definite section of DNA (the gene) determines, via m-RNA, the synthesis of one protein, and the whole

DNA of the given organism (its chromosomes) determines the entire enormous set of its proteins.

Ribosomes, as well as chromosomes, viruses etc., belong to another important class of biological objects, nucleoproteins (last line of Table 2.2), in which nucleic acids, combining with the proteins, form complicated structures.

The ribosome is an extremely complex molecular aggregate consisting of two subparticles (Fig. 2.161a, b, c). It contains molecules of globular proteins (about 70−80 in all) and of still another ribosomal variety of RNA. Individual ribosomal RNA segments have the secondary structure of a double helix and form complicated loops (Fig. 2.162).

Genetic information is stored in chromosomes. According to present-day concepts based on electron microscopy and neutron diffraction analysis the simplest component of the chromosome is a so-called chromatin filament consisting of double-helical DNA with nucleoprotein globules, or nucleosomes, which are regularly repeated along it.

Nucleosomes have been successfully crystallized, and the principal features of their structure studied. It has been found (Fig. 2.163) that the double helix of DNA (a section of 140 pairs of nucleotides) itself bends, and its chain forms a gently sloping superhelix with $1\frac{3}{4}$ turns around the protein core.

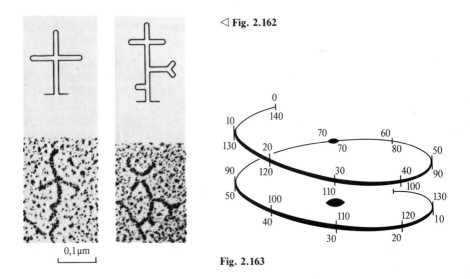

◁ Fig. 2.162

0,1 μm

Fig. 2.163

Fig. 2.162. Electron micrographs of parts of the RNA molecule, and corresponding schemes of the folding of the precursor of ribosomal RNA, on the basis of which the ribosome is further formed, showing the double helical segments [2.110]

Fig. 2.163. Structure of the nucleosome core consisting of $1\frac{3}{4}$ turns of the DNA superhelix. Numbers of nucleotide pairs are indicated [2.111]

2.9.6 Structure of Viruses

Regular virus particles are made up of nucleic acid and globular protein molecules. A virus has no reproduction mechanism of its own, but by penetrating into a cell of the host organism, it makes its protein-synthesizing apparatus work for it according to the program of its own nucleic acid, rather than that of the host cell.

Let us consider the structure of the so-called "small" regular viruses which consist of nucleic acid and one or two varieties of protein. The function of protein is the formation of a coat or framework containing and protecting nucleic acid, the infection agent of the virus. Two types of such viruses are known, rod-shaped and spherical.

The rod-shaped viruses show a helical symmetry, according to which their constituent protein subunits are packed. The classical, best-studied representative of this class of virus is the tobacco mosaic virus TMV; the other viruses of this type (the punctate mosaic virus of barley, Adonis, etc.) also cause plant diseases. Figure 2.164 shows an electron micrograph of TMV. The virus contains an RNA chain. It is possible to break the virus down into separate protein subunits and RNA, and to repolymerize the former into the helical packing typical of the virus, but already without RNA. X-ray structure

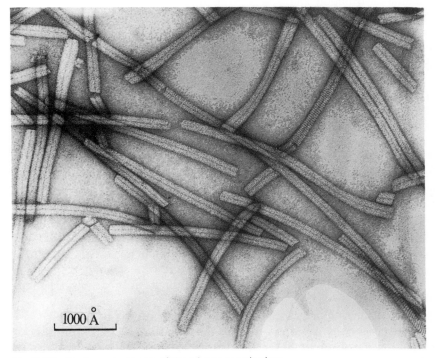

1000 Å

Fig. 2.164. Electron micrograph of the tobacco mosaic virus

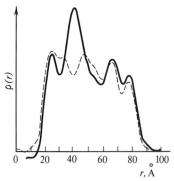

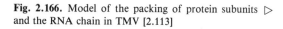

RNA

Fig. 2.165. Radial electron density distribution of TMV (solid line) and of a repolymerized protein without RNA. Maxima at 40 Å on the TMV curve show the position of RNA [2.112]

Fig. 2.166. Model of the packing of protein subunits ▷ and the RNA chain in TMV [2.113]

analysis of the initial virus and the repolymerized particles made it possible to construct the radial distribution of the electron density for them (Fig. 2.165). The maximum at a distance of 40 Å from the particle axis, which is present on the first distribution and absent from the second, evidently localizes the position of RNA. A TMV model shown in Fig. 2.166 was constructed on the basis of x-ray diffraction and also electron microscopic data. TMV has the shape of a rod, 3000 Å long and 170 Å in diameter, with a 40-Å-diameter channel inside it. The molecular weight of each protein subunit is 17,420; all of them are identical. There are about 2140 such subunits in TMV. The TMV symmetry is s_M, where $M = 49/3$, i.e., there are 49 subunits per three turns of the helix. The pitch of the helix is 23 Å, and its period, corresponding to three turns, is 69 Å. The distance between the nucleotides in RNA is about 5.1 Å, i.e., the chain is not extended. Obligatory intermediates in the assembly of TMV are TMV disks. A disk consists of two layer rings of 17 protein subunits each. A stack of disks serves as a precursor of the virus, which is assembled in a helical structure with the help of viral RNA.

It is possible to crystallize disks in three-dimensional crystals under specific conditions. X-ray structure analysis of disk crystals [2.114] has established the structure of the protein subunits and the details of their packing. The electron density map of a protein subunit is given in Fig. 2.167a. Figure 2.167b shows schematically the structure of a subunit; the chain is folded in four approximately parallel α helices. Figure 2.167c shows the shape of the polypeptide chain in one subunit in side projection.

The packing of the subunits in rod-shaped viruses is predetermined by their shape (they narrow down towards the cylinder axis) and by the disposition of the active groups on the surface. So, self-assembly is ensured, first in disks and, then, in a stable helical structure.

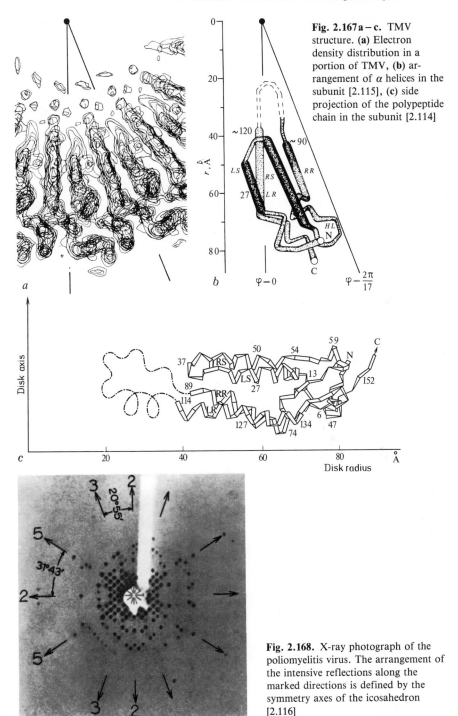

Fig. 2.167 a – c. TMV structure. **(a)** Electron density distribution in a portion of TMV, **(b)** arrangement of α helices in the subunit [2.115], **(c)** side projection of the polypeptide chain in the subunit [2.114]

Fig. 2.168. X-ray photograph of the poliomyelitis virus. The arrangement of the intensive reflections along the marked directions is defined by the symmetry axes of the icosahedron [2.116]

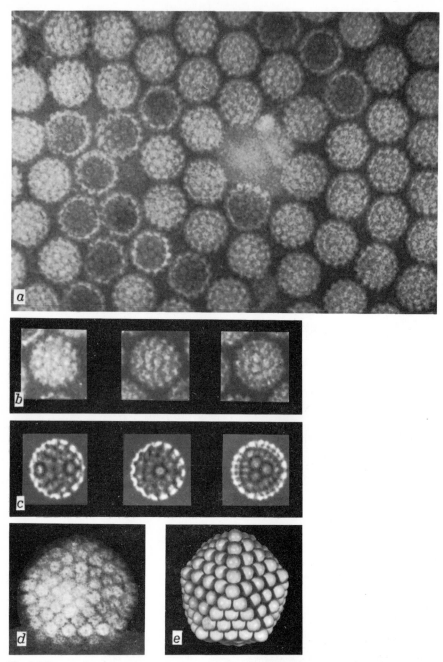

Fig. 2.169 a – e. Some spherical viruses. (a) Human wart virus ($\times$ 300,000) [2.117], (b) particles of this virus in different orientations, (c) computer simulation of the image of these particles which establishes precisely the emergences of the axes 2, 3 and 5 [2.118]; (d) herpes virus; (e) its model made up of 162 balls

In another type of viruses, spherical, or, more precisely, icosahedral, the nucleic acid is located in a closed protein shell. Because of their regular structure and approximately spherical shape many viruses of this class form good crystals, which often show a cubic symmetry (see Figs. 1.1, 19). The periods of such crystals may reach 2000 Å. The arrangement of strong reflections on x-ray photographs indicates the presence of fivefold axes in the virus particles (Fig. 2.168) forming such a crystal (not in the crystal itself, of course). The structure of spherical viruses was studied by means of electron microscopy (Fig. 2.169).

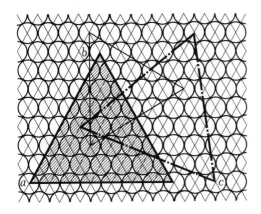

Fig. 2.170. Close packing of balls on a plane. Triangles depicted in the figure illustrate possible variants of the formation of protein capside faces: (*a*) the face passes through the centers of the balls limiting the morphological units, (*b*) through the centers of the balls and between them, and (*c*) skew face

Let us consider a geometric scheme illustrating the possibility of the formation of icosahedral closed shells, capsides, from certain protein units. A close-packed plane layer of such units with a sixfold coordination (Fig. 2.170) may serve as a flat two-dimensional model of the material for such a shell. The triangles shown in Fig. 2.170 may be used as the "pattern" elements; the bends in the protein shell would correspond to their sides. A polyhedron whose faces are equal-sided triangles and shape is closest to spherical is the icosahedron, its volume-to-surface ratio being the most advantageous (Fig. 2.171 a). Thus we arrive at the icosahedral symmetry inherent in "spherical" viruses and other pseudospherical shells (Fig. 2.171b, see also [Ref. 1.6, Fig. 2.50]). An individual protein molecule is asymmetric; therefore, the morphological protein unit, which is conventionally depicted as a ball in Fig. 2.170, must consist of six protein molecules in a flat layer, i.e., it must be a hexamer (Fig. 2.172a). At the same time, the explanation of the formation of a closed icosahedral shell requires the assumption that the same molecules are aggregated into pentamers, as can be seen from Fig. 2.172b. The pentamers correspond to the points where axes 5 emerge at the surfaces of icosahedral symmetry (Fig. 2.173, cf. Fig. 2.171b and [Ref. 1.6, Fig. 2.50])..

Pentamers and hexamers are morphological units of the virus shell readily distinguishable in electron micrographs (Fig. 2.169). (Trimer or dimers are also found in certain viruses.) The possible schemes of the packing of

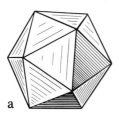

Fig. 2.171. (a) Icosahedron. **(b)** Football sewn according to icosahedral symmetry

pentamers and hexamers in the shell of a virus particle are given in Fig. 2.173. In some cases one can observe the intrinsic structure of pentamers and hexamers in high-resolution electron micrographs (Fig. 2.169d). Figure 2.174 illustrates the shell structure of the turnip yellow mosaic virus (TYMV) based on electron-microscopy and x-ray data. This virus is about 300 Å in diameter; the capside is composed of 180 subunits joined into 32 morphological units (12 pentamers and 20 hexamers). The molecular weight of the virus is 5.5 mln. All the subunits are identical, their molecular weight is about 20,000. The virus crystallizes in a cubic lattice with "diamond" packing, a = 700 Å. The thickness of the protein shell is about 35 Å. The RNA chain inside it is so folded that it forms 32 dense packets closely connected with the morphological units.

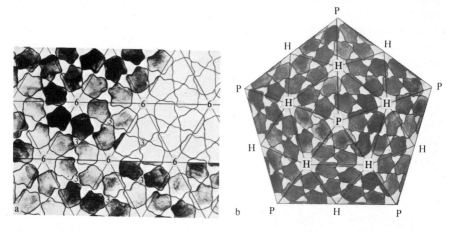

Fig. 2.172a, b. Hexagonal packing of symmetric combination of asymmetric subunits. **(a)** Joining of subunits into hexamers (H) (each hexamer corresponds to one ball in Fig. 2.170); **(b)** during the formation of an icosahedron face, pentamers (P) also arise in addition to hexamers (H)

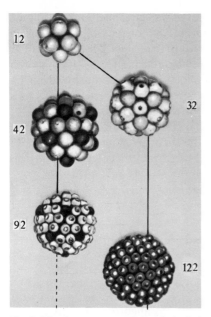

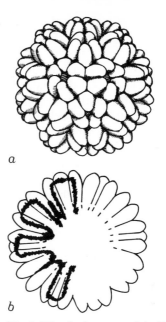

Fig. 2.173. Arrangement of morphological units – hexamers and pentamers – in the lower members of the two icosahedral classes $P = 1$ (on the left) and $P = 3$ (on the right). The numbers of morphological units in a shell are indicated [2.113]

Fig. 2.174a, b. Structure of the TYMV. (a) External surface of the virus: 180 protein molecules form pentamers and hexamers [2.119], (b) scheme illustrating the possible arrangement of the RNA [2.120]

Fig. 2.175. Different deltahedra with an icosahedral symmetry ("icosadeltahedra"). Each deltahedron has on its surface $20\,T$ equilateral triangles [2.113]

The general regularities in the structure of icosahedral shells are as follows [2.113]. The icosahedron faces cut out of the flat surface of close-packed morphological units can be divided into unit triangles (Fig. 2.170). The packing units corresponds to the vertices of these triangles. An icosahedron has 20 faces, and if T is the triangulation number, i.e., the number of triangles per face, then the total number of triangles is 20 T. The icosahedral polygons thus obtained are called deltahedra; some of them are shown in Fig. 2.175. The triangulation number T depends on the different possibilities of selecting the triangular face of the icosahedron on the surface depicted in Fig. 2.170: $T = pf^2$ (f is an integer), $P = h^2 + hk + k^2$ (h and k integers having no common multiplier). A list of possible classes is: $P = 1, 3, 7, 13, 19, 21, \ldots$. The following classes of deltahedra exist:

Class	T values							
$P = 1$	1	4	9		16		25	$\ldots$
$P = 3$		3		12				27 $\ldots$
Skew classes			7		13	19 21		$\ldots$

In the class $P = 1$ the generating face of the icosahedron runs through the centers of all the units of the flat layer (triangle a in Fig. 2.17 and the first structure in Fig. 2.175). In the class $P = 3$ the bend goes in another convenient direction, also through the centers of the units (triangle b in Fig. 2.170 and the second structure in 2.175). In the general case of a skew bend $P = 7, 13, 19 \ldots$ (triangle c in Fig. 2.170 and the last six structures in Fig. 2.175). Here, the planes of the triangles of the deltahedron are slightly curved; it is not a "polyhedron" in the strict sense of the word. The number M of protein molecules (subunits) in the shell is 60 T. They are joined into 12 pentamers and 10 ($T - 1$) hexamers. The total number of morphological units is $V = 12$ pentamers $+ 10 (T - 1)$ hexamers $= 10 T + 2, M = 60 T = 12 \times 5 + (V - 12)6$. For the class $P = 1$, $V = 12, 42, 92, 162, 262, \ldots$, for the class $P = 3$, $V = 32, 122, 272 \ldots$. Figure 2.173 shows models made up of morphological units which correspond to the vertices in the deltahedra, for $P = 1$ and $P = 3$.

Most of the viruses belong to the first class, for instance, phase $\phi X 174$: $T = 1$, i.e., $V = 12, M = 60$; the polyoma virus: $T = 7$, i.e., $V = 72, M = 420$; viruses of herpes and varicella: $T = 16$, i.e., $V = 162, M = 960$; the virus of the rabbit hepatitis: $T = 25$, i.e., $V = 252, M = 1500$. Representatives of the second class are the TYMV and the tomato bushy stunt virus (TBSV): $T = 3, M = 32, B = 180$. Figure 2.176 shows the result of three-dimensional reconstruction of the space structure of certain viruses according to electron microscopy data.

Incidentally, the above-mentioned term "bend" of the flat layer has only a formally geometric meaning and indicates the relationship between the structure of deltahedron surface and the flat surface divided into triangles, rather than the actual character of self-assembly of the shell of spherical

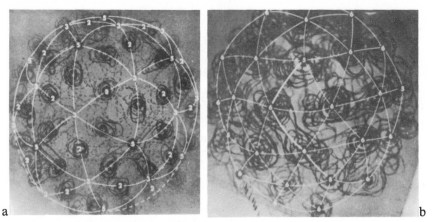

a   b

Fig. 2.176a, b. Three-dimensional reconstruction of spherical viruses. (a) TBSV (surface lattice T = 3 is drawn); (b) human wart virus (surface lattice T = 7 is drawn) [2.121]

viruses. This self-assembly is evidently effected by successive joining and crystallization of the packing units. According to the presently adopted point of view the icosahedral protein shells of some small viruses, for instance TYMV, are formed according to the "all or nothing" principle as a result of self-assembly. The larger icosahedral shells are formed step-by-step. For the adenovirus, for instance, it has been shown that the faces of its shell are formed first and can exist independently. Faces are capable of joining into icosahedral shells similar to virus shells, but without the "pentamers", i.e., those subunits which are located at the fivefold axes. There are evidently viruses where a small internal protein capsule is first formed, on which the nucleic acid is condensed, with the outer shell formed subsequently.

X-ray structure investigations into virus crystals, carried out in recent years, have revealed the packing of the polypeptide chain in shell subunits. Thus, the tomato bushy stant virus (TBSV) crystallizes in the cubic space group $I23$, a = 383.2 Å, and the unit cell contains two virus particles about 330 Å in diameter. The virus shell consists of 180 subunits (T = 3) with a molecular weight of 41,000 daltons. To obtain a resolution of 2.9 Å, about 200,000 reflections of native crystals were measured, as well as many for two heavy-atom derivatives [2.122]. The subunits are built of two domains (Fig. 2.177a) domain P (projection) and domain S connected by a "hinge", and of the N terminal arm. Besides the elements of icosahedral symmetry (axes 5, 3, and 2), the packing has additional quasi-symmetry axes 3^q and 2^q, by which the subunits of the shell are related (Fig. 2.177d). Domains S form the virus surface (Fig. 2.177b, c), while domains P stick over this surface, grouping together in pairs. The S domains of subunits denoted by A, B, and C are related by quasi-axis 3^q, subunits A are disposed around axes 5, and subunits B and C, around axes 3. The schematic course of the polypeptide chain in the subunit is given in Fig. 2.177d; in both domains it forms an antiparallel β

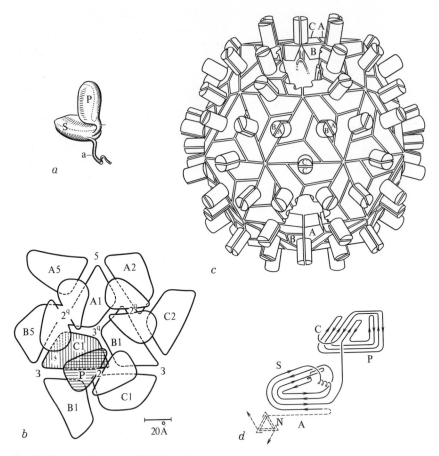

Fig. 2.177 a – d. Structure of TBSV. (a) Shape of a protein subunit consisting of P and S domains and an N-terminal arm. (b) Scheme of the packing of the subunits and their domains in one asymmetric region of icosahedral packing; P domains protrude outwards; 2,3,5: regions where the respective axes emerge, 2^q, 3^q: regions where the quasi-axes emerge. (c) General scheme of the virus structure. (d) Scheme of the folding of the polypeptide chain in subunit domains

structure. N terminals of 60 C subunits have been clearly revealed on a Fourier synthesis; they folded together around axes 3, connecting three subunits C. The N terminals of the other 120 subunits (B and C) are not clearly identified; they are evidently disordered and penetrate deep into the virus particle. Their function is, possibly, to connect the protein shell with the RNA located inside the virus.

It is believed that the presence of RNA is necessary for self-assembly of certain spherical viruses, since it can serve as a kind of a "matrix" for protein crystallization. Yet it is possible to repolymerize the protein molecules of certain spherical viruses in the absence of RNA as well, which points to a high adaptability of the shape of the subunits to the formation of icosahedral

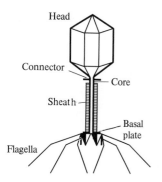

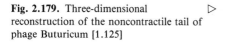

Fig. 2.178. General scheme of the structure of phages with a contractile tail

Fig. 2.179. Three-dimensional   ▷
reconstruction of the noncontractile tail of phage Buturicum [1.125]

shells. At the same time, under different conditions the protein molecules of some spherical viruses can be crystallized into tubes.

The spherical viruses are the most complicated biological structures which obey exact symmetry laws. Still more intricate systems are, for instance, bacteriophages and cell organelles; they consist of many varieties of different molecules (DNA, proteins, lipids, etc.), and some of their elements show strict structural regularities.

Of great interest are the structures of bacteriophages. They consist of several functional elements (Fig. 2.178). The head, containing nucleic acid, is joined to the tail by a connector. The head capside has the form of a poly-hedron, with a pseudoicosahedral symmetry; it is built up of hexamers and pentamers. In phages with a contractile tail the latter consists of a narrow tubular rodlike core surrounded by a sheath. In phages with a noncontractile tail the latter is built of protein molecules of one sort. In both cases the tail structure can be described as a stack of disks which are laid on one another with a rotation determined by the helical parameters p and q. At its end, the tail has a basal plate with branching off fibrils.

The most comprehensive information on the phage structure is provided by electron microscopy and the three-dimensional reconstruction method [2.123 – 125] (see also [Ref. 1.6, Sect. 4.10.3]).

Figure 2.178 shows the structure of the phages (see also [Ref. 1.6, Figs. 4.11, 112]). Figure 2.179 is a model of the tail of the phage Buturicum with a noncontractile tail. The shell disk has a 6-fold symmetry, and the helical symmetry is $S_{3(1/6)}6$. Most of the phages have contractile tails; these are the so-called T-even phages ($T2$, $T4$, and $T6$), $DD6$ phage, phage Phy-1 E.Coli K-12, etc. [2.126, 127]. The structure of the latter and the changes in its structure on

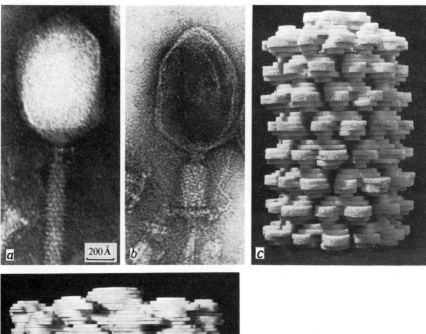

Fig. 2.180. (a, b) Electron micrographs
of the phage Phy-1 E.coli *K*-12;
(a) intact state, (b) contracted state;
(c, d) three-dimensional reconstruction
of the tail in the intact (c) and the
contracted (d) state
(Courtesy of A. M. Michailov)

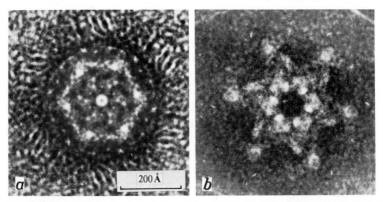

Fig. 2.181a, b. Electron micrographs of the basal plate of the Phy-1 E. Coly *K*-12 phage. (a)
Intact hexagon configuration, (b) star configuration after contraction of the tail (Courtesy of
A. M. Michailov)

contraction are shown in Fig. 2.180a – d. The diameter of the intact tail is equal to 210 Å, the number of molecules in the disk being 6; the symmetry of the tail is $D_{7/2} \cdot 6$. Three-dimensional reconstruction reveals two families of helical grooves on the surface of the tail and the dimeric structure of the sheath molecules (Fig. 2.180c).

The contractile tail of the phages is one of the simplest propulsion devices in nature. During the process of contraction of the tail (Fig. 2.180b, d) its diameter increases from 210 to 280Å, apparently owing to the rearrangement of protein subunits. The symmetry of the contracted tail is $S_{11/1} 6$. In the intact structure, the sheath subunits are arranged with their long axes being approximately tangential to the circumference, whereas in the contracted state they are rotated in the horizontal plane and are found to be close to the radial direction (Fig. 2.180d). The rotation of the subunits and their penetration into the gaps between the subunits of the neighboring "disks" occur simultaneously [2.128, 129]. The process of contraction is initiated by a contact of baseplate fibrils with bacteria. The rearrangement of the protein subunits of the baseplate takes place during this process, as shown in Fig. 2.181. The contraction of the phage tail enables the rod to penetrate the wall of the bacterial cell and "inject" the nucleic acid of the phage into it.

Symmetry plays no part in the cell structure, but at the macroscopic level it manifests itself in the structural organization and the shape of plants and animals.

3. Band Energy Structure of Crystals

A number of physical phenomena in crystals are determined by their electron energy spectrum. Some phenomena are associated with the motion of electrons in the periodic field of the lattice and with their scattering on lattice vibrations. The optical, electrical, magnetic, galvanomagnetic, and other properties of crystals of dielectrics, semiconductors, and metals are intimately connected with the nature of the electron energy spectrum and the geometry of the isoenergetic surfaces of the electrons in the crystal, the peculiarities of vibrations of the lattice atoms, and the dispersion of the frequencies of these vibrations. This chapter discusses the energy spectrum of the electrons in a crystal.

The most clear-cut physical interpretation of the energy spectrum of electrons and the shape of the electron isoenergetic surfaces can be obtained by viewing them in the space of the reciprocal lattice of the crystal, rather than in the real space. Like all other properties, of course, the energy spectrum of the electrons and the shape of the isoenergetic surfaces intrinsically depend on the crystal symmetry. Taking into account the space symmetry of the crystal and the wave nature of the electron results in the division of the reciprocal-lattice space into energy bands whose boundaries satisfy the conditions of interference of short waves in the crystal. Thus, the energy-band structure of the crystal, on the one hand, and the scattering of external electron beams or short waves (for instance, x rays) in the crystal, on the other, have a common physical nature. In the final analysis, the band spectrum depends on the nature of the chemical bond of the atoms in the crystal. The concept of electron energy bands is of fundamental importance in modern crystalphysics and quantum electronics and is widely used in the study of electron phenomena in solids. This calls for consideration of the band-energy structure within the framework of crystallography. It should also be emphasized that we discuss the ideal crystal, neglecting the defects of the crystal lattice.

3.1 Electron Motion in the Ideal Crystal

3.1.1 Schrödinger Equation and Born-Karman Boundary Conditions

The steady state of an electron in a crystal is described by the Schrödinger equation, which is reduced to the following form with the aid of the Hartree self-consistent-field method:

$$\mathscr{H} \psi_k = E \psi_k. \tag{3.1}$$

Here, the Hamiltonian of the system

$$\mathscr{H} = - \frac{\hbar^2}{2m} \nabla^2 + U(r), \qquad \hbar = \frac{h}{2\pi},$$

where $U(r)$ is the potential energy of the electron in the periodic field of the crystal lattice, ψ_k is the wave function, which describes the state of an electron with a wave vector k, and $E = \varepsilon$, denotes the eigenvalues of the electron energy. The square of the modulus of the wave function $\psi(r) \psi^*(r) = |\psi|^2$ is equal to the probability of finding the electron at a point of the crystal with a radius vector r. Equation (3.1) is often called the one-electron approximation, since here the extremely complex problem of interaction of the system of electrons and nuclei in the crystal reduces to determining the state of a single electron in an averaged periodic external field. It can be demonstrated that if $U(r)$ has the translation symmetry of the lattice, the steady-state solution of (3.1) can be represented as

$$\psi_k(r) = U_k(r) \exp(ikr), \tag{3.2}$$

where amplitude $U_k(r)$ also shows the translation symmetry of the crystal lattice. This statement is called Bloch's theorem. In the case of a free electron, $U_k(r) = \text{const}$, and (3.2) transforms into the expression for the wave function of a free electron

$$\psi_k(r) = \text{const} \exp(ikr). \tag{3.3}$$

Let us denote by

$$t = p_1 a_1 + p_2 a_2 + p_3 a_3, \quad H = h_1 a_1^* + h_2 a_2^* + h_3 a_3^*,$$

the arbitrary vectors of the direct and reciprocal lattices, respectively, where a_1, a_2, a_3 and a_1^*, a_2^*, a_3^* are the basic vectors of the direct and reciprocal crystal lattices. Substituting $k_1 = k + 2\pi H$ for k in (3.2) and taking into account that, by definition (see [Ref. 1.6, Chap. 3]), Ht is equal to an arbitrary integer, we find that $\psi_k(r) = \psi_{k_1}(r)$. Thus, in the reciprocal-lattice space the wave function of the electron $\psi_k(r)$, as well as its energy $\varepsilon(k)$, is periodic, with periods a_1^*, a_2^*, and a_3^*. The region of the reciprocal space (or the k space) where the functions of the wave vector are defined and which satisfies the condition

$$- \pi \leqslant ka_i \leqslant \pi, \tag{3.4}$$

is called the first Brillouin zone (a_i is the basis vector of the direct lattice, and $i = 1, 2, 3$). For a cubic lattice, the first Brillouin zone is a cube, whose edges are equal to $2\pi/a$, and according to (3.4)

$$- \frac{\pi}{a} \leqslant k_i \leqslant + \frac{\pi}{a}. \tag{3.5}$$

In Sect. 3.2 we shall analyze the general rule for constructing Brillouin zones, for which definition (3.4) is a particular case.

In solving the Schrödinger equation (3.1) one uses the Born – von Karman conditions for boundary conditions; accordingly, the wave function of the electron (3.2) is unaltered on displacement of the radius vector by a whole number N of unit cells of the direct lattice, i.e., by Na_i. Using expression (3.2) and the translation symmetry of amplitude $U_k(r)$, the Born – von Karman conditions can be represented in the form

$$ka_i = \frac{2\pi}{N} g_i. \tag{3.6}$$

It is easy to see that condition (3.6) is equivalent to the statement that the projection of the wave vector within a Brillouin zone can take only N discrete values. Indeed, using (3.4), we get from (3.6)

$$- \frac{N}{2} \leqslant g_i \leqslant + \frac{N}{2}. \tag{3.7}$$

Since the number N can be chosen arbitrarily large, a Brillouin zone can be regarded as quasicontinuous.

Substituting wave function (3.3) into (3.1) yields the following expressions for the energy and momentum of a free electron:

$$E = \frac{\hbar^2 k^2}{2m}, \quad p = \hbar k. \tag{3.8}$$

Thus, the momentum of a free electron coincides in direction with the wave vector and is related to it by the simple expression (3.8). A free electron corresponding to a wave with a wave vector k moves in space at a constant velocity

$$v = \frac{\hbar k}{m} = \frac{1}{\hbar}\frac{d\varepsilon}{dk}. \tag{3.9}$$

According to (3.8), the energy of a free electron satisfies the classic relation $\varepsilon = mv^2/2$. In the space of the wave vector, the isoenergetic surface of a free electron is a sphere.

For an electron moving in a crystal, the dependence of energy ε on wave vector k is generally more complicated. Since the Schrödinger equation (3.1) is invariant to reversal of the time sign, $\varepsilon(k) = \varepsilon(-k)$; this means that the iso-energetic surface of the electron $\varepsilon = \varepsilon(k)$, defined in the corresponding Bril-louin zone, has a center of symmetry. It can be shown that the isoenergetic surfaces of an electron in a crystal $\varepsilon = \varepsilon(k)$ have all the elements of the point symmetry of the crystal and, besides, a center of symmetry as well. In the vicinity of a point of minimum or maximum $k = k_0$, which is not a point of degeneracy, energy ε can be expanded into a series in the powers of the projection of the wave vector, accurate up to the quadratic term

$$\varepsilon(k) = \varepsilon(k_0) + \frac{1}{2}\sum_{ij}\left(\frac{\partial^2\varepsilon}{\partial k_i\partial k_j}\right)_{k=k_0}(k_i - k_{i0})(k_j - k_{j0}). \tag{3.10}$$

The symmetric second-rank tensor $\partial^2\varepsilon/\partial k_i\partial k_j$ is called the inverse-effective-mass tensor

$$(m_{ij}^*)^{-1} = \frac{1}{\hbar^2}\left(\frac{\partial^2\varepsilon}{\partial k_i\partial k_j}\right)_{k=k_0}. \tag{3.11}$$

If we reduce tensor (3.11) to the diagonal form with diagonal components

$$(m_i^*)^{-1} = \frac{1}{\hbar^2}\left(\frac{\partial^2\varepsilon}{\partial k_i^2}\right)$$

and introduce the quasimomentum

$$p = \hbar(k - k_0), \tag{3.12}$$

(3.10) transforms to

$$\varepsilon(k) = \varepsilon(k_0) + \sum_i\frac{\hbar^2 k_i^2}{2m_i^*} = \varepsilon(k_0) + \sum_i\frac{p_i^2}{2m_i^*}. \tag{3.13}$$

Expression (3.13) is similar to (3.8), from which it can be seen that m_i^* and p_i

have the dimensions of mass and momentum, respectively. Thus (3.13) represents the kinetic energy of an electron possessing an "anisotropic" mass. This mass "anisotropy" is naturally the direct result of the crystal anisotropy and is just a convenient method for describing the motion of an electron in the periodic field of the crystal lattice. In the case of a free electron, all the three components of the inverse-effective-mass tensor coincide, and the effective mass of a free electron

$$m^{-1} = \frac{1}{\hbar^2} \frac{\partial^2 \varepsilon}{\partial k_x^2} = \frac{1}{\hbar^2} \frac{\partial^2 \varepsilon}{\partial k_y^2} = \frac{1}{\hbar^2} \frac{\partial^2 \varepsilon}{\partial k_z^2}$$

is equal to the true mass of an electron and is independent of the wave vector, i.e., of the direction of motion. It is thus obvious that the effective mass of an electron moving in a crystal does not generally coincide with its true mass. The concept of the effective mass of an electron is of fundamental importance in semiconductor physics, for instance, in electrical conductivity and other phenomena associated with electron transport in a crystal.

From (3.13) it follows that in the extreme point region the isoenergetic surfaces of an electron in the reciprocal space (in the k space) are closed and are ellipsoids near the extreme points ($k = k_0$). If the extreme point does not coincide with the symmetry element of the crystal (i.e., in the general position), some more such ellipsoids must be located in the Brillouin zone in accordance with the symmetry of the crystal. Their maximum number for a cubic crystal is 48. As already noted, $\varepsilon = \varepsilon(k)$ is a periodic function in the reciprocal-lattice space. Thus the system of the above-described closed isoenergetic surfaces (ellipsoids in particular) must be repeated periodically throughout the reciprocal-lattice space. It is self-evident that the closed surfaces surrounding the points of energy minimum and maximum must alternate with open surfaces passing thorughout the reciprocal-lattice space. In Sect. 3.3 we shall consider in more detail some simple types of closed and open isoenergetic surfaces.

The movement of an electron in a crystal is effected by the periodic field of the lattice, which alternately decelerates and accelerates it. Therefore, the instantaneous velocity of an electron, which is equal to $v^* = p/m$ in accordance with (3.9), does not remain constant. If we, however, introduce the concept of the average electron velocity, which is equal numerically to the velocity of the center of gravity of a wave packet, corresponding to wave function (3.2), this velocity will be constant. Indeed, according to the principles of quantum mechanics the motion of the center of gravity of a wave packet is equivalent to that of a classical particle, which is constantly subjected to the action of a force equal to the average value of the force field. Since the average value of the lattice field is zero, the average electron velocity thus introduced and corresponding to the steady state (3.2) is constant. If we assume that average electron velocity v thus introduced is equal numerically to the group velocity of a packet of plane waves, then, according to the well-known formula

$$v = \frac{\partial \omega}{\partial k} = \frac{1}{\hbar} \frac{\partial \varepsilon}{\partial k}, \tag{3.14}$$

because $\varepsilon = \hbar\omega$. The rigorous derivation of (3.14) is based on quantum-mechanical averaging of the value of p/m corresponding to the steady state of the electron (3.2). A comparison of (3.9) and (3.14) shows that for an electron in a crystal the direction of wave vector k does not generally coincide with that of velocity v because of the difference between the phase and group velocities of the wave (3.2). From the definition of the average velocity of an electron (3.14), it is clear that near the energy minimum the direction of velocity v coincides with the direction of the external normal to the isoenergetic surface, while near the maximum, with the direction of the internal normal, the velocity is directed oppositely. Since all the isoenergetic surfaces and their set in the Brillouin zone have centers of symmetry, $\varepsilon(-k) = \varepsilon(k)$, then, according to (3.14), $v(-k) = -v(k)$ and, hence, the sum of the electron velocities for all the values of the wave vector in the Brillouin zone is equal to zero. Thus, is the absence of an external field the summary electric current through the crystal is zero.

Suppose now that an electron in a crystal is subjected to an external force F, which changes wave vector k and, hence, average velocity $v(k)$, energy $\varepsilon(k)$, and the state of the electron in the zone, which is described by the wave function (3.2). A force F is so small that it does not lead to transitions between the bands. The change in the average velocity of the electron, i.e., its acceleration, is

$$\frac{dv}{dt} = \frac{1}{\hbar} \frac{d}{dt} \frac{\partial \varepsilon(k)}{\partial k} \tag{3.15}$$

or

$$\frac{dv_i}{dt} = \frac{1}{\hbar} \frac{d}{dt} \left(\frac{\partial \varepsilon}{\partial k_i} \right) = \frac{1}{\hbar} \sum \frac{\partial^2 \varepsilon}{\partial k_i \partial k_j} \frac{dk_j}{dt}. \tag{3.16}$$

On the other hand, from the law of conservation of energy

$$\frac{d\varepsilon(k)}{dt} = \frac{\partial \varepsilon}{\partial k} \frac{dk}{dt} = vF; \tag{3.17}$$

and from (3.14) we obtain the equation of motion in the form

$$\hbar \frac{dk}{dt} = F. \tag{3.18}$$

A comparison of (3.16) and (3.18) brings us to the equation of motion in its final form

$$\frac{dv}{dt} = (m^*)^{-1}F, \tag{3.19}$$

where the tensor of the inverse effective mass $(m^*)^{-1}$ coincides with the above-introduced tensor (3.11), provided the dependence of the electron energy on the quasimomentum is quadratic (3.10). Thus, it is possible to impart the classical form to the equation of motion of an electron in a crystal by replacing the inverse scalar mass of the electron by tensor of the inverse effective mass $(m^*)^{-1}$, and the momentum of the free electron by quasimomentum $\hbar k$. This replacement actually represents consistent inclusion of the interaction of the moving electron with the crystal-lattice field. From (3.19) it is clear that, because of this interaction, the direction of electron acceleration does not generally coincide with that of the acting external force (it may be, for instance, an electric field applied to the crystal).

3.1.2 Energy Spectrum of an Electron

The periodicity of the wave function (3.2) leads to the division of the reciprocal-lattice space into quasi-continuous bands, to which corresponds the band nature of the energy spectrum of the electron in the periodic field of the crystal. We shall go into more detail on this further on; here we only note that the formation of energy bands, which are generally separated by energy-gap bands, is the result of the quantum-mechanical uncertainty relation

$$\Delta\varepsilon\Delta t \simeq \hbar, \tag{3.20}$$

where Δt is the time of localization of the electron near a fixed lattice point, and $\Delta\varepsilon$ is the band of the possible energy values. In an isolated atom, the electron localization time is infinitely large, which results in discrete energy values in accordance with (3.20). As atoms draw closer together and a crystal lattice is formed, the electron localization time in the atom is limited by the tunneling of the electron through the potential barrier separating neighboring atoms. The probability of tunneling W can be roughly estimated assuming the barrier to be rectangular

$$W \simeq 10^{16} \exp\left(-\frac{2}{\hbar}\sqrt{2mul}\right). \tag{3.21}$$

Here, u and l are the barrier height and width, respectively, and m is the mass of the free electron. For valence electrons, $u \simeq 10$ eV, $l \simeq 10^{-8}$ cm, which leads to $W \simeq 10^{15}$ s^{-1} and, hence, to the time of electron localization in the atom $\Delta t \simeq W^{-1} \simeq 10^{-15}$ s in conformity with (3.21). According to (3.20) this causes extension of the level of the valence electron into an energy band of width $\Delta \varepsilon \simeq h/\Delta t \simeq 1$ eV. It can similarly be shown that the electrons of the inner atomic shells are delocalized. In the crystal they are associated with an energy band of finite width; also, the bands narrow down on transition from the outer to the inner shells. Thus, the steady state corresponds to an equi-probable distribution of the electrons over all the lattice points, which in turn results in the splitting of the electron levels into quasi-continuous energy bands. If a level in an atom is filled, all the N^3 (neglecting the spin) electron states corresponding to this band level are also occupied by electrons, i.e., the band is filled completely. If the atomic level is vacant or only partly filled, the band arising from it is also vacant of partly filled.

It is easy to see that the electrons of a completely filled band make no con-tribution to the electrical conductivity of the crystal. Indeed, in accordance with (3.18) the electric field applied to the crystal changes the state of the electron, i.e., its wave vector. If all the states in the band are occupied by electrons, then, in accordance with the Pauli principle, the field cannot trans-fer the electron from one state to another, i.e., accelerate it. Thus, only elec-trons belonging to partly filled energy bands take part in electrical conduc-tivity.

By way of example, Fig. 3.1 shows how energy bands 1s and 2s arise from the corresponding electron levels during the formation of the body-centered cubic lattice of lithium. Since there are two electrons in the lithium atom at level 1s, and one electron at level 2s, band 1s is filled completely, and band 2s only partly. (The filled part of band 2s is shaded in Fig. 3.1). In accordance with (3.20), band 2s is broad, and band 1s narrow. Thus, the electrical con-ductivity of the metal lithium is due to the electrons of band 2s. The possible energy-band structure of the metals is not restricted to the case just discussed. The metallic conductivity of a crystal may be due to the overlapping of energy bands, one of which is vacant, and the other completely filled. The case where a vacant and a completely filled band do not overlap corresponds to a dielectric or a semiconductor. Then, the electrical conductivity of the crystal is

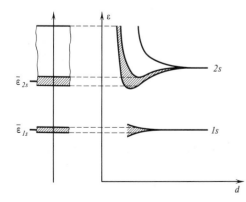

Fig. 3.1. Formation of energy bands in lithium

due to interband electron transitions, i.e., transitions of the electrons from a completely filled to a vacant band under the effect of heat or light (internal photoeffect). On transfer of electrons to a vacant band from a completely filled one, the latter acquires vacant states, which are called holes. Here, the electrical conductivity of the crystal may be due, not only to the electrons which have passed to the vacant band, but also to a redistribution of the electrons over the states in the nearly completed band. It can be rigorously demonstrated that the electron motion in a nearly completed band is equivalent to that of positively charged vacancies, i.e., holes, whose effective mass is generally different from that of the electrons. The division of the crystal electrical conductivity into the electron and hole components (n- and p-type conductivity, respectively) and the determination of the sign of the main current carriers are achieved by measuring (see [Ref. 1.7., Chap. 7]) the Hall effect and also the galvano- and thermomagnetic phenomena.

3.2 Brillouin Zones

3.2.1 Energy Spectrum of an Electron in the Weak-Bond Approximation

It has been shown above that the energy of an electron in a crystal ε is a periodic function of wave vector k. This results in the division of the reciprocal-lattice space into regions in which energy ε has the same values. These regions, which are called Brillouin zones, can be obtained on the basis of solution of the Schrödinger equation (3.1) in the nearly free-electron approximation.

Substituting wave function (3.2) into the Schrödinger equation (3.1) and reducing the expression obtained by $\exp(ikr)$, we arrive at the following equation for amplitude $U_k(r)$:

$$\left[\varepsilon_k - \frac{\hbar^2 k^2}{2m} - U(r) \right] U_k = -\frac{\hbar^2}{2m} \nabla^2 U_k - \frac{i\hbar^2}{m} (k \nabla U_k) . \qquad (3.22)$$

Since the potential energy of an electron $U(r)$ is a periodic function, it can be expanded into a Fourier series

$$U(r) = \sum_{H \neq 0} U_H \exp[2\pi i(Hr)] . \qquad (3.23)$$

As the summation in (3.23) is performed both in the positive and negative direction of the basis vectors of the reciprocal lattice, and $U(r)$ is a real function, we have $U_{-H} = U_H$. Besides, in (3.23) $U_{000} = 0$. We similarly expand periodic amplitude $U_k(r)$ into a Fourier series

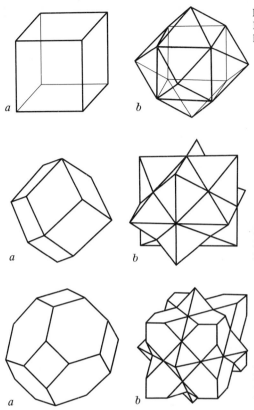

Fig. 3.4a, b. The first (**a**) and second (**b**) Brillouin zone for a primitive cubic lattice

Fig. 3.5a, b. The first (**a**) and second (**b**) Brillouin zone for a body-centered-cubic lattice

Fig. 3.6a, b. The first (**a**) and second (**b**) Brillouin zone for a face-centered-cubic lattice

For the body-centered-cubic lattice, the reciprocal lattice is face centered. Here, each point of the reciprocal lattice is surrounded by twelve nearest points, while the first Brillouin zone is accordingly bounded by the twelve faces of the rhombododecahedron (Fig. 3.5a). The second Brillouin zone for the body-centered-cubic lattice is sandwiched between the rhombododecahedron and the cubooctahedron embracing it (Fig. 3.5b). Figure 3.6 portrays the first and second Brillouin zones for the face-centered-cubic lattice.

3.2.3 Band Boundaries and the Structure Factor

As mentioned above, on the plane bounding a Brillouin zone and satisfying condition (3.29) the electron energy undergoes a jump. In an individual case, this jump may not take place. Then, the electron energy is continuous in transition from one band to the next and, hence, two neighboring Brillouin zones overlap. Since (3.29) can be regarded as the condition for the interference of an electron wave similar to the Laue condition, we can introduce the concept of the scattering factor Φ_H for electrons in the crystal. We shall now demonstrate that in this case the disappearance of Φ_H is equivalent to the absence of the corresponding boundaries of the energy band.

If n identical atoms are located in each unit cell of a crystal, the energy of an electron can be represented as a sum of energies of its interaction with each atom of the cell,

$$U(r) = \sum_{m=1}^{n} U'(r - r_m).$$ (3.34)

Expanding periodic potential U' into a Fourier series, we obtain, by analogy with (3.23),

$$U(r) = \sum_{H} U_H \exp(2\pi i H r),$$ (3.35)

where

$$U_H = f \hat{\Phi}_H,$$ (3.36)

$$\hat{\Phi}_H = \sum_{m=1}^{n} \exp(-2\pi i H r_m)$$ (3.37)

and f is the atomic-scattering factor. From (3.36) it is seen that if $\hat{\Phi}_H = 0$, the coefficient U_H in expansion (3.35) or (3.23) reduces to zero and, in accordance with (3.30), electron energy ε_k does not experience a discontinuity on the surface of the Brillouin zone. Thus, the absence of a reflection hkl consisting of identical atoms also means that the hkl plane is not the boundary of the Brillouin zone. Broadened Brillouin zones, from which the fictitious boundaries are removed, are called Jones zones.

3.3 Isoenergetic Surfaces. Fermi Surface and Band Structure

The solution of the Schrödinger equation (3.1) is, as a rule, considered for two limiting cases, the weak and the strong interaction of an electron with the crystal lattice. The weak-interaction case (the nearly free-electron approximation) has been considered above. In accordance with (3.30) it is associated with the spherical isoenergetic surfaces of the electron. In actual fact, the isoenergetic surfaces of an electron in a crystal have a more complex form. Qualitatively, this can be illustrated as follows. The normal component of the electron velocity at the boundary of the Brillouin zone is zero. According to (3.14) this means that in the nondegenerate case the derivative $\partial \varepsilon / \partial k$ with respect to the direction perpendicular to the band boundary vanishes at the boundary. Thus, the isoenergetic surfaces intersect the boundary of the Brillouin zone at right angles. It follows that at a sufficient distance from the zone center the isoenergetic surfaces are nonspherical. An analysis of a more general shape of isoenergetic surfaces can conveniently be performed by solving the Schrödinger equation (3.1) corresponding to the other limiting case of strong bond of the electron with the lattice atoms (the strongly bonded electron approximation).

3.3.1 Energy Spectrum of an Electron in the Strong-Bond Approximation

Here, the solution of (3.1) can be represented as a sum of atomic functions each of which describes the state of the electron in an isolated atom, the overlapping of the wave functions taking place only within a single coordination sphere. The wave function of the electron then has the form (Bloch's solution)

$$\psi_k(r) = \sum_n \exp(i k a_n) \varphi_a(r - a_n),\tag{3.38}$$

where $\varphi_a(r - a_n)$ is the atomic-wave function describing the behavior of the electron in the atom with radius vector a_n. The expression for the eigenvalue of energy ε_k is

$$\varepsilon_k = \varepsilon_a - C - \sum_n \varepsilon(a_n) \exp(i k a_n),\tag{3.39}$$

where ε_a is the energy of the assigned state of the electron in the isolated atom, and $\varepsilon(a_n)$ and C are the overlapping integrals

$$\varepsilon(a_n) = - \int \psi_a^*(r - a_n)[U(r) - U_a(r)]\,\psi_a(r)\,dr\tag{3.40}$$

and

$$C = - \int |\psi_a(r)|^2 [U(r) - U_a(r)]\,dr.\tag{3.41}$$

Here, $U_a(r)$ is the electron potential in the field of the isolated atom, and $U(r)$ is the periodic potential of the lattice. If $\varepsilon(a_n)$ is identical for all the atoms of the coordination sphere, then (3.39) can be represented as

$$\varepsilon_k = \varepsilon_a - C - \varepsilon \sum_n \exp(i k a_n).\tag{3.42}$$

In this case the calculation of the energy spectrum of the electron $\varepsilon = \varepsilon_k(k)$ reduces to the calculation, in (3.42), of the coordination sum $\sum_n \exp(i k a_n)$, where n is the coordination number.

By way of example we consider the primitive cubic lattice ($n = 6$). Calculation by (3.42) yields:

$$\varepsilon_k = \varepsilon_a - C - 2\varepsilon(\cos a k_x + \cos a k_y + \cos a k_z).\tag{3.43}$$

Recall that in this case the first Brillouin zone is a cube with edge $2\pi/a$. From (3.43) it is seen that, on formation of a primitive cubic lattice of N atoms, the level of an electron in an isolated atom ε splits up into a quasi-continuous energy band of N levels of widths 12ε. Indeed, assuming that the additive constant in (3.43) $\varepsilon_a - C = 0$, we find that the minimum of ε_k occurs at the center of the band at $k_x = k_y = k_z = 0$, and the maximum, at the eight vertices of the cube at $k_x = k_y = k_z = \pm\pi/a$; $\varepsilon_{min} = -6\varepsilon$, and $\varepsilon_{max} = +6\varepsilon$. Thus, the width of the whole band is $\varepsilon_{max} - \varepsilon_{min} = 12\varepsilon$. Expansion of ε_k into a series (3.43) in the range of low k values with an accuracy to the quadratic term leads to the expression

$$\varepsilon_k \simeq -6\varepsilon + \varepsilon a^2 k^2. \tag{3.44}$$

Consequently, at the center of the Brillouin zone the isoenergetic surface is a sphere. This permits introducing the scalar effective mass of the electron in accordance with (3.11)

$$m_i^* = \frac{\hbar^2}{\left(\dfrac{\partial^2 \varepsilon_k}{\partial k^2}\right)_{k=0}} = \frac{\hbar^2}{2\varepsilon a^2}. \tag{3.45}$$

Near the cube vertices the isoenergetic surfaces are also spheres. This can be demonstrated by expanding ε_k into a series near the upper edge of the band. The scalar effective mass of the hole, introduced for these spheres by analogy with (3.45), is negative, its modulus being equal to that of the effective mass of the electron, i.e. $m_h^* = -m_e^*$. Isoenergetic surfaces of more intricate shape are located between the spherical surfaces corresponding to the extreme points of ε_k. All of them satisfy (2.43), where $-6\varepsilon < \varepsilon_k < +6\varepsilon$. Some of these surfaces are open. Since $\varepsilon = \varepsilon(k)$ is a periodic function, closed isoenergetic surfaces repeat periodically in all the unit cells of the reciprocal lattice, and the open ones, also repeating, traverse the entire space of the reciprocal lattice. Figure 3.7 presents an open isoenergetic surface, for which the dependence of energy ε_k on wave vector k is given by (3.43). Figure 3.8 shows the section of

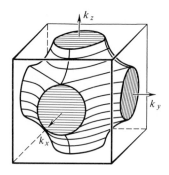

Fig. 3.7. Open isoenergetic surface according to (3.43)

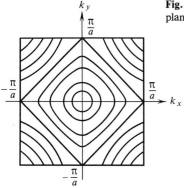

Fig. 3.8. Section of the isoenergetic surface (Fig. 3.7) by plane $k_z = 0$

isoenergetic surfaces (3.43) by plane $k_z = 0$. Here, the open "surfaces" are the isoenergetic lines $k_x + k_y = \pi/a$, corresponding to $\varepsilon_k = -2\varepsilon$ in (3.43). Isoenergetic surfaces for a body-centered and face-centered cubic lattice can be constructed in a similar way. Figure 3.9 exhibits an element of an open isoenergetic surface or a face-centered cubic lattice located inside a cubooctahedron, i.e., inside the corresponding first Brillouin zone.

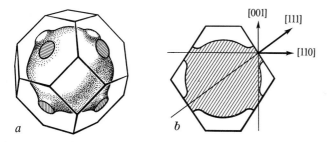

Fig. 3.9a,b. Open Fermi surface for fcc metals (**a**) and its section (**b**) by the (110) plane

3.3.2 Fermi Surfaces

As indicated in Sect. 3.1, in metals the energy bands are filled only partly. An isoenergetic surface which serves as the boundary between the occupied and vacant states in the reciprocal-lattice space is called the Fermi surface. Since, in conformity with the Pauli principle, only electrons situated near the Fermi surface take part in electrical conductivity (and other charge-transfer) phenomena, the topology of this surface essentially determines the electrical, galvanomagnetic, and other properties of metals. It can be demonstrated, for instance, that in a magnetic field the electron trajectory in the reciprocal-lattice space depends on the section of the Fermi surface transversed by a plane perpendicular to the direction of the magnetic field. If the Fermi surface is open, the trajectory may also be open, while period T and the rotation frequency ω of the electron are equal to infinity and zero, respectively. If the section of the Fermi surface gives a closed trajectory, the following expressions hold true for period T and the rotation frequency of the electron:

$$T = \frac{c}{eH}\frac{\partial S}{\partial \varepsilon}, \qquad \omega = 2\pi\frac{eH}{c}\left(\frac{\partial S}{\partial \varepsilon}\right)^{-1}. \tag{3.46}$$

$$m^* = \frac{1}{2\pi}\frac{\partial S}{\partial \varepsilon}. \tag{3.47}$$

Here, S is the cross-sectional area of the Fermi surface, ε is the energy of the electron, and H is the magnetic-field. Comparing (3.46) with the corresponding expressions for a free electron, we arrive at the expression relating the effective mass m^* of the electron to the shape of the Fermi surface

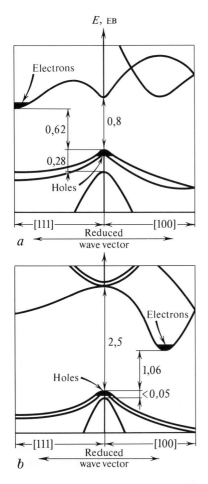

Fig. 3.10a, b. Band structure of germanium (a) and silicon (b)

According to (3.47) the sign of the effective mass depends on whether the Fermi surface contains the energy minimum or maximum. Figure 3.9 depicts the open Fermi surface for gold, copper, and silver and its section by the (110) plane. The existence of open directions [111], [110], and [001], to which the open trajectories correspond, can be observed.

In semiconductors and dielectrics, the zones are vacant or completely filled. Here, it is important to study the structure of the isoenergetic surfaces corresponding to the minimum and maximum energy of the bands separated by the energy gap, called the forbidden energy band. The band structure is usually characterized by the dependence of the boundary energy on the reduced wave vector. The scheme of reduced bands in Fig. 3.10 illustrates a particular case of this dependence for elemental semiconductors, e.g., germanium and silicon.

4. Lattice Dynamics and Phase Transitions

Vibrations of atoms about their equilibrium position is one of the fundamental properties of the crystal lattice. The set of phenomena associated with such vibrations and describing their laws is called lattice dynamics. Lattice dynamics lies at the basis of the theory of thermal properties of crystals and the present-day concepts of the electrical and magnetic properties of crystals, light scattering in them, etc. For instance, the anharmonicity of atomic vibrations in the crystal lattice determines the ratio between the heat capacity, compressibility, and the coefficient of linear thermal expansion (the Grüneisen ratio). The concept of the thermal motion of atoms and the vibration anharmonicity is the foundation of the modern theory of phase transitions in crystals (ferroelectric in particular, see [1.7]). Here we shall only state the principal conclusions of the theory of crystal lattice dynamics and consider, on their basis, the heat capacity, thermal conductivity, and thermal expansion of crystals.

4.1 Atomic Vibrations in a Crystal

4.1.1 Vibrations of a Linear Atomic Chain

At not-too-low temperatures, when the amplitude of atomic vibrations in the lattice greatly exceeds the de Broglie wavelength corresponding to the atoms, the atomic vibrations obey the laws of classical mechanics. The main features of the vibrations can be understood simply by considering the vibrations of a linear atomic chain (one-dimensional lattice model). Assuming that the spacing of the unit cell of such a one-dimensional lattice is the shortest distance a between neighboring atoms of the same sort, we shall consider the case where a one-dimensional unit cell contains two atoms. The crystal lattices of alkali-halide crystals and a number of semiconductors can serve as a three-dimensional analog of such a lattice.

Figure 4.1 presents a linear atomic chain consisting of atoms of two sorts with serial numbers m' and m'' and their nearest neighbors. We denote the masses of the atoms by m_1 and m_2 and the coefficients of elasticity for the neighboring pairs m', m'', and m', $m'' - 1$ by β_1 and β_2, respectively. In the approximation of elastic forces acting only between neighboring atoms, the

$m'-1\ \ m''-1\quad m'm''\quad m'+1\ \ m''+1$

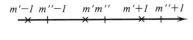

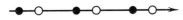

Fig. 4.1. Analysis of the vibrations of a linear atomic chain

equations of motion of the atoms have the form

$$m_1 \ddot{u}'_m = -\beta_1(u'_m - u''_m) - \beta_2(u'_m - u''_{m-1}),$$
$$m_2 \ddot{u}''_m = -\beta_1(u''_m - u'_m) - \beta_2(u''_m - u'_{m+1}),$$

(4.1)

where u'_m and u''_m are the coordinates of the atoms with numbers m' and m'', respectively. We seek the solution of (4.1) in the form of a running wave

$$u'_m = A' \exp[\mathrm{i}(kam - \omega t)], \quad u''_m = A'' \exp[\mathrm{i}(kam - \omega t)], \quad (4.2)$$

where k is the modulus of the wave vector of the atom ($k = 2\pi/\lambda$), amplitudes A' and A'' are independent of m, and the role of the modulus of the radius vector is played by term am (a is the basis vector of the lattice). Substituting (4.2) into (4.1) and cancelling the factor $\exp[\mathrm{i}(kam - \omega t)]$, we obtain the following set of linear equations for amplitudes A' and A'':

$$\left(\omega^2 - \frac{\beta_1 + \beta_2}{m_1}\right)A' + \left[\frac{\beta_1 + \beta_2 \exp(-\mathrm{i}ak)}{m_1}\right]A'' = 0,$$

(4.3)

$$\left[\frac{\beta_1 + \beta_2 \exp(\mathrm{i}ak)}{m_2}\right]A' + \left(\omega^2 - \frac{\beta_1 + \beta_2}{m_2}\right)A'' = 0.$$

Set (4.3) gives nonzero solutions for A' and A'', the corresponding determinant being equal to zero. This condition, in turn, leads to a quadratic equation for ω^2, which is satisfied by the following solutions:

$$\omega^2_{\mathrm{ac}} = \frac{1}{2}\omega^2_0\left(1 - \sqrt{1 - \gamma^2 \sin^2\frac{ak}{2}}\right),$$

(4.4)

$$\omega^2_{\mathrm{opt}} = \frac{1}{2}\omega^2_0\left(1 + \sqrt{1 - \gamma^2 \sin^2\frac{ak}{2}}\right),$$

where

$$\omega^2_0 = \frac{(\beta_1 + \beta_2)(m_1 + m_2)}{m_1 m_2}, \quad \gamma^2 = 16\frac{\beta_1 \beta_2}{(\beta_1 + \beta_2)^2}\frac{m_1 m_2}{(m_1 + m_2)^2}. \quad (4.5)$$

4.1.2 Vibration Branches

Solutions (4.2,4) show that elastic vibrations of atoms can be described with the aid of a running monochromatic wave, provided the frequencies of these vibrations satisfy the laws, or branches, of dispersion $\omega = \omega(k)$, one of which $\omega = \omega_{ac}(k)$ is usually called acoustic, and the other $\omega = \omega_{opt}(k)$, optical. Similar to the Bloch function (3.2), solution (4.2) is periodic in the reciprocal-lattice space. Therefore, all the features of atomic vibrations can be understood by considering wave (4.2) as a function of wave vector k within the first Brillouin zone (3.4) or

$$- \pi/a \leqslant k \leqslant + \pi/a . \tag{4.6}$$

Applying the Born–von Karman boundary conditions (3.6) and in accordance with (3.7), we conclude that for the volume of a lattice with N cells the projection of wave vector k can have N discrete values within the Brillouin zone. This discreteness or quasicontinuity of the values of the wave vector and of the corresponding vibration frequencies is the result of the Born–von Karman boundary conditions (3.6).

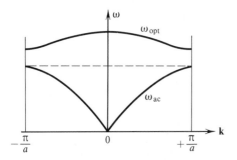

Fig. 4.2. Dispersion of the optical and acoustic vibration branches

Figure 4.2 illustrates the dependence of ω_{ac} and ω_{opt} on k, defined by (4.4) within the first Brillouin zone or, in other words, the dispersion of the acoustic and optical vibration branches (for $\gamma^2 < 1$ and $m_1 \neq m_2$). With small k (long waves), expansion of (4.4) into a series with respect to the small parameter $ak \ll 1$ yields.

$$\omega_{ac} = vk , \quad v \simeq \frac{1}{4}\omega_0 \gamma a , \quad \omega_{opt} \simeq \omega_0 \left(1 - \frac{\gamma^2 a^2}{32} k^2\right), \tag{4.7}$$

where v is the velocity of sound. This corresponds, as seen from Fig. 4.2, to the different nature of dispersion of the acoustic and optical branches at $k \simeq 0$, and also to the principal property of these vibrations, namely $\omega_{ac}(0) = 0$, $\omega_{opt}(0) \neq 0$. To elucidate another fundamental property of these vibrations we analyze the ratio

$$\frac{u'_m}{u''_m} = \frac{A'}{A''} = \frac{\beta_1 + \beta_2 \exp(-ika)}{(\beta_1 + \beta_2) - m_1 \omega^2}.$$

For long waves ($k \to 0$) we have, with due regard for (4.7),

$$\left(\frac{u'_m}{u''_m}\right)_{ac} = 1, \qquad \left(\frac{u'_m}{u''_m}\right)_{opt} = -\frac{m_2}{m_1}. \qquad (4.8)$$

It is seen that the acoustic branch is characterized by vibration of the atoms in phase, and for the optical, in counterphase. The same result can be obtained for the shortest waves ($k \to \pi/a$ or $\lambda \to 2a$). If atoms with masses m_1 and m_2 are oppositely charged ions, the optical vibrations are naturally associated with variation in the dipole momentum of the unit cell, which manifests itself, for instance, in the additional absorption of infrared light. Figure 4.2 shows that for all the k in the Brillouin zone $\omega_{ac} < \omega_{opt}$. Energywise, this means that at sufficiently low temperatures only acoustic vibrations are excited in the crystal, while at higher temperatures optical vibrations play the determining role. If we denote the limiting frequency of acoustic vibrations by $\omega^m_{ac} = \omega_{ac}(\pi/a)$ and introduce the characteristic (Debye) temperature

$$T_D = \hbar\omega^m_{ac}/k_0, \qquad (4.9)$$

then at $T \leqslant T_D$ the contribution of optical vibrations can be neglected (k_0 is the Boltzmann constant).

The same features characterize, in principle, the vibrations of atoms in a three-dimensional crystal lattice with a unit cell containing S different atoms. An analysis of the vibrations of atoms in the crystal volume $N^3(a_1[a_2a_3])$ indicates that there are generally $3S$ different vibration branches, for each of which the projections of wave vector k can take N discrete values within the Brillouin zone. Three vibration branches are acoustic and $(3S - 3)$ optical; $\omega^j_{ac}(0) = 0$ ($j = 1, 2, 3$) and $\omega^j_{opt}(0) \neq 0$ ($j = 3, \ldots S$). In this general case, too, the character of the dispersion of the acoustic and optical branches is reflected by Fig. 4.2. In a three-dimensional lattice, the dispersion of all the $3S$ vibration branches can be characterized by surfaces $\omega = \omega(k)$ in the reciprocal-lattice space. The symmetry of these surfaces depends on that of the direct lattice; besides, surfaces $\omega = \omega(k)$ have a center of symmetry.

4.1.3 Phonons

In solid-state physics, elementary excitations of the crystal lattice associated with atomic vibrations are called phonons. Thus, the phonons can be regarded as quasiparticles with a quasimomentum $\hbar k$ and energy $\hbar \omega_k$. This approach is convenient in considering a number of phenomena, such as electron scattering on lattice vibrations, thermal conduction, etc.

Below Debye temperatures $T < T_D$ the phonons obey Bose–Einstein quantum statistics, and their average number at thermal equilibrium is determined by the Planck function

$$n = \frac{1}{\exp(\hbar \omega / k_0 T) - 1}. \tag{4.10}$$

Here, n is the equilibrium number of phonons with energy $\hbar \omega$ in a phase space cell of volume $(2\pi \hbar)^3$. The number of phase space cells per interval dk is equal to

$$dn_q = \frac{4\pi k^2 dk}{(2\pi \hbar)^3} V, \tag{4.11}$$

where V is the volume of the crystal.

Taking into account, at $T < T_D$, only the acoustic-vibration branches and assuming, according to (4.7), the dispersion of the acoustic frequencies to be linear for all k, i.e., $k \approx \omega / v$, we transform (4.11) to

$$dn_q = \frac{3V}{2\pi^2 v^3} \omega^2 d\omega. \tag{4.12}$$

Here, factor 3 corresponds to three acoustic modes (one longitudinal and two transverse), and v is the average velocity of sound.

Thus, the total number of phonons in the volume V of the crystal is

$$n\, dn_q = \frac{3V}{2\pi^2 v^3} \frac{\omega^2 d\omega}{\exp(\hbar \omega / k_0 T) - 1}, \tag{4.13}$$

and, accordingly, the total energy of the phonons in volume V

$$E = \frac{3V\hbar}{2\pi^2 v^3} \int_0^{\omega_{ac}^m} \frac{\omega^3 d\omega}{\exp(\hbar \omega / k_0 T) - 1}, \tag{4.14}$$

where ω_{ac}^m is the maximum frequency of acoustic vibrations corresponding to the boundary of the Brillouin zone. The value of ω_{ac}^m is determined from the condition of equality of the total number of vibrations in the three acoustic branches to the value of $3N^3$

$$\frac{3V}{2\pi^2 v^3} \int_0^{\omega_{ac}^m} \omega^2 d\omega = \frac{V(\omega_{ac}^m)^3}{2\pi^2 v^3} = 3N^3. \tag{4.15}$$

Therefore

$$\omega_{ac}^m = v \left(\frac{6\pi^2 N^3}{V} \right)^{1/3} = v \left(\frac{6\pi^2}{\Omega_0} \right)^{1/3}, \tag{4.16}$$

where Ω_0 is the unit cell volume. Using (4.16) and (4.9), we obtain the following expression for the Debye temperature:

$$T_D = v \left(\frac{6\pi^2}{\Omega_0} \right)^{1/3} \frac{\hbar}{k_0}. \tag{4.17}$$

At high temperatures, the contribution of the optical vibrations to phonon energy E becomes substantial.

4.2 Heat Capacity, Thermal Expansion, and Thermal Conductivity of Crystals

4.2.1 Heat Capacity

It is well known that the heat capacity of crystals at high temperatures is constant $c_v \simeq 6$ cal/deg mol and independent of the type of the crystal (the Dulong-Petit law). Below Debye temperatures the heat capacity strongly depends on the temperature; at $T \to 0$, $c_v \to 0$. The temperature dependence of the heat capacity can be obtained on the basis of the above-described concepts of atomic vibrations in the crystal lattice. By definition, the heat capacity of a crystal at a constant volume is

$$c_v = \partial E / \partial T, \tag{4.18}$$

where E is the total internal energy of the crystal. It is expedient to consider two temperature ranges below and above Debye temperature T_D.

At temperatures $T < T_D$, the expression for E is given by (4.14). Expansion of the integrand with respect to the small parameter $\hbar \omega / k_0 T$ and subsequent integration in (4.14) yields

$$E \simeq \frac{\pi^2 V (k_0 T)^4}{10 \hbar^3 v^3}, \tag{4.19}$$

from which, according to (4.18), we arrive at the Debye equation

$$c_v = \frac{12 \pi^4 k_0}{5} \left(\frac{T}{T_D} \right)^3. \tag{4.20}$$

This equation fairly well describes the temperature dependence of the heat capacity in the temperature range of $10-50$ K for a number of crystals with a simple structure, for instance, for alkali-halide crystals and most of the elements. For crystals with a complex structure and, hence, an intricate vibration spectrum there is a range of characteristic (Debye) temperatures. In addition, the temperature dependence $c_v = c_v(T)$ is more complicated, although within a sufficiently small vicinity of absolute zero the law T^3 is fulfilled in this case as well.

At high temperatures $T > T_D$, the energy of optical vibrations can be calculated on the basis of the classical model of a set of linear harmonic oscillators. Since the average energy of an oscillator is equal to $k_0 T$, and the total number of oscillators in volume V is $3SN^3$,

$$E = 3SN^3 k_0 T;$$ (4.21)

thus we have, according to (4.18),

$$c_v = 3SN^3 k_0.$$ (4.22)

For a gram molecule of a substance $SN^3 = N_0 \simeq 6 \times 10^{23}$ (Avogadro constant), from which the Dulong – Petit law follows.

4.2.2 Linear Thermal Expansion

We have so far been considering harmonic vibrations of atoms in a crystal. Consequently we have restricted ourselves to the linear terms in the right-hand side of the equations of motion (4.1), which corresponds to the quadratic terms in the potential energy equation. Let us now discuss the interaction of two neighboring atoms under anharmonic conditions. Then the expression for interaction force F and potential energy U as functions of the displacement x of the atoms from their equilibrium position has the form

$$F = -\frac{dU}{dx} = -2\beta x + 3\gamma x^2,$$ (4.23)

$$U(x) = \beta x^2 - \gamma x^3,$$ (4.24)

where coefficient γ is called the coefficient of anharmonicity. Let us calculate the average displacement x with the aid of Boltzmann's distribution function

$$\bar{x} = \frac{\int\limits_{-\infty}^{+\infty} x \exp\left[-U(x)/k_0 T\right] dx}{\int\limits_{-\infty}^{+\infty} \exp\left[-U(x)/k_0 T\right] dx}.$$ (4.25)

Substituting the expression for $U(x)$ (4.24) into (4.25), expanding the integrands in the approximation of the smallness of the anharmonic term, and integrating lead to an average displacement

$$\bar{x} = \frac{3k_0 T\gamma}{4\beta^2} \tag{4.26}$$

and, accordingly, to the coefficient of linear thermal expansion α

$$\alpha = \frac{\bar{x}}{aT} = \frac{3k_0\gamma}{4\beta^2 a}, \tag{4.27}$$

where a is the interatomic distance. From (4.27) it can be seen that the coefficient of linear thermal expansion α is directly proportional to the coefficient of anharmonicity γ and, hence, in the absence of vibration anharmonicity $\alpha = 0$. This follows from the fact that for a linear oscillator in the harmonic approximation $\bar{x} = 0$. By calculating $\bar{x}$ for an oscillator in the quantum-mechanical approximation it is possible to obtain the theoretical dependence $\alpha = \alpha(T)$; α decreases as $T \to 0$, which corresponds to the Nernst theorem and agrees with the experimental data. The coefficient of linear thermal expansion of most substances lies in the range of $(10-100) \times 10^{-6}$ deg^{-1} and shows a strong anisotropy in crystals.

4.2.3 Thermal Conductivity

Another thermal property of crystals, which is essentially associated with the anharmonicity of atomic vibrations, is thermal conductivity. By definition, the coefficient of thermal conductivity K relates thermal flux j to the temperature gradient in a definite direction

$$j = K \operatorname{grad} T. \tag{4.28}$$

Debye used for the coefficient of thermal conductivity the following expression borrowed from the kinetic theory of gases:

$$K = \tfrac{1}{3}cv\lambda, \tag{4.29}$$

where c is the heat capacity, v is the velocity of sound, and λ is the mean free path of phonons, which depends on phonon–phonon interaction. It can be shown that phonon–phonon interaction does not exist in the harmonic approximation. This is understandable if one takes into account that the solution of linear equations (4.1) consists of the superposition of harmonic waves and, hence, their independent propagation in the crystal. Under these conditions the thermal resistance of the crystal is zero and, accordingly,

$K = \infty$. The final value of the coefficient of thermal conductivity is thus determined by the vibration anharmonicity. The foregoing naturally refers to the ideal crystal. In a real crystal, an additional mechanism of phonon scattering on lattice defects operates, which makes an additional contribution to the thermal resistance of the crystal.

Debye showed that at high temperatures $T > T_D$, $\lambda \sim T^{-1}$. In the low-temperature range $T < T_D$, the exponential dependence $\lambda \sim \exp(-T_D/2T)$ holds true. Experimental values of K at room temperature vary over a wide range from 0.9 for metals to 10^{-3} cal cm^{-1} s^{-1} deg^{-1} for dielectrics (the above-mentioned mechanism of thermal conductivity is not valid for metals). The anisotropy of thermal conductivity in crystals is described by a second-rank tensor.

4.3 Polymorphism. Phase Transitions

It has been indicated in Sect. 1.3.2, that the equilibrium crystal structure corresponds to the minimum of the free energy F of the crystals. But there may be several such minima within a wide range of temperatures and pressures. Each of them may be associated with its own crystal structure, which sometimes also differ in the nature of the chemical bond. Such structures are called polymorphous modifications, and the transitions from one modification to another are called polymorphous transformations or phase transitions.

It was also emphasized above that in explaining polymorphous transformations the energy of the thermal motion of the lattice atoms cannot be neglected, as in the calculations of the crystal-lattice energy. The reason is that the phase-transition mechanism involves changes in the frequencies of atomic vibrations in the lattice and sometimes the appearance of an unstable vibration mode at a particular temperature or pressure. Thus, ferroelectric phase transitions [1.7] are due to the instability of one of the transverse optical vibrations, i.e., to the appearance of the so-called soft mode.

Apart from a change in temperature, phase transitions may be caused by changes in pressure, by external fields, or by combinations of these effects.

We shall consider only phase transitions in the solid state, although this concept is equally well applied in describing solid – liquid and liquid – gas transitions, as well as transitions within liquid-crystal phases (see Sect. 2.8) or liquids. For instance, the transition of helium to the superfluid state is a phase transition.

A phase transition is a finite change in the microscopic structure and macroproperties of the medium due to continuous small change in the external conditions (thermodynamic parameters). The changes which structures undergo in phase transitions are usually changes in the arrangement or the nature of odering of the atoms (their centers), but there are also phase transitions related only to the state of the electron subsystem. Magnetic transitions, for instance, are associated with a change in spin ordering, while the transition of some metals to the superconducting state is related to a change in the type of interaction between conduction electrons and phonons.

Polymorphism in the crystalline state was discovered by Mitcherlich in 1822 for crystals of sulphur and potassium carbonate. This phenomenon is widespread. As has been mentioned above, the structures of almost all the elements, as well as those of many inorganic (Sect. 2.3.4) and organic (Sect. 1.6) compounds, have polymorphous modifications (Sect. 2.1). For instance, the cubic modification of tin with a diamond-type structure (grey tin) is stable below $+13.3°C$. Above $+13.3°C$, another modification of tin − white tin with a body-centered tetragonal lattice − is stable (Fig. 2.5a, e). The physical properties of these two polymorphous modifications of tin are essentially different; for instance, white tin is plastic, while grey tin is brittle. Other classical examples are diamond and graphite (Fig. 2.5a, c). Quartz has several polymorphous modifications (Fig. 2.23). The hexagonal crystals of the semiconductor CdS undergo a phase transition into the cubic modification at room temperature and a pressure of ~ 20 kbar. Besides, the cubic phase of CdS is nonphotosensitive, as distinct from the hexagonal phase. Another important example is the phase transition in the crystal of $BaTiO_3$; this crystal is ferroelectric, has a tetragonal lattice (in the temperature range of $+5°C < T < +120°C$), and transforms to the cubic paraelectric phase above $+120°C$.

The phase equilibrium and phase composition of a substance are usually characterized by a phase, or state, diagram. The simplest example of a phase diagram is the p, T diagram (p being the pressure, and T the temperature). Here, each point with coordinates p and T, which is called a figurative point, characterizes the state of a substance at a given temperature and pressure. Curves $T = T(p)$ in this diagram separate the possible phases of the substance, which include, in particular, the gas, liquid, and various crystal phases (Fig. 2.2). By way of example, Fig. 4.3 shows the simplified phase diagram of sulphur. Curve OD separates the T and p ranges, respectively, where the rhombic and monoclinic phases of sulphur are stable. At atmospheric

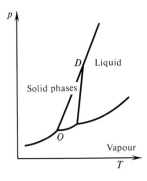

Fig. 4.3. Simplified phase diagram of sulphur

pressure the transition from the rhombic to the monoclinic phase occurs at a temperature of 368.5 K. As seen from the phase diagram, the phase-transition temperature increases with the pressure. Many crystalline solid phases are, however, metastable, i.e., they may exist out of their equilibrium regions on the phase diagram (see below).

4.3.1 Phase Transitions of the First and Second Order

We distinguish phase transitions of the first and second order. First-order phase transitions are accompanied by a jump in such thermodynamic functions as the entropy, volume, etc., and hence the latent heat of transition. Accordingly, the crystal structure also changes jumpwise. Thus, for the first-order phase transitions, curves $T = T(p)$ in the phase diagram satisfy the Klausius – Clapeyron equation

$$dT/dp = T(\Delta V)/Q, \qquad\qquad (4.30)$$

where ΔV is the volume jump, and Q is the latent heat of transition. In the second-order phase transitions, it is the derivatives of the thermodynamic functions that experience a jump (for instance, the heat capacity, compressibility, etc., change jumpwise). In a second-order phase transition the crystal structure changes continuously. Since a first-order phase transition is, irrespective of its structural mechanism, associated with the nucleation process, it is attended by a temperature hysteresis. This means the non-coincidence of the phase-transition temperatures during heating and cooling, and implies that each first-order phase transition involves superheating or supercooling. A typical example is the crystallization process, which is a particular case of a first-order phase transition. On second-order phase transitions no temperature hysteresis is observed.

For phase transitions of both the first and second order the crystal symmetry changes jumpwise at the phase-transition point. There is, however, a substantial difference between the change in symmetry on first- and second-order phase transitions. In second-order phase transitions the symmetry of one of the phases is a subgroup of the symmetry of the other phase (see Sect. 4.8). In most cases (but not necessarily) the high temperature phase is the more symmetric, and the low temperature phase is the phase less symmetric. On first-order phase transitions the crystal symmetry generally changes arbitrarily, and the two phases may have no symmetry elements in common.

4.3.2 Phase Transitions and the Structure

From the standpoint of the changes in the crystal structure upon phase transition it is customary to distinguish reconstruction phase transitions of the displacement type and of the order – disorder type.

In a reconstruction phase transition the crystal structures of the initial and the resulting phase differ essentially; the coordination numbers change and the displacements of the atoms to the new equilibrium positions are commensurate with the interatomic distances. For instance, in graphite the c.n. of the carbon is 3, in diamond 4, in α-Fe $8 \div 6$, in γ-Fe 12. In one of the transitions in NH_4Cl the c.n. changes from 6 to 8, etc. In some cases, however, one can find some crystallographic agreement between neighboring, reconstructively differing phases. The phase structures in reconstruction transformations differ greatly at specific volumes. Sometimes the term "polymorphous modifications", in the narrow sense, is applied exclusively to reconstructively differing phases. Reconstruction phase transitions are always of the first order.

Sometimes a polymorphous transition is accompanied by a change in the type of closest packing without a change in coordination number; this is called polytypy. Polytypy is known for zinc sulphide, rutile, SiC, CdI_2, MoS_2, etc.

In displacement-type phase transitions, which will be considered below, the changes in atomic positions are moderate. In the transitions intermediate between those of the reconstructive and displacement types the centers of the structural groupings do not change their positions, but they (molecules and radicals, in particular) turn about or begin to rotate. Thus, the structures of α- and β-quartz differ by a mutual rotation of the SiO_4 tetrahedra, i.e., by an angle $Si - O - Si$. In a number of crystalline paraffins and alcohols, polymorphous transitions are attended by the rotation of molecules about their long axis. In other organic compounds a polymorphous transition involves the simultaneous rotation of molecules and change in their angle of inclination to the base. In many cases a polymorphous transition is associated, not with free rotation of molecules, but with their torsional vibrations. Here, the molecules have a lower degree of relative orientation at temperatures above the transition point. This mechanism was first pointed out by Frenkel; it evidently operates in a number of ammonium salts. In crystals with polar molecules (for instance, in solid hydrogen chloride or iodide, hydrogen sulphide) a change in the orientation of polar molecules results in a polymorphous transformation, which often has the nature of a ferroelectric phase transition. An example is the phase transition in HCl at 98.8 K. Above this temperature hydrogen chloride has a cubic structure, and the chlorine atoms form a phase-centered cubic lattice which corresponds to a statistically random orientation of the dipoles. At the phase-transition temperature the hydrogen chloride shifts to the tetragonal phase jumpwise, with simultaneous orientation of the dipoles. Phase transitions involving molecule rotation or dipole ordering (in particular, ferroelectric) are characterized by temperature dependences of the ordering parameter $\eta = \eta(T)$ which are the same type as those presented in Fig. 4.5.

Let us now consider displacement-type phase transitions. An example of such a transition is the above-mentioned phase transition in $BaTiO_3$ (Fig. 4.4). At temperatures above the Curie point ($T_C \simeq +120°C$), barium titanate has a cubic lattice. The barium atoms are located at the cube vertices, titanium at the cube center, and oxygen at the face centers, thus forming an oxygen octahedron. The resultant dipole moment of the lattice is zero, and therefore the crystal is in the paraelectric region. At the Curie point, the Ti and O atoms shift jumpwise with respect to the Ba atoms in the direction of one of the cube edges. The lattice then transforms from cubic to tetragonal and a ferroelectric polarization arises in the direction of the atomic shift.

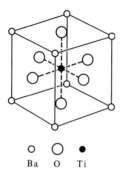

○ ○ ●
Ba O Ti

Fig. 4.4. Crystalline structure of barium titanate (perovskite type, see Fig. 2.16)

An example of the displacive transformation is the so-called martensite transformation associated with diffusionless crystal lattice rearrangement. One of the well-known examples of martensite transformation is that in iron-carbon alloys. It occurs at the quenching of the high-temperature cubic face-centered phase of an alloy which, by slight distortion, transforms into a tetragonal body-centered martensite phase. Martensite transformations occur in many other alloys, such as copper – zinc, nickel – titanium, etc.

There is a correlation between the crystal lattice of the martensite and the parent phase. The latter is characterized by orientation relationships between the crystal lattice planes and the crystallographic directions of both phases. The crystal lattice of a martensite phase can be obtained from the parent one either by a homogeneous strain or by a combination of a homogeneous strain and heterogeneous modes of atomic displacements. Owing to these factors the typical rate of transformation is very high. It is commensurate with the typical rate of propagation of elastic strain perturbation, i. e., with the sound velocity.

Another feature is that in many cases a martensite transformation does not provide the complete transition of the parent phase into a new one. Moreover, some alloys can be described by relationships between the volume fraction of the martensite phase and the temperature of supercooling. This effect results from the so-called thermoelastic equilibrium which arises when a transformation is not accompanied by irreversible plastic deformation.

All these distinctive features of martensite transformations are caused by strong constraints, which necessitate a good crystal lattice correspondence on interphase boundaries.

We shall now consider an order-disorder phase transition. An example is the ferroelectric phase transition due to the ordering of the hydrogen bonds. The difference between the paraelectric and ferroelectric phase is that in the former the probabilities of the positions of the H atoms in the hydrogen bond $AH...B$ and $A...HB$ are equal, while in the latter they are different.

Another example of order – disorder phase transition is a phase transition in a binary alloy consisting of atoms of two sorts, A and B. Above the transition temperature, such an alloy is disordered, which corresponds to a statistically random distribution of the A and B atoms at the lattice points. Below the phase-transition temperature we observe ordering when regions consisting entirely of A or B atoms or of two sublattices A and B appear in the lattice. In Ising's model, two opposite spin directions are associated with A and B atoms. Thus, this model describes the ferromagnetic or antiferromagnetic phase transitions in binary alloys associated with the ordering of their structure.

Let us denote by $P_a(A)$ the probability that an atom A will occupy its own point in the lattice, and by $P_a(B)$, the probability that this point will be occupied by an atom B. Then the following ordering parameter η can be introduced for such a binary alloy as CuZn (β-brass) with equal numbers of a- and b-type points:

$$\eta = P_a(A) - P_a(B).$$

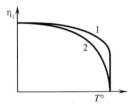

Fig. 4.5. Temperature dependence of the order parameter for (1) first- and (2) second-order phase transition

During the transition from the ordered to the disordered phase parameter, η varies from unity to zero. Figure 4.5 presents two types of temperature dependence of η, corresponding to a first- and a second-order transition, respectively. β-brass experiences a second-type phase transition at $T_C = 480°C$. The mechanism of ordering in β-brass is clear from Fig. 4.6. In the high-temperature phase each point of a body-centered cubic lattice is equiprobably occupied by a Zn or a Cu atom ($\eta = 0$). At $T < T_C$, copper atoms settle predominantly at the center of the cube, and the zinc atoms, at its vertices ($\eta = 1$). At the phase-transition point itself, parameter η varies continuously with the temperature. Alloy Cu_3Au also experiences an ordering-type transition, but of the first order, at $T = 380°C$ (Fig. 2.9a, b). Here, parameter η changes jumpwise at $T = T_C$.

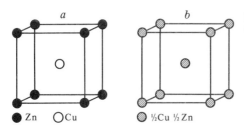

Fig. 4.6a, b. (a) Low- and (b) high-temperature phases of β-brass

When discussing the phase equilibrium and phase diagram we dealt with thermodynamically equilibrium phases corresponding to the minimum of free energy F. But very often, some crystal phase, thermodynamically disadvantageous at a given figurative point, may exist at that point indefinitely because of the low mobility of atoms in a solid. An example is diamond, which forms from graphite and is stable at a pressure of $\sim 10^5$ kgf/cm^2 and a temperature of $\sim 2 \times 10^3$°C. If a diamond is cooled to room temperature, it may exist indefinitely at atmospheric pressure, although this phase is metastable. Note that near the transformation points the phases show various instabilities (such as critical opalescence in quartz).

The polymorphous transformations are often classified into monotropic and enantiotropic. Monotropic transitions proceed only in one direction, and enantiotropic, in two opposite directions. Examples of monotropic transitions are irreversible transitions in arsenic, antimony, and other elements of the fifth group, and also the above-mentioned graphite – diamond transition. In actuality, the reverse transition from the metastable to the equilibrium phase often proceeds very slowly for the reason stated above. One of the examples is metal phase quenching.

One of the relatively rare cases is a so-called homeomorphous transition, when the polymorphous transformation and the corresponding change in physical properties is not accompanied by a change in crystal structure (for instance, the α and β phases of iron have identical body-centered cubic lattices, but different magnetic structures).

4.4 Atomic Vibrations and Polymorphous Transitions

It is natural to use the thermodynamic approach for a quantitative description of polymorphous transformations. According to the Boltzmann theorem the probability that at a temperature T the crystal is in the α phase with an energy E_α is equal to W_α

$$W_\alpha = \exp\left(-\frac{F_\alpha}{k_0 T}\right) = \exp\left[-\frac{E_\alpha - TS(E_\alpha)}{k_0 T}\right], \tag{4.31}$$

where $F_\alpha = E_\alpha - TS_\alpha$ is the free energy, and S is the entropy. The probability W_α is maximum at values of E_α and S_α satisfying the condition

$$dE_\alpha/dS_\alpha = T. \tag{4.32}$$

Figure 4.7 exhibits the dependence of crystal energy E on entropy S. According to (4.32), at a temperature T the equilibrium state of the crystal corresponds to a point with coordinates E_α, S_α, where a line tangent to curve $E = E(S)$ makes an angle with the x axis, whose tangent is numerically equal to T. The intercept made by the tangent line on the y axis is equal numerically to free energy $F_\alpha = E_\alpha - TS_\alpha$. If the crystal reveals polymorphism and, hence, the existence of two phases, α and β, the phase-transition temperature $T = T_0$ can, according to (4.31), be found from the condition $W_\alpha = W_\beta$ or from the equality of the free energies $F_\alpha = F_\beta$.

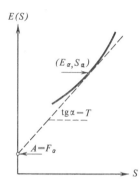

Fig. 4.7. Dependence of the internal energy of a crystal, E, on the entropy S

In the approximation of equal frequencies of atomic vibrations the total internal energy of the crystal E has the form

$$E = E' + \hbar\omega n, \tag{4.33}$$

where E' is the internal energy of the crystal at $T = 0$, and n is the phonon concentration. Entropy S can be expressed as the configurational part of the energy

$$S = k_0 \ln P, \tag{4.34}$$

where P is the number of possible distributions of n phonons over $3N$ degrees of freedom (N is the number of projections of the wave vector within the first Brillouin zone)

$$P = \frac{(3N + n - 1)!}{(3N - 1)!\, n!}. \tag{4.35}$$

Substituting (4.35), (4.34), and (4.33) into the expression for free energy

$F = E - TS$ and using the condition for the minimum of free energy $dF/dn = 0$ and the Stirling formula $\ln n! \simeq n \ln n$, we arrive at the following expressions for the phonon concentration and free energy F:

$$n = 3N \frac{1}{\exp(\hbar\omega/k_0 T) - 1}, \tag{4.36}$$

$$F = E - TS = E' + 3Nk_0 T \ln[1 - \exp(-\hbar\omega/k_0 T)]. \tag{4.37}$$

According to (4.37) the free energies of the α and β phases at temperature T satisfy the following expressions:

$$F_\alpha(T) = E'_\alpha + 3Nk_0 T \ln[1 - \exp(-\hbar\omega_\alpha/k_0 T)],$$
$$F_\beta(T) = E'_\beta + 3Nk_0 T \ln[1 - \exp(-\hbar\omega_\beta/k_0 T)]. \tag{4.38}$$

Equating energies F_α and F_β, we can determine the phase-transition temperature $T = T_0$ from the following equation

$$\exp\left(-\frac{E'_\alpha - E'_\beta}{3Nk_0 T_0}\right) = \frac{1 - \exp(-\hbar\omega_\alpha/k_0 T_0)}{1 - \exp(-\hbar\omega_\beta/k_0 T_0)}. \tag{4.39}$$

It is seen that the polymorphous transformation is intimately connected with the jumpwise change in the frequency of atomic vibrations. If $E'_\beta > E'_\alpha$, then (4.39) has a solution at $\omega_\alpha > \omega_\beta$, i.e., phase transition takes place when the β phase is more "friable" (with respect to the vibrations of the lattice atoms) than the α phase. Figure 4.8 demonstrates curves $E = E(S)$ for both phases.

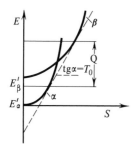

Fig. 4.8 Dependence $E = E(S)$ for the α and β phases

The temperature of transformation from the α to the β phase $T = T_0$ is equal numerically to the slope of the common tangent line to the curves, and the energy difference at the points of tangency is numerically equal to the latent heat of transition Q. At temperatures $T > T_0$ the β phase is stable, while at temperatures $T < T_0$, the α phase is.

A similar description of the phase transition can be obtained by taking into account the dispersion of frequencies in the two phases $\omega_\alpha = \omega_\alpha(k)$ and

$\omega_\beta = \omega_\beta(k)$. Here, the expressions for the free energy of the α and β phase have the form

$$F_\alpha(T) = E'_\alpha + k_0 T \sum_{k,s} \ln \{1 - \exp[-\hbar\omega^s_\alpha(k)/k_0 T]\},$$

$$F_\beta(T) = E'_\beta + k_0 T \sum_{k,s} \ln \{1 - \exp[-\hbar\omega^s_\beta(k)/k_0 T]\}, \qquad (4.40)$$

where $\omega^s(k)$ is the frequency of a phonon with a wave vector k and a polarization $s = 1, 2, 3$. Summation in (4.40) is performed over all the discrete values of wave vector k in the first Brillouin zone and over all the branches of vibrations s. Equating the values of F_α and F_β at phase transition point $T = T_0$, we have

$$E'_\alpha - E'_\beta = k_0 T \sum_{k,s} \ln \frac{1 - \exp[-\hbar\omega^s_\alpha(k)/k_0 T]}{1 - \exp[-\hbar\omega^s_\beta(k)/k_0 T]}. \qquad (4.41)$$

The right-hand side of (4.41) is a function of the temperature $E = E(T)$. Figure 4.9 presents function $E = E(T)$ and its section by straight line $E = E'_\alpha - E'_\beta$, which determines phase transition temperature $T = T_0$. Above Debye temperatures $T > T_D$, the right-hand side of (4.41) is a linear function of the temperature

$$\varepsilon = k_0 T \sum_{k,s} \ln \frac{\omega^s_\alpha(k)}{\omega^s_\beta(k)} = k_0 T \ln \frac{\prod\limits_{k,s} \omega^2_\alpha(k)}{\prod\limits_{k,s} \omega^s_\beta(k)}. \qquad (4.42)$$

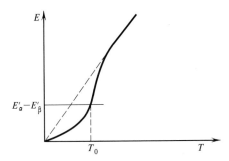

Fig. 4.9. The determination of the phase transition temperature T_0

The right-hand side can be crudely estimated as follows. We can assume that the frequency ratio $\omega^s_\alpha/\omega^s_\beta$ is equal to the ratio of the sound velocities in a wave with a polarization s, i.e.,

$$\omega^s_\alpha/\omega^s_\beta \simeq v^s_\alpha/v^s_\beta. \qquad (4.43)$$

Then, substituting (4.43) into (4.42), we have

$$\varepsilon(T) \simeq k_0 T \ln \prod_s \prod_k \left(\frac{v_\alpha^s}{v_\beta^s}\right) = k_0 T \ln \prod_s \left(\frac{v_\alpha^s}{v_\beta^s}\right)^N = k_0 T N \ln \frac{v_\alpha^l v_\alpha^{t_1} v_\alpha^{t_2}}{v_\beta^l v_\beta^{t_1} v_\beta^{t_2}} . \quad (4.44)$$

Here v^l is the longitudinal sound velocity, and v^{t_1} and v^{t_2}, respectively, are the transverse sound velocities referring to the α or β phase. The phase transition temperature $T = T_0$ can be obtained by graphical solution of (4.41) using the construction illustrated in Fig. 4.9. But, as indicated in Sect. 4.2, the Debye approximation is not always good for crystal phases with a complex structure; this restricts the possibilities of quantitative description of polymorphous transformations on the basis of (4.41). Besides, (4.41) does not take into account the effects related to the vibration anharmonicity.

We have discussed the effect of the temperature T on the possibility of the appearance of some phase or other, which is taken into account by the thermodynamic function of the free energy $F = E - TS$. If the pressure changes, the thermodynamic potential $\Phi = E - TS + pV$ is considered. Qualitatively, the consideration is based on to an analysis of the behavior of functions similar to those depicted in Figs. 4.7, 8. Yet the pressure may shift the phase equilibrium points quite considerably, and it can be utilized to obtain phases which do not arise on changes in temperature alone.

In the $p - T$ diagram, the line of first-order phase transitions can merge into the line of second-order phase transitions. Such a merger was first predicted by *Landau* in 1935 [4.1], and the transition point itself is called critical. The pressure dependence of the phase transition temperature has no kink at the critical point, and only the second derivative $d^2 T/dp^2$ experiences a jump. According to *Landau* the heat capacity, the compressibility, and the coefficient of thermal expansion of a crystal near the critical point T_{cr} show a temperature dependence of the type $(T_{cr} - T)^\alpha$, where $\alpha = 1/2$ (in the current literature the Landau point is often called tricritical). Near the critical point, light scattering similar to critical opalescence is observed. As shown by *Ginzburg* [4.2], ferroelectric crystals exhibit dynamic light scattering near the critical point; this scattering is associated with the Rayleigh scattering on polarization fluctuations. The Landau critical point was first revealed experimentally in a ferroelectric crystal SbSI by *Volk* et al. [4.3]. Later on, critical points were revealed in NH_4Cl by *Garland* and *Weiner* [4.4] and in a number of other crystals (KH_2PO_4, $BaTiO_3$).

4.5 Ordering-Type Phase Transitions

In the preceding section we considered the mechanism of first-order phase transitions associated with the predominant contribution of the vibration energy of the atoms to the entropy. We shall now discuss phase transitions of the ordering type, when the increase in crystal entropy in order – disorder transition is mainly due to the configurational energy directly related to the ordering mechanism. We assume that the contribution of the atomic-vibration energy can be neglected. Consider the mechanism of ordering in a binary alloy. The results obtained can be generalized for the other cases of phase transitions of the order − disorder type.

Let us take a look at the simplest case, where the binary alloy consists of two sorts of atoms A and B arranged in the ordered phase at points "a" and "b" with a definite degree of probability, the number of a and b points being identical and equal to N. Then, by virtue of the definition given in Sect. 4.3, we can introduce the ordering parameter

$$\eta = P_a(A) - P_a(B) .$$

Proceeding from the definition of $P_a(A)$ and $P_a(B)$, we find that $P_a(A) + P_a(B) = 1$ and, consequently,

$$\eta = 2P_a(A) - 1 = 2p - 1 , \tag{4.45}$$

where p stands for $P_a(A)$.

Let us determine the entropy of the ordered phase as a configurational energy associated with the motion of atoms A and B. We take into account that N_p atoms of type A can be distributed over N points of type a in the following number of independent ways

$$n = \frac{N!}{(N - N_p)! \, (N_p)!} . \tag{4.46}$$

It is easy to see that the remaining $(N - N_p)$ atoms of type A can be distributed over N points of type B ("foreign" points) in the same number of independent ways n. Therefore the total number of independent ways for distributing atoms A is equal to n^2 and, hence, according to (4.34) the entropy of the ordered phase is equal to

$$S = C - 2Nk_0 \left[(1 - p) \ln (1 - p) + p \ln p - \tfrac{1}{2} \ln 2 \right] , \tag{4.47}$$

where constant C is independent of p. The energy of the unordered phase with respect to the ordered is thus

$$E = UN(1 - p) , \tag{4.48}$$

where $N(1 - p)$ is the number of atoms transferred from proper to improper points, and U stands for the energy required for the transfer of atom A to point b, or of atom B to point a. Expressing now, with the aid of (4.47) and (4.48), the free energy $F = E - TS$ from the free-energy-minimum condition, we obtain the final equation relating U, p, and T. Replacing parameter p in it by ordering parameter η according to (4.45), we arrive at the equation

$$\eta = \text{th}\,(U/k_0 T)\,. \tag{4.49}$$

Now we recall that energy U depends on η. Since in the unordered phase the difference between points a and b is lost, it is clear that $U = 0$ at $\eta = 0$. At the same time, $U \neq 0$ at $\eta \neq 0$. In the theory of unordered distributions of atoms, the dependence between U and η is assumed linear

$$U = U_0 \eta\,. \tag{4.50}$$

This leads to the temperature dependence of the ordering parameter

$$\eta = \text{th}\,(U_0 \eta / k_0 T)\,. \tag{4.51}$$

From (4.51) it follows that $\eta = 1$ at $T = 0$ and $\eta = 0$ at $T = T_0$, where

$$T_0 = U_0/2 k_0\,. \tag{4.52}$$

As follows from (4.51), in the range of the phase transition temperature $T_0 = T$, parameter η varies continuously and, hence, solution (4.51) describes a second-order phase transition, for instance, an ordering-type transition in β-brass (Figs. 4.5, 6). When deriving (4.51) the number of atoms of type A and B was assumed equal. It can be shown that precisely this assumption leads to a second-order phase transition and, conversely, at any other ratio between the numbers of atoms of the two types the theory of the ordering of binary alloys leads to first-order transitions. Thus one can describe phase transitions of the type observed in alloy Cu_3Au.

Figure 4.8 shows the curves for the dependence of the energy E of the crystal on its entropy S, which illustrate a first-order transition from one polymorphous phase to another. It is easy to imagine what the dependence $E = E(S)$ in order – disorder phase transition must look like. If we assume that the ordered phase corresponds to lower temperatures and that a rise in temperature increases the system energy, dependence $E = E(S)$ is a monotonically increasing function (Fig. 4.10). In the case of a first-order

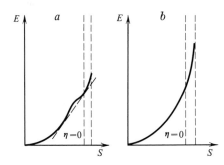

Fig. 4.10a, b. Dependence $E = E(S)$ for (**a**) first- and (**b**) second-order transitions of the order-disorder type

phase transition, curve $E = E(S)$ has an inflection point. This enables one to draw a tangent to the curve, which has two points of tangency. Hence, for one and the same temperature $T = dE/dS$ at the points of tangency there are two different corresponding values of entropy. An entropy jump corresponds to a release of latent heat and, consequently, to a first-order phase transition (Fig. 4.10a). The absence of an inflection point on curve $E = E(S)$ corresponds to a second-order phase transition (Fig. 4.10b). In the former case, the ordering parameter η experiences a jump at the inflection point. In the latter case, the parameter η varies continuously in the phase transition region, and for the phase-transition temperature and for $\eta = 0$, there is a corresponding point on curve $E = E(S)$, at which the curvature changes its sign.

The real mechanism of an order – disorder phase transition includes both a change in entropy due to the mixing of atoms and a change in entropy associated with the energy of atomic vibrations. Therefore the real mechanism is much more complicated than the one considered above. Allowance for the vibration energy in the mechanism of an ordering-type phase transition may change not only the dependence $\eta = \eta(T)$, but the very nature of the transition.

In conclusion we also note that a similar approach is possible with respect to ordering-type phase transitions whose mechanism involves a change, the freezing of the free rotation of molecules, or the ordering of polar molecules. For these cases *Pauling* and *Fowler* developed a quantitative theory, which leads to a dependence $\eta = \eta(T)$ similar to (4.51) and to analogous solutions corresponding to first-order phase transitions. Finally, it should be emphasized that we do not consider here the theory of second-order phase transitions according to *Landau*, because this theory describes, among other things, pyroelectric or ferroelectric phase transitions, where the ordering parameter η has the meaning of spontaneous electric polarization of the crystal. This theory is discussed in [1.7], which deals with the electric and, in particular, the ferroelectric properties of crystals.

4.6 Phase Transitions and Electron – Phonon Interaction

When considering the phase transition mechanism one usually ignores the contribution of the electron subsystem (or the free energy of the electrons) to the total free energy of the crystal and thus assumes that the phase transition mechanism has no bearing on electron excitations in the crystal. This assumption is based on the fact that at above-Debye temperatures the contribution of the electron heat capacity to the total heat capacity of the crystal can be ignored. Let us discuss this point in more detail.

4.6.1 Contribution of Electrons to the Free Energy of the Crystal

From the free-energy expression (4.37), it is possible to obtain the Gibbs –
Helmholtz equation by taking into account that $S = -\partial F/\partial T$,

$$F = E + T(\partial F/\partial T). \tag{4.53}$$

On integration this can be represented as

$$F = -T\int_0^T (E/T^2)\,dT. \tag{4.54}$$

By definition (4.18), integral energy E can be expressed via heat capacity c_v

$$E(T) = E(0) + \int_0^T c_v(T)\,dT. \tag{4.55}$$

Substituting (4.55) into (4.54), we obtain the final expression for free energy F

$$F = E(0) + T\int_0^T \frac{dT}{T^2} \int_0^T c_v(\tau)\,d\tau. \tag{4.56}$$

It can be seen that the free energy of the crystal is completely determined by
the temperature dependence of the specific heat capacity. It is known from
solid-state physics that for a nondegenerate semiconductor or dielectric, at an
above-Debye temperature $T > T_D$ the ratio of electron heat capacity c_v^{el} to
lattice heat capacity c_v^L is

$$c_v^{el}/c_v^L = N_c/n_0 \ll 1, \tag{4.57}$$

where $N_c = (2\pi m k_0 T/h^3)^{3/2}$ is the density of electron or hole states in the
band; m is the effective mass of the electron or hole, and n_0 is the number of
atoms in 1 cm^3. For metals at $T > T_D$

$$c_v^{el}/c_v^L \simeq k_0 T/E_f \ll 1, \tag{4.58}$$

where the Fermi energy E_f is of the order of several electronvolts, and
$k_0 T \simeq 0.025$ eV at room temperature. Thus, both for metals and for semiconductors and dielectrics at $T > T_D$ the contribution of the electrons to the
free energy of the crystal should be neglected. The electron contribution may
become substantial only near $T = 0$, where, according to (4.20), c_v tends to
zero as $\sim T^3$.

Nevertheless, it turns out that in addition to the absolute-zero range the
temperatures in the vicinity of the phase transition temperature also form the

special temperature range, where the contribution of the electron subsystem to the free energy of the crystal may be substantial. The physical meaning of this statement consists in the fact that near the phase transition temperature the electrons may make an appreciable contribution, not to the heat capacity itself, but to its anomaly at phase transition. This conclusion was drawn for the first time for ferroelectric phase transitions within the framework of the phenomenological theory of Landau – Ginsburg [4.6]. The effect of electron excitations on phase transitions of a different nature was investigated in a number of independent papers. Later it was shown that, at least for ferroelectric phase transitions, interband electron-phonon interaction is the microscopic mechanism responsible for the effect of electrons on phase transitions. Since the theory of ferroelectric phase transitions based both on the lattice dynamics and on the phenomenological theory of Landau – Ginsburg will be considered separately in [1.7], here we shall only briefly consider the role of interband electron – phonon interaction in the phase transition mechanism. As a corollary, we shall discuss some new effects due to the influence of electron excitations on phase transitions.

4.6.2 Interband Electron – Phonon Interaction

In Sect. 4.4 of this chapter the mechanism of polymorphous transformations in crystals was related to the change in atomic vibration frequencies. This change was introduced in the form $\omega_\alpha > \omega_\beta$, and the question of the vibration anharmonicity and the relevant instability mechanism remained open. The mechanism of ferroelectric phase transitions in modern dynamic theory is usually associated with the instability of one of the transverse optical vibrations at the center of the Brillouin zone $k = 0$, and the anharmonicity of the corresponding vibration branch is considered to be responsible for the instability (see [1.7]). One of the mechanisms possibly responsible for the vibration instability and, hence, for the phase transition is interband electron – phonon interaction [4.5]. This mechanism consists of interaction of the electrons of two neighboring bands of the crystal with one of the optical vibrations; one of the bands is vacant (or almost vacant), while the other is completely filled with electrons. Such an interaction, or "mixing" of neighboring energy bands, leads, on the one hand, to a change in the frequency of the interacting optical phonon and, on the other, to a change in the electron spectrum (the energy gap width). This interaction of the filled and vacant bands results in the instability of the "mixing" optical vibrations and, accordingly, in phase transitions from the more symmetric to the less symmetric phase. It is known from quantum chemistry that the disappearance of the orbital electron degeneracy in a molecule leads to its transition from the symmetric to the nonsymmetric configuration (Jahn – Teller effect). Hence, the phase transition due to the interband electron – phonon interaction is called the Jahn – Teller pseudoeffect.

Let us consider two neighboring energy bands of a crystal denoted by the indices $\sigma = 1, 2$, with boundary energies ε_1 and ε_2 and with a corresponding energy gap width $E_{g0} = \varepsilon_2 - \varepsilon_1$. Suppose some active optical vibration with a coordinate u and a frequency ω interacts with the electrons in the bands. Neglecting the dependence of ε and u on k, i.e., the band dispersion, we write the Hamiltonian for the crystal

$$\mathcal{H} = \sum_{\sigma} \varepsilon_{\sigma} a_{\sigma}^{+} a_{\sigma} + \frac{1}{2}\left(-\frac{\hbar^2 \partial^2}{M \partial u^2} + M\omega^2 u^2\right) + \sum_{\sigma, \sigma'} \frac{\tilde{V}_{\sigma\sigma'}}{\sqrt{N}} a_{\sigma}^{+} a_{\sigma} u . \quad (4.59)$$

Here, a_{σ}^{+} and a_{σ} are the operators of electron production and annihilation, $V_{\sigma\sigma'}$ are the constants of interband electron–phonon interaction, M is the corresponding mass factor, and N is the number of electrons in the lower band (equal to the number of unit cells by the order of magnitude).

An analysis of this solution shows that taking interband interaction into account leads to the renormalization of the electron spectrum

$$\varepsilon_{1,2}^{*} = \frac{\varepsilon_1 + \varepsilon_2}{2} \mp \sqrt{E_{g0} + \frac{4\tilde{V}^2}{N} u^2} , \quad (4.60)$$

where $\tilde{V} = \tilde{V}_{12}$. From (4.60) it follows immediately that interband electron–phonon interaction changes the energy gap width

$$E_g = 2\sqrt{\frac{E_{g0}^2}{4} + \frac{\tilde{V}^2}{N} u_0^2(T)} . \quad (4.61)$$

Here, E_g is the energy gap width in the presence of the Jahn–Teller pseudo-effect, and $u_0(T)$ is the active-vibration coordinate corresponding to the minimum of the free energy of the crystal. The latter can be defined as the sum of the vibration energy F_ω, corresponding to the active optical vibration, and the energy of electron subsystem F_e. In the approximation of a completely filled and a vacant band we obtain the following expression for free energy F

$$F(T, u) = \frac{N E_{g0}}{2} - N k_0 T \ln \left\{ 2 \left[1 + \cosh\left(\frac{1}{k_0 T} \sqrt{\frac{E_{g0}^2}{4} + \frac{\tilde{V}^2}{N} u^2}\right) \right] \right\}$$
$$+ \frac{M\omega^2}{2} u^2 . \quad (4.62)$$

It is seen that taking into account interband electron–phonon interaction leads to anharmonicity of the vibration branch under consideration, since, along with quadratic term u^2, (4.62) contains terms with higher powers of u [cf., for instance, (4.24)]. We denote by $u = u_0$ the coordinate of the vibration corresponding to the minimum of free energy F

$$\left.\frac{\partial F}{\partial u}\right|_{u=u_0} = 0 . \tag{4.63}$$

Then we obtain the following expression for u_0 from (4.62):

$$u_0^2 = N\left\{\frac{[f_1(u_0) - f_2(u_0)]^2}{M^2\omega^4}\tilde{V}^2 - \frac{E_{g0}^2}{4\tilde{V}^2}\right\}, \tag{4.64}$$

where the amount of filling of the bands by electrons f_1 and f_2 are Fermi functions

$$f_{1,2}(u_0) = \left\{\exp\left[\mp\frac{1}{k_0 T}\left(\frac{E_{g0}^2}{4} + \frac{\tilde{V}^2}{N}u_0^2\right)^{1/2}\right] + 1\right\}^{-1} . \tag{4.65}$$

Solutions (4.64) and (4.65) define the temperature dependence $u_0 = u_0(T)$, which is presented in Fig. 4.11. At $T = 0, f_1 = 1$ and $f_2 = 0$, and from (4.64) it follows that at $T = 0$

$$u_0^2(0) = N\left[\frac{\tilde{V}^2}{(M\omega^2)^2} - \frac{E_{g0}^2}{4\tilde{V}^2}\right] . \tag{4.66}$$

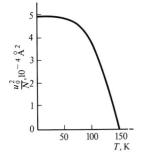

Fig. 4.11. Temperature dependence of the active-vibrations coordinate in the quasi-Jahn–Teller model of a ferroelectric phase transition (BaTiO₃)

With increasing temperature, u_0^2 decreases monotonically from $u_0^2(0)$ to zero. Function $u_0^2(T)$ vanishes at temperature $T = T_C$, which plays the role of the phase transition temperature

$$k_0 T_C = \frac{E_{g0}}{4}(\text{arcth }\tau)^{-1}, \tag{4.67}$$

where

$$\tau = 2\tilde{V}^2/M\omega^2 E_{g0} . \tag{4.68}$$

The nonzero solution (4.63) has the corresponding condition

$$\tau > 1 . \tag{4.69}$$

The nature of the phase transition based on the mechanism discussed above depends on the type of active vibration. If we take into account the dispersion, the solution $u_0 = u_0(k)$, which is related to an arbitrary point of the Brillouin zone, may correspond to condition (4.63). If $u_0 \neq 0$ for $k = 0$, this corresponds to a change in crystal structure and symmetry due to the relative shift of the sublattices. If the crystal is ionic and the shift of the sublattices results in spontaneous polarization, the phase transition is ferroelectric, and point $T = T_C$ is the Curie point. If $u_0 \neq 0$ at the boundary of the Brillouin zone, then the transition to the antiferroelectric phase takes place at $T = T_C$. If the phonon inside the Brillouin zone is active, the phase transition may be different in nature; in particular, it may not be ferroelectric.

For ferroelectric transitions, expansion of free energy F, determined in (4.62), into a series in even powers of u near $u = 0$ leads to a similar series, which is used in the phenomenological theory of Landau – Ginsburg (see [1.7]). Then the relationship between the spontaneous polarization P_s and u_0 is given by the simple expression

$$P_s = \frac{\bar{e}}{\Omega_0} \frac{u_0}{\sqrt{N}}, \tag{4.70}$$

where $\bar{e}$ is the effective charge corresponding to active optical vibration, and Ω_0 is the unit-cell volume. Thus, the temperature dependence shown in Fig. 4.11 is, in the case of a ferroelectric phase transition, the well-known temperature dependence of the square of spontaneous polarization (see [1.7]). The coefficients in the expansion of free energy F with respect to polarization P (or to u) can be expressed in terms of the parameters of microscopic theory, in particular, via the constant of the interband electron – phonon interaction $\tilde{V}$.

Renormalization of the coefficient at u^2 in the free-energy expression (4.62) leads to a change in the frequency of the active-vibration mode ω. For a high-symmetry phase

$$(\omega')^2 = \omega^2 + \frac{2\tilde{V}^2}{ME_{g0}} [f_2(0) - f_1(0)] . \tag{4.71}$$

From (4.71) and (4.67) it follows that at the phase transition temperature $T = T_C$ the frequency ω' reduces to zero. The temperature dependence of ω' is presented in Fig. 4.12.

Thus the interband electron – phonon interaction results, on fulfilment of condition (4.69), in the instability of the vibration interacting with the bands. In the case of a ferroelectric phase transition the active optical mode with a frequency ω' at $k = 0$ is called "soft" (see [1.7]). We thus see that the presence of the "soft" vibration mode is due to the anharmonicity near the phase transition, but the anharmonicity itself is, in the model at hand, the consequence of electron – phonon interaction.

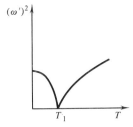

Fig. 4.12. Temperature dependence of the "soft" vibration mode due to interband electron – phonon interaction

The above-described mechanism makes it possible to investigate, on the basis of (4.62, 64, 65), all the basic properties (ferroelectric, in particular) of the crystal near the phase transition. Since the ferroelectric properties are specially considered in [Ref. 1.7, Chap. 13] we restrict ourselves to two new effects, which are the direct consequence of interband electron-phonon interaction.

4.6.3 Photostimulated Phase Transitions

The model just discussed explains the experimentally observed effect of non-equilibrium electrons on the phase transition temperature. Let us consider a photoconducting crystal undergoing a phase transition at $T = T_C$; a high-symmetry phase corresponds to temperature $T > T_C$. It is known from experiment that illumination of a crystal in the spectral region where it shows photoconductivity reduces the phase transition temperature. Suppose the illumination of a crystal transfers the electrons from the lower to the upper energy band, and the electron concentration in the upper band increases by Δn. Then, according to (4.87), the phase transition temperature shifts towards the lower temperatures by

$$\Delta T_c = \frac{E_g}{4k_0} \left[\left(\text{arcth} \frac{\tau}{1 + 2\Delta n \tau} \right)^{-1} - (\text{arcth } \tau)^{-1} \right]. \tag{4.72}$$

The physical meaning of this effect is that photoactive absorption of light changes the filling of the band with electrons and, hence, the contribution to the free energy due to interband electron – phonon interaction. Note that photoactive illumination changes not only the concentration of free electrons in the band, but also the filling of all the levels in the forbidden band due to impurities or defects. In many cases the photosensitive shift in phase transition temperature must be of an "impurity" nature, because the concentration of electrons on impurity levels and its change under illumination may be several orders of magnitude higher than in the band. Photostimulated phase transitions have been observed in a number of independent investigations [4.6].

Figure 4.13 illustrates the shift in Curie temperature on illumination of $BaTiO_3$ crystals in the intrinsic absorption field. These data indicate that il-

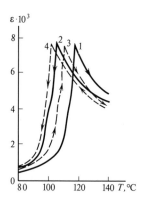

Fig. 4.13. Photostimulated phase transition in barium titanate. (1, 2) Temperature dependences of the dielectric constant in the dark under heating and cooling conditions, respectively. (3, 4) Same dependences, but under illumination of the crystal in the spectral region of its photosensitivity [4.7]

lumination affects the phase transition in $BaTiO_3$ from the tetragonal to the cubic phase; the shift occurs towards the lower temperatures and is equal to several degrees.

Figure 4.14 illustrates the same phenomenon in photosensitive crystals of HgI_2 experiencing a nonferroelectric phase transition from the tetragonal to the orthorhombic phase. Since in this case the low-temperature phase is more symmetric, in accordance with theory the photoactive illumination increases the phase transition temperature. A considerable body of experimental material on photostimulated phase transitions in the crystals of aromatic hydrocarbons and elemental semiconductors of group V has been accumulated.

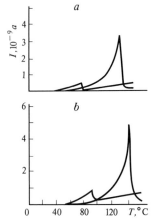

Fig. 4.14a, b. Photostimulated nature of the transition of HgI_2 from the tetragonal D_{4h}^{15} to the orthorhombic C_{2v}^{12} phase at 400 K. Pseudopyrocurrent at the phase transition point in the dark (a) and under illumination at the spectral photosensitivity maximum (b). The larger current maximum corresponds to heating, the smaller, to cooling [4.7]

4.6.4 Curie Temperature and the Energy Gap Width

It should also be noted that (4.67) relates the Curie temperature of a ferroelectric T_C to the energy gap width E_{g0}. The results agree qualitatively with experiment. It will suffice, for instance, to compare the Curie temperature of the narrow-band ferroelectric semiconductor GeTe and the broad-band dielectric $BaTiO_3$.

Proceeding from (4.61), we can expand the energy gap width E_g of the crystal into a series in even powers of u_0 or P_s. Restricting ourselves to the quadratic term, we have

$$E_g \simeq E_{g0} + \frac{a}{2} P_s^2 ,$$ (4.73)

where the constant

$$a = 4 \frac{\Omega_0^2}{\bar{e}^2} \frac{\tilde{V}^2}{E_{g0}}$$ (4.74)

is proportional to the square of the constant of electron–phonon interaction $\tilde{V}$. Equation (4.73) enables one to predict the nature of the temperature anomaly of E_g in the range of first- and second-order ferroelectric phase transitions. Indeed, in first-order phase transitions P_s^2 experiences a jump and, accordingly, the energy gap width undergoes a finite jump $\Delta E_g = E_g - E_{g0} \simeq \frac{a}{2} P_s^2$. In second-order phase transitions $P_s \sim (T - T_C)^{1/2}$, and therefore it is the temperature coefficient of the energy gap width that experiences a finite jump. The experimental dependence $E_g = E_g(T)$ near the phase transition, for instance, for $BaTiO_3$ (first-order phase transition) and triglycinesulphate (second-order phase transition), agrees with these conclusions. As we see, measurements of $E_g = E_g(T)$ directly confirm the existence of the Jahn-Teller pseudoeffect in ferroelectrics. From the jump ΔE_g and with the aid of (4.73, 74) it is possible to estimate the constant of electron-phonon interaction for $BaTiO_3$ $\tilde{V} \simeq 0.6$ eV/Å ($\Delta E_g \simeq 0.02$ eV $P_s \simeq 18 \times 10^{-6}$ C/cm^2, $\Omega_0 \simeq 64$ Å^3, $\bar{e} \simeq 2.4$ e).

4.7 Debye's Equation of State and Grüneisen's Formula

The "equation of state" means the relationship between the volume V, the pressure p, and the temperature T of a solid. Here, one proceeds from the thermodynamics equation

$$p = -(\partial F/\partial V)_T .$$ (4.75)

To express free energy F we use (4.40) for the sum of free oscillator energies

$$F(T) = E' + k_0 T \sum_{k,s} \ln [1 - \exp(-h\omega^s(k)/k_0 T)] .$$

With due allowance for the Debye frequency distribution we replace the sum in (4.40) by the integral

$$F = E_0 + k_0 T \frac{3V}{2\pi^2 v^3} \int_0^{\omega_m} \ln[1 - \exp(-\hbar\omega/k_0 T)] \omega^2 d\omega$$

$$= E_0 + 9Nk_0 T \left(\frac{T}{T_D}\right)^3 \int_0^{T_D/T} \ln[1 - \exp(-x)] x^2 dx,$$

(4.76)

where, according to (4.9), the relationship between limiting vibration frequency ω_m and Debye temperature T_D is given by the expression $T_D = \hbar\omega_m/k_0$. Taking a derivative of (4.75), we assume that the Debye temperature or the limiting frequency is a function of volume V

$$p = -\frac{\partial E_0}{\partial V} - 3Nk_0 TD\left(\frac{T_D}{T}\right) \frac{1}{T_D} \frac{\partial T_D}{\partial V}.$$

(4.77)

Here, $D = D(z)$ is the Debye function

$$D(z) = \frac{3}{z^3} \int_0^z \frac{x^3}{\exp(x) - 1} dx.$$

(4.78)

It can be shown that in the harmonic approximation $dT_D/dV = 0$ and the vibration anharmonicity leads to $dT_D/dV < 0$. The Grüneisen constant is the temperature-independent relation

$$\gamma_G = -\frac{V}{T_D} \frac{dT_D}{dV} = -\frac{d\omega_m/\omega_m}{dV/V} = -\frac{d\ln\omega_m}{d\ln V} > 0.$$

(4.79)

In the harmonic approximation constant $\gamma_G = 0$. Since the temperature-dependent part of the internal energy is $E_T = 3Nk_0 TD(T_D/T)$ (see Sect. 4.2), the final Debye equation of state takes the form

$$p = -\frac{\partial E_0}{\partial V} + \gamma_G \frac{1}{V} E_T,$$

(4.80)

where $\partial E_0/\partial V$ is temperature independent.

From (4.80) we can obtain Grüneisen's formula relating the linear coefficient of expansion α with isothermal compressibility k. By differentiating with respect to the temperature and allowing for (4.18) we obtain from (4.80)

$$\left(\frac{\partial p}{\partial T}\right)_V = \gamma_G \left(\frac{c_v}{V}\right).$$

(4.81)

Introducing the linear coefficient of expansion

$$\alpha = \frac{1}{3V}\left(\frac{\partial V}{\partial T}\right)_p = -\frac{1}{3V}\frac{(\partial p/\partial T)_V}{(\partial p/\partial V)_T} = -\frac{1}{3}\left(\frac{\partial V}{\partial p}\right)_T \frac{1}{V}\left(\frac{\partial p}{\partial T}\right)_V \quad (4.82)$$

and the isothermal compressibility

$$k = -\frac{1}{V}\left(\frac{\partial V}{\partial p}\right)_T,$$

we get Grüneisen's formula in its final form

$$\alpha = \frac{1}{3}\frac{k\gamma_G c_v}{V}. \quad (4.83)$$

By investigating the compressibility of crystals at high pressures it is possible to determine Grüneisen's constant γ_G and compare it with the value of γ_G calculated by (4.83). For cubic crystals, the agreement is particularly good. The values of Grüneisen's constant for some substances are given in Table 4.1.

Table 4.1. Grüneisen's constant

Substance	Calculated value	Experimental value
Na	1.25	1.50
K	1.34	2.32
Fe	1.6	1.4
Co	1.87	1.8
Ni	1.88	1.9
NaCl	1.63	1.52
KCl	1.60	1.26

4.8 Phase Transitions and Crystal Symmetry [1]

4.8.1 Second-Order Phase Transitions

We have so far considered the basic features of phase transitions and their relationship with the thermodynamic characteristics and the vibration spectrum of crystals. We focused our attention on first-order phase transitions, when a change in thermodynamic parameters results in a considerable

---

[1] This section was written by E. B. Loginov.

rearrangement of the structures and changes in phase properties. A certain correlation between the structures of the initial and the newly formed phase may, or may not, take place. One cannot generally predict the relationship between the structure and the symmetry of the phases lying on the different sides of the phase-equilibrium line.

This section treats transitions accompanied by such a small change in the atomic structure of the crystal that uniform description of the structure and thermodynamic potential of the two phases is possible. The most prominent examples of this are second-order phase transitions, when the atomic structure of the crystal varies continuously, but at the transition point the crystal symmetry changes jumpwise. Indeed, the symmetry can be either the one or the other; continuous change in symmetry is impossible. For instance, on very small displacements of atoms in the cubic structure, tetragonal or rhombohedral distortion may take place, i.e., the cubic symmetry is lost at once. Thus, at the point of second-order phase transition the structures and states of the two phases coincide. As we have already noted, this is not so for first-order transitions, when two phases with different structures and properties are in equilibrium.

A characteristic feature of a second-order phase transition is the fact that the symmetry group of one of the phases is a subgroup of the symmetry group of the other phase, because during the displacement of atoms only some symmetry elements are lost, while the others remain, and they form a subgroup. The more symmetric phase corresponds, as a rule, to the high-temperature modification. There is always a certain quantity (the transition or order parameter) equal to zero in the high-symmetry phase which increases continuously from zero to the final value as it recedes to the lower-symmetry phase; such a change in the transition parameter is sufficient for a complete description of the change in symmetry on phase transition. By taking into account the dependence of the thermodynamic potential on the transition parameter one can achieve uniform description of the two phases. The structure and thermodynamic potential of the equilibrium phase are obtained as a result of finding the minimum of the thermodynamic potential.

By way of example we consider the possible ferroelectric phase transitions in triglycinesulphate[2]. The transition parameter is the vector of electric polarization P. The crystal symmetry imposes certain limitations on the character of the dependence of thermodynamic potential Φ on the components of vector P. Since Φ must remain unaltered on all transformations of the symmetry group of the crystal, it is a function of invariant combinations of components P_i.

The point symmetry group of the high-temperature phase is C_{2h}-$2/m$. The z axis is directed along the twofold axis. Then there are four invariant combinations of the components of the polarization vector, P_x^2, P_y^2, P_xP_y, and P_z^2. Thus, from the crystal symmetry it follows that the dependence of the thermo-

---

[2] The thermodynamic theory of ferroelectric phase transitions is considered in [1.7].

dynamic potential on the polarization must have the form

$$\Phi = \Phi(P_x^2, P_y^2, P_x P_y, P_z^2, T, p), \tag{4.84}$$

where T is the temperature, and p is the pressure (it is assumed that the dependence of Φ on the other variables, for instance, deformations, is excluded by finding the minimum with respect to these variables). Expression (4.84) contains all the information that can be obtained from the symmetry. We must now find the minimum of expression (4.84). To do this, we take advantage of the fact that in the high-temperature phase the minimum falls to the value $P = 0$. We also assume that there is a second-order phase transition; hence, near the transition point all the components of vector P are small, so that we can use an expansion of Φ in powers of their invariant combinations. To begin with, we restrict ourselves to consideration of invariant-linear (quadratic in P_i) terms in expansion (4.84)

$$\begin{aligned} \Phi = \Phi_0(T,p) + A_{11}(T,p)P_x^2 + 2A_{12}(T,p)P_x P_y + A_{22}(T,p)P_y^2 \\ + A_{33}(T,p)P_z^2. \end{aligned} \tag{4.85}$$

In the high-symmetry phase (and at the transition point) the minimum of (4.85) falls to the value $P_i = 0$, i.e., the quadratic form of P_x, P_y, and P_z must be positive valued. Consequently, the following inequalities hold true in the high-temperature phase and at the transition point:

$$A_{11} \geqslant 0; \qquad A_{11} A_{22} - A_{12}^2 \geqslant 0; \qquad A_{33} \geqslant 0. \tag{4.86}$$

At the transition point $T = T_c$ one of the inequalities (4.86) turns into an equality (otherwise all three inequalities would have held true in the vicinity of the transition point as well, i.e., no transition would have taken place). In the low-symmetry phase this inequality is violated, and the minimum of the thermodynamic potential falls to the nonzero values of P_x, P_y (if the second inequality is violated) or to P_z (if the third inequality is violated).

Let us first consider the case of the change of sign of coefficient A_{33}. Since the second inequality (4.86) is fulfilled for the low-symmetry phase, the minimum of the thermodynamic potential falls to the values $P_x = P_y = 0$. Substituting these values into (4.84) and expanding Φ in powers of P_z^2 with an accuracy to the second-order terms, we obtain

$$\Phi = \Phi_0 + \alpha(T - T_c)P_z^2 + \frac{1}{2}\beta P_z^4. \tag{4.87}$$

Here, $A_{33} = \alpha(T - T_c)$. The values of α, β, and Φ_0 only slightly depend on the temperature, and this dependence can be neglected. For definiteness sake we put $\alpha > 0$ and assume that $\beta > 0$ (the case $\beta < 0$ corresponds to a first-order phase transition). Then, at $T > T_c$, the minimum of expression (4.87) falls to the value $P_z = 0$ (high-symmetry phase), and at $T < T_c$, to the value

$$P_z = \sqrt{\frac{\alpha(T_c - T)}{\beta}}, \qquad (4.88)$$

i.e., the polarization appears beginning with $T = T_c$ and increases continuously with a further decrease in temperature. Thus, at $T = T_c$, a second-order phase transition actually takes place. The point symmetry of the low-temperature phase is described by group C_2-2, and the thermodynamic potential of the equilibrium phase is given by the expression

$$\Phi = \Phi_0 - \frac{\alpha^2(T_c - T)^2}{2\beta}. \qquad (4.89)$$

At the transition point entropy $S = \partial \Phi / \partial T$ is continuous (i.e., the heat of a second-order transition is equal to zero), while the heat capacity experiences a jump $\Delta C = \alpha^2 T_c / \beta$, the heat capacity being higher in the low-symmetry than in the high-symmetry phase. The dielectric susceptibility (polarizability) is $\chi = (\partial^2 \Phi / \partial p^2)^{-1} = 2 [\alpha \cdot (T - T_c) + \beta P_z^2]^{-1}$. In the high-temperature phase $\chi = 2/\alpha(T - T_c)$ (the Curie–Weiss law), and in the low-temperature phase, $\chi = 4/\alpha(T_c - T)$, i.e., the susceptibility goes to infinity at point T_c. Precisely such a phase transition occurs in triglycinesulphate at temperature $T_c = 49\,°C$.

Investigation into the case of the violation of the second inequality in (4.86) (which does not take place in triglycinesulphate) after diagonalization of the quadratic form from P_x, P_y in (4.85) is carried out similar to that considered above. The symmetry of the low-temperature phase would, in this case, be described by group C_s-m.

The example just discussed reveals many features of second-order phase transitions, although in transitions with multicomponent order parameters there may also be qualitative differences, which are indicated below.

4.8.2 Description of Second-Order Transitions with an Allowance for the Symmetry [4.8]

Before we proceed to the general description of second-order phase transitions it is necessary to introduce a more rigorously unified description of the thermodynamic potential of both phases. To do this, we consider the potential as a function of the crystal structure which can be characterized by the density function $\rho(x, y, z)$. In a crystal consisting of atoms of the same sort $\rho(x, y, z)$ can be considered the probability distribution of the positions of the atomic (nuclear) centers; in a multiatomic crystal it is more convenient to use the electron density for this purpose; additionally, in magnetic crystals one should take into account the distribution of the current $j(x, y, z)$ or the spin orientation. As we know, the symmetry group of a crystal [Ref. 1.6, Chap. 2] is a set of operations (transformations of the coordinates) leaving $\rho(x, y, z)$ unaltered.

Let G_0 be the group which the crystal possesses at the transition point, and $\rho(x, y, z)$ is the density function in some phase (generally speaking, this functions depends on the temperature and pressure). If we perform all the transformations $g_i \in G_0$ over $\rho(x, y, z)$, we obtain a set of functions which transform linearly through each other as a result of the operations g_i, i.e., the representation of group G_0. Expanding this representation into irreducible ones [Ref. 1.6, Sects. 2.6.7, 6.8, 8, 15], we obtain the following expansion of function $\rho(x, y, z)$ with repsect to the basis functions $\psi_i^{(n)}(x, y, z)$ of irreducible representations:

$$\rho = \sum_n \sum_i c_i^{(n)} \psi_i^{(n)}, \tag{4.90}$$

where n is the number of the irreducible representation, and i is the number of the function at its basis. Hereafter we assume functions $\psi_i^{(n)}$ to be normalized in some definite way.

Among the functions $\psi_i^{(n)}$ there is always one which is in itself invariant to all the transformations of group G_0 (it realizes the identity representation of the group). Denoting this function by ρ_0 and the remaining part of ρ by $\Delta\rho$, we can write

$$\rho = \rho_0 + \Delta\rho, \; \Delta\rho = \sum_n' \sum_i c_i^{(n)} \psi_i^{(n)}. \tag{4.91}$$

The prime after the symbol of the sum means that the identity representation is excluded from the summation. Note that if $\Delta\rho \neq 0$, the symmetry group G_1 of the crystal does not coincide with G_0, but is its subgroup $G_0 \in G_0$.

Since function $\rho(x, y, z)$ is real, the sum (4.91) must contain a conjugate complex representation along with the complex one. A pair of conjugate complex irreducible representations can be regarded as a single physically irreducible representation of a double dimension. Accordingly, functions $\psi_i^{(n)}$ can be selected as real.

The thermodynamic potential of a crystal with a density function $\rho(91)$ is a function of the temperature, pressure, and coefficients $c_i^{(n)}$ (and naturally depends on the concrete form of functions $\psi_i^{(n)}$ themselves). The values of $c_i^{(n)}$ which are actually realizable (as a function of p and T) are determined thermodynamically from the equilibrium conditions, i.e., from the conditions of the minimum of Φ. This also assigns crystal symmetry G_0, since it is clear that the symmetry of the function ρ, described by functions $\psi_i^{(n)}$, whose laws of transformation are known, depends on the values of the coefficients $c_i^{(n)}$ in expansion (4.91).

Since the expansion (4.91) of function $\Delta\rho$ in irreducible representations contains no identity representation, either $\Delta\rho = 0$ (i.e., all the $c_i^{(n)} = 0$) or $\Delta\rho$

changes on certain transformations from group G_0, and hence the symmetry of function $\Delta\rho$ (and ρ) is lower than G_0. Consequently, at the transition point itself all the $c_i^{(n)}$ are zero. Since the state changes continuously during a second-order phase transition, all the $c_i^{(n)}$ also reduce to zero continuously and take infinitely low values near the transition point. Accordingly, let us expand potential $\Phi(T, p, c_i^{(n)})$ near the transition point into a series in powers of $c_i^{(n)}$.

It should be noted that since in transformations of group G_0 functions $\psi_i^{(n)}$ transform through each other (within the basis of each irreducible representation), it can be imagined that coefficients $c_i^{(n)}$, rather than functions $\psi_i^{(n)}$, are transformed (by the same law). Further, since the thermodynamic potential of a body evidently cannot depend on the choice of a reference system, it must be invariant to transformations of group G_0. Therefore, each term of expansion of Φ in powers of $c_i^{(n)}$ may only contain an invariant combination of the values of $c_i^{(n)}$ of the corresponding power.

A linear invariant cannot be compiled from values which transform according to an irreducible representation of a group (except for the identity representation). A second-order invariant always exists; this is the sum of the squares of coefficients $c_i^{(n)}$. Thus, in the approximation quadratic with respect to $c_i^{(n)}$ the expansion of Φ has the form

$$\Phi = \Phi_0 + \sum_n {}' A^{(n)} \sum_i (c_i^{(n)})^2, \tag{4.92}$$

where $A^{(n)}$ are functions of p, T.

Since at the transition point itself the minimum of (4.92) falls to $c_i^{(n)} = 0$, all the $A^{(n)}$ at this point are nonnegative. If all the $A^{(n)}$ at the transition point were positive, they would also be positive in some vicinity of this point, i.e., we could have $c_i^{(n)} = 0$ on both sides of the transition, and the symmetry would remain unaltered. It is obvious that at the transition point one of the coefficients $A^{(n)}$ must reverse its sign (two coefficients can vanish at the same point only if it is an isolated point on the $p-T$ diagram which corresponds to the intersection of several lines of a second-order phase transition).

Thus, on one side of the transition point all the $A^{(n)} > 0$, and on the other one of the coefficients $A^{(n)}$ is negative. Accordingly, on one side of the transition point all the $c_i^{(n)} = 0$, and on the other side nonzero $c_i^{(n)}$ appear, which correspond to one and the same irreducible representation. Here, a phase with a symmetry group $G_1 \subset G_0$ is formed, where G_0 is the symmetry of the initial phase. Hereafter we omit index n denoting the number of the representation.

We introduce the notation $\eta^2 = \sum_i (c_i)$, $c_i = \eta\gamma_i$, so that $\sum_i \gamma_i^2 = 1$, and expand Φ as

$$\Phi = \Phi_0(p, T) + \eta^2 A(p, T) + \eta^3 \sum_\alpha B_\alpha(p, T) f_\alpha^{(3)}(\gamma_i)$$
$$+ \eta^4 \sum_\alpha C_\alpha(p, T) f_\alpha^{(4)}(\gamma_i) + \cdots, \tag{4.93}$$

where $f^{(3)}$ and $f^{(4)}$ are invariants of the third, fourth, etc., orders compiled from values γ_i; sums over α have as many terms as there are independent invariants of the corresponding order which can be made up from γ_i.

Since at the transition point $A = 0$ and the potential must have a minimum at $\eta = 0$, a second-order phase transition requires that no third-order invariance be present and that the minimum of fourth-order invariants with respect to γ_i give positive coefficients at η^4.

Assuming these (so-called Landau) conditions to be fulfilled, we write the expansion of the thermodynamic potential with an accuracy to the fourth-order terms

$$\Phi = \Phi_0 + A(p, T)\eta^2 + \eta^4 \sum_\alpha c_\alpha(p, T) f_\alpha^{(4)}(\gamma_i) . \tag{4.94}$$

The second-order term does not contain γ_i; these values are determined from the condition of the minimum of the coefficient at η^4. The obtained values of γ_i define the symmetry of the function

$$\Delta\rho = \eta \sum_i \gamma_i \psi_i , \tag{4.95}$$

i.e., the symmetry of the density function and the symmetry group G_1 of the phase arising from the phase with symmetry G_0 as a result of a second-order transition. The temperature variation in the parameter of the order $\eta = \eta(p, T)$ is found by minimizing equation (4.94) with respect to η (with given γ_i) precisely as we did in considering the example with the ferroelectric phase transition in triglycinesulphate.

4.8.3 Phase Transitions Without Changing the Number of Atoms in the Unit Cell of a Crystal

It is known [Ref. 1.6, Sect. 2.8.15] that in the basis functions of irreducible representations of space groups it is possible to isolate factor $\exp(i\mathbf{k}\mathbf{r})$, where $\mathbf{k}$ is the wave vector lying in the first Brillouin zone. It is easy to see that the functions ψ with $\mathbf{k} \neq 0$ in the expansion (4.91) correspond to an increase in lattice spacings by two or by any multiple number of times in general, and that the initial unit cells subsequently turn into a multiple number of distinct "subcells" of a new, large cell. In other words, in the lattice being formed, it is possible to isolate sublattices corresponding to the initial lattice.

Indeed, at $\mathbf{k} \neq 0$ the initial lattice retains only those points for which $\mathbf{k}\mathbf{r} = 2\pi n$, where n is an intenger. Figure 4.15 illustrates a wave vector lying at the boundary of the Brillouin zone. The lattice splits up into two sublattices. The principal spacing along the x axis doubles, and so do the volume of the unit cell and the number of atoms in it.

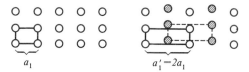

Fig. 4.15. The simplest scheme of a phase transition with a change in the number of atoms in the unit cell of the crystal $k = b_{1/2}$, $a_1' = 2a$, [4.9]

If $k = 0$, all the translations are preserved. This means that all the transitions with the preservation of the number of atoms in the unit cell must be described by representations with $k = 0$ [4.9]. Then, both a change in cell spacings and a mutual shift of the sublattices may take place. Figure 4.16a depicts a case where the lattice undergoes shear deformation accompanied by the disappearance of two symmetry planes perpendicular to the drawing. In Fig. 4.16b, two sublattices shift relative to each other so that the center of symmetry is lost and the lattice becomes polar. In Fig. 4.16c, three sublattices merging into each other on rotation through 120° shift by equal distances so that the sixfold axis disappears.

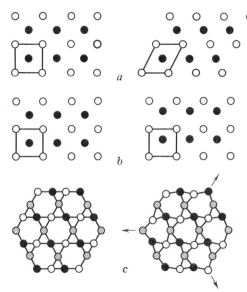

Fig. 4.16a – c. Phase transitions with retention of the number of atoms in the unit cell. (a) Shear deformation of the unit cell; (b) mutual displacement of the two sublattics without a change in cell parameters; (c) same for three sublattices. In all cases $k = 0$ [4.9]

It is worth noting that under the effect of symmetry operations of space group $\Phi = G_3^3$ the functions with $k = 0$ transform precisely in the same way as do the basis functions of the representations of the corresponding point symmetry group $K = G_0^3$. Indeed, the basis functions of the representations of groups k depend exclusively on direction and transform under the effect of the rotational components of the operations of group G_3^3 in the same way as they do under the effect of operations of group G_0^3. In other words, although group K is not a subgroup of group $\Phi = G_3^3$ (being its homomorphous mapping, [Ref. 1.6, Sect. 2.8.4], all the representations of group K are rep-

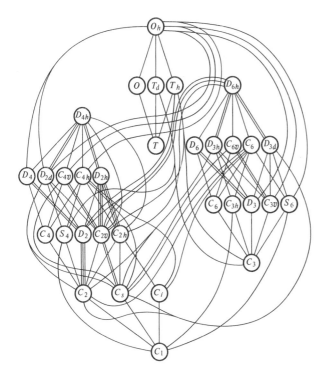

Fig. 4.17. General scheme of all possible second-order phase transitions with retention of the number of atoms in the unit cell of the crystal [4.9]

resentations of the given group Φ (and all the other Φ homomorphous to K, i.e., belonging to the given crystal class) corresponding to $k = 0$.

Thus, in order to investigate structural changes not involving a change in the number of atoms in the unit cell of the crystal it will suffice to take into account, in expansions of the type (4.91), the irreducible representations of point groups K [Ref. 1.6, Table 2.8]. An analysis of these representations leads to the scheme given in Fig. 4.17 for all the possible second-order phase transitions between crystal classes K (corresponding transitions are possible in all the space groups Φ of the given class K). Each line corresponds to some representation. The straight lines denote one-dimensional representations, and the curves, multidimensional. The branching of the lines means that the given representation can (depending on the ratio between the factors at the invariants in the expansion of the thermodynamic potential) give a transition to different crystal classes. For instance, in group $O_h^1\text{-}Pm3m$ a ferroelectric phase transition of the second order (representation F_{1u}) can give a tetragonal group $C_{4v}^1\text{-}P4mm$ (polarization along a cube edge) or a rhombohedral group $C_{3v}^1\text{-}P3m1$ (polarization along the body diagonal of the cube).

4.8.4 Changes in Crystal Properties on Phase Transitions

Let us assume that a phase transformation involves the introduction of additional degrees of freedom C_i corresponding to an irreducible representation Γ, and consider the physical implications of such a change in structure. To do this we have to find out which invariants can be composed from coefficients C_i and the external parameters of the system. If we restrict ourselves to the invariants linear in C_i (and hence to the physical characteristics depending linearly on these coefficients), it will be sufficient to find all the physical quantities which transform according to representation Γ. To find quantities depending on coefficients C_i in a more complex way, we must consider the symmetric degrees of representation Γ.

Ferroelectric phase transitions. Suppose that representation Γ belongs to the vector representations V of a given group, i.e., that one or several components of the polar vector transform according to representation Γ. Then the crystal symmetry permits us to compose invariant $\Sigma c_i E_i$, where E_i are the components of the vector of the electric field tension; the phase transition with inclusion of the degrees of freedom c_i causes the appearance of spontaneous polarization

$$P_i = -\frac{1}{4\pi}\frac{\partial \Phi}{\partial E_i}, \tag{4.96}$$

i.e., is ferroelectric, the polarization depending linearly on coefficients c_i. If, however, some symmetric degree (for instance, the second) of representation Γ contains a vector representation, there is invariant $\sum_{ijk} \alpha_{ijk} c_i c_j P_k$, and the transition results in a polarization proportional to $P_k \sim \sum \alpha_{ijk} c_i c_j$. Such ferroelectrics are called improper [4.10, 11]. In particular, all ferroelectrics which form as a result of transitions according to representations with $k = 0$ are improper. In contrast to the proper ferroelectrics, the improper ones do not obey the Curie – Weiss law, while the dielectric constant experiences a jump at the transition point.

Ferromagnetic phase transitions. If vector representations induce phase transitions with the appearance of a polar vector of electric polarization, pseudovector representations $\tilde{V}$ induce transitions with the appearance of an axial vector, for instance a magnetization vector, i.e., ferromagnetic phase transitions.

Ferroelastic phase transitions. Expansion of symmtric square $[V]^2$ of a vector representation in irreducible representations gives a set of all representations according to which the components of the deformation tensor (as well as of the stress tensor, the dielectric-constant tensor, etc.) are transformed. The relevant phase transitions involve spontaneous deformation (namely, the component which is transformed according to the given representation) and a change in optic indicatrix. Such transitions are called ferroelastic, and the substances, ferroelastics. Ferroelastic phase transitions are accompanied by the vanishing of the elastic moduli corresponding to the deformation components which are transformed according to the given representation.

Elastic constants. Similarly, expansion of $[[V]^2]^2$ (i.e., the symmetric square of the symmetric square of a vector representation) in irreducible representations gives a set of all the representations according to which the tensor components of elastic constants are transformed. The corresponding phase transitions will involve a change in the elastic constants of the crystal.

The list of physical properties can be supplemented by the piezoelectric (representation $V \times [V]^2$), piezomagnetic (representation $\tilde{V} \times [V]^2$), piezo-optical, electro- and magnetostrictive (representation $[[V]^2]^2$) characteristics, etc.

By way of illustration we note that the transition depicted in Fig. 4.16a is associated with representation Γ, which appears in $[V]^2$, i.e., is ferroelastic; and in Fig. 4.16c, with representation Γ belonging to $[[V]^2]^2$, i.e., the transition involves a change in elastic constants.

4.8.5 Properties of Twins (Domains) Forming on Phase Transformations

The values of the quantities γ_i determining the symmetry of the low-temperature phase are found from the condition of the minimum of the factor at η^4 in (4.94). But to the minimum of this factor there correspond several solutions for γ_i (at least replacement of γ_i by $-\gamma_i$ does not affect its value), and since a phase transition begins independently in different regions of a crystal, it results in the formation of regions with different values of γ_i. These regions are called domains or twins. We shall indicate the properties distinguishing twins which form as a result of different phase transitions. If a phase transition is ferroelectric (representation Γ belongs to vector representation V) or ferromagnetic (Γ belongs to pseudovector representation $\tilde{V}$), the domains formed as a result of the transition differ in the direction of the electric polarization or magnetization. If a transition is ferroelastic (Γ belongs to $[V]^2$), the twins formed undergo different spontaneous deformation (those components are different which transform according to representation Γ) and have different (differently oriented) optic indicatrices. These are ordinary twins, which can be characterized by the twinning planes (the planes along which the ellipsoids of spontaneous deformation of the twin components intersect) and the twinning directions (directions lying in the twinning planes and perpendicular to the line of intersection of the adjacent twinning planes).

If Γ does not belong to representation $[V]^2$, the twins differ neither in spontaneous deformation, nor in optic indicatrices. The concepts of a twinning ellipsoid and of twinning planes and directions are inapplicable to such twins.

Two cases are possible depending on whether representation Γ appears in representation $[[V]^2]^2$. If it does, the twin components differ in elastic constants. Under the effect of stress such twins experience only second-order effects; for instance, the direction of forced displacement of the twin boundary remains unaltered on change in the sign of the stress. An example are Dophinais twins, which form on $\alpha \rightarrow \beta$ transformation of quartz (Fig. 4.16 c).

If Γ does not appear, not only in $[V]^2$, but also in $[[V]^2]^2$, the twin components differ neither in optic indicatrices, nor in elastic constants. Such "masked" twins may not be noticed at all unless the structure or the physical properties of the crystal are thoroughly investigated. The twins of the described type include all the inversion twins and some types of rotation and reflection twins. If representation Γ corresponds to a nonzero value of wave vector k, translation twins form as a result of the corresponding phase transition.

4.8.6 Stability of the Homogeneous State of the Low-Symmetry Phase

Until now we assumed that at neighboring points on the line of second-order phase transition the transition occurs according to one and the same representation. This problem requires more detailed consideration. The fact is that irreducible representations of space groups are classified not only by discreteness (the number of small representations), but also by the continuous value of wave vector k. Therefore coefficients $A^{(n)}$ in expansion (4.92) must depend, not only on discrete number n, but also on continuous variable k.

Suppose the phase transition involves the vanishing (as a function of p and T) of coefficient $A^{(n)}(k)$ with a definite number n and a definite value of $k = k_0$. Then the coefficient $A^{(n)}(k)$ must have a minimum with respect to k at $k = k_0$, i.e., the first derivatives with respect to k must be equal to zero at this point. If the vanishing of $\partial A^{(n)}/dk$ is not due to the crystal symmetry, these derivatives are nonzero (although small) at a neighboring point, and the minimum of $A^{(n)}(k)$ falls on a value of k different from k_0, although close to it. Here, the value of k changes continuously along the line of second-order phase transitions.

If, however, the crystal symmetry requires that $\partial A^{(n)}/\partial k = 0$, then the transition occurs according to one and the same representation at different points on the phase-transition line. We shall indicate the group-theoretical criterion of the vanishing of $\partial A^{(n)}/\partial k$. To do this we allow for the fact that the value k of the wave vector defines the translation symmetry (periodicity) of a new phase. Then a structure with a vector k close to k_0 can be regarded as a structure with a wave vector k_0 and with coefficients $C_i(r)$ slowly (by virtue of the smallness of $k - k_c$) varying in space. The required absence of linear terms in the expansion of $A^{(n)}(k)$ in powers of $k - k_0$ leads to the absence of invariants in the products of the values c_i and their gradients in the expansion of the thermodynamic potential. Since terms of the type $c_i \nabla c_j + c_j \nabla c_i$ reduce to the divergency of some vector, they make no contribution to the volume energy. Hence, for the transition according to the given representation to occur along the entire line on the phase diagram it is necessary that no invariants be composed of the combinations $c_i \nabla c_j - c_j \nabla c_i$ [4.12]. In the language of representation theory that means that the antisymmetric square of representations Γ, responsible for the phase transition, may not contain a vector representation.

If, however, Lifshits invariants do exist, the value of k varies along the phase-transition line, and, as noted above, this variation can be represented as a variation of coefficients c_i in space, which means that the so-called incommensurate, or modulated, structure arises as a result of the corresponding phase transition. It is easiest to describe such a structure in the case of a ferromagnetic or ferroelectric phase transition, when the incommensurate structure is due to the change of the vector of electric polarization or magnetization in space; this structure is ususally helicoidal, and the helicoid period is not necessarily a multiple of (commensurate with) the lattice spacing. Structures of this type are observed in the most diverse materials: in quasi-one-dimensional and two-dimensional conductors, magnetics, ferroelectrics, alloys, liquid crystals, etc.

We remark in conclusion that the predictions of the phenomenological theory as to the nature of the temperature dependence of physical quantities are valid for regions which are neither very close to the phase transition point (in the vicinity of a phase transition an essential role is played by fluctuations of the transition parameter), nor very far from this point (since the theory uses expansion of the thermodynamic potential into a power series). The fluctuation range may be either very narrow (for instance, in superconductors and magnetics) or rather wide (for instance, in quartz and most of the ferroelectrics). The predictions relating to changes in symmetry evidently hold true for second-order phase transitions.

5. The Structure of Real Crystals

The regular, strictly periodic structure of the crystal discussed in the preceding chapters is just an idealized picture. In nature, even under conditions of ideal thermodynamic equilibrium, crystals must show various deviations from this structure, which are called crystal lattice defects. Equilibrium lattice defects should by no means be interpreted as crystal defects. They can be regarded as elementary excitations of the ground state of the crystal, being just as inherent in the crystal as phonons or electrons, etc. While phonons and electrons are elementary excitations in the phonon and electron subsystems of a crystal, which were considered in Chaps. 3, 4, lattice defects are elementary excitations in the atomic subsystem of a crystal, whose ground state was described in Chap. 1.

Apart from equilibrium lattice defects, real crystals exhibit nonequilibrium defects due to the nonideal conditions of the origin and life of the crystal. These defects do not disappear completely, even after a very long time, from thermal motion alone; they are in a "frozen" state. Nonequilibrium lattice defects are often stabilized by electric, magnetic, or elastic fields arising in the course of crystal growth, phase transformations, or under external influences. The density of nonequilibrium lattice defects can be considerably reduced by improving the methods for the preparation and treatment of crystals.

All the so-called structure-sensitive properties of crystalline materials are due to the presence of equilibrium and nonequilibrium lattice defects. A crystal responds to external influences by changing its real structure through the generation, rearrangement, motion, and annihilation of lattice defects. The plastic deformation of crystals, for instance, consists completely in the motion of various lattice defects. The thermal expansion of crystals is caused, not only by the anharmonism of atomic vibrations, but also by an increase in the density of the lattice defects. An electric current in ionic crystals is mainly due to the migration of charged lattice defects, while the most important properties of semiconductors depend on the number of electrically active defects of the lattice, etc.

5.1 Classification of Crystal Lattice Defects

Because of their low mobility and long life time lattice defects lend themselves to a pictorial geometric description (the only exception being quantum crystals, such as those of helium, where zero quantum vibrations are so intensive that the localization of the lattice defects is disturbed and they behave as quasiparticles, i.e., similar to phonons and electrons). By "lattice sites" we mean the positions which the atoms must occupy in a crystal having an ideal atomic structure. Strictly speaking, not a single atom in a real crystal is located at a lattice site. The description of the atomic structure of real crystals is, however, greatly facilitated by the fact that it does not differ substantially from the ideal structure; the arrangement of most of the atoms can, therefore, be described with reference to the network of the lattice sites. It is convenient to classify the lattice defects according to a purely geometric feature − their dimensionality, i.e., the number of dimensions in which the qualitative disturbances of the ideal structure of the crystal (the absence or anomalous disposition of neighboring atoms) extend to macroscopic distances.

Zero-dimensional (point) defects are vacant sites of the lattice, interstitial atoms, atoms at lattice sites of a "foreign" sublattice, impurity atoms in various positions, etc.

One-dimensional (linear) defects are chains of point defects, and also dislocations, i.e., specific defects disturbing the regular succession of atomic planes.

Two-dimensional (surface) defects are the crystal surface, stacking faults (irregularly packed atom layers), grain and twin boundaries, domain walls, etc.

Three-dimensional (volume) defects are pores, inclusions, precipitations, and similar macroscopic formations.

Local distortions of the atomic structure of a crystal, similar to lattice defects, are sometimes caused by electron-type elementary excitations, which strongly interact with the lattice. In semiconductors, electrons and holes distort (polarize) the lattice around them, forming *polarons*. In ionic crystals, local excitation of the electron state may be transmitted from one ion to another, thus migrating through the crystal in the form of an *exciton*. Theory also predicts the existence of *fluctuons*, i.e., local fluctuations of the density, electric polarization, or magnitization, which are stabilized by electrons. On the other hand, lattice defects distort the electron and phonon structure of the crystal; the levels of the electron and phonon spectra are displaced, new levels and local vibrations arise. Accumulation of lattice defects may result in their coalescence leading to precipitation of a new phase inside the crystal (condensation of vacancies, excitons, etc.) and in overall instability throughout the crystal, which leads to a phase transformation of the whole crystal (ordering with respect to vacancies, stacking faults, etc.).

5.2 Point Defects of the Crystal Lattice

5.2.1 Vacancies and Interstitial Atoms

Vacancies, or unoccupied lattice sites, and *interstitial atoms*, or the atoms implanted into the interstices, are defects antipodes: the filling of a vacancy or the removal of an interstitial atom restore the regularity of the crystal lattice. The energy necessary for the formation of a vacancy is defined as the work necessary to shift an atom from a lattice site to the crystal surface and is usually of the order of one electron volt. The energy of formation of an interstitial atom is defined as the work done in shifting an atom from the crystal surface to an interstice and may be as high as several electron volts because of the large contribution of the energy of local distortions arising on incorporation of an atom into an interstice[1].

The possibility of the existence of defects with such a high energy and under conditions of thermodynamic equilibrium is due to the fact that the formation of point defects greatly increases the crystal entropy. From a crystal containing N identical atomes one can remove n atoms in

$$C_N^n = \frac{N!}{n!\,(N-n)!}$$

different ways. According to the Boltzmann equation, the corresponding increment of the configurational entropy is

$$\Delta S = k \ln \frac{N!}{n!\,(N-n)!}. \tag{5.1}$$

If the energy of formation of one defect is E, the formation of n defects at a temperature T changes the free energy of the crystal by

$$\Delta F = nE - T\Delta S. \tag{5.2}$$

Minimization of free energy (5.2) taking into account only configurational entropy (5.1) and using Stirling's formula $\ln m! \approx m \ln m$ for evaluating the factorials of large numbers gives the following estimate for the equilibrium number of point defects:

$$n = N \exp(-E/kT). \tag{5.3}$$

---

1 In calculating the energy of formation of vacancies and interstitial atoms, possible changes in the surface energy of the crystal must be excluded. Therefore, only those surface atoms which are located at kinks of atomic steps on the crystal surface can take part in the point defect formation

For copper, for example, the energy of formation of a vacancy is about 1 eV, and that of an interstitial atom, 3.4 eV. According to (5.3) the concentration of vacancies at the melting point ($T = 1356$ K) has the magnitude 2×10^{-4}, and of interstitial atoms, only 2×10^{-13}.

It should be noted that in deriving (5.3) the change in the vibration entropy of the crystal was neglected. Near point defects the atoms vibrate with changed frequencies and amplitudes, which increases the crystal entropy by some value $n \Delta S'$, proportional to the number of defects. Therefore, (5.3) must have a correction factor $\exp(\Delta S'/k)$, which, however, does not affect the results as regards the order of magnitude. As an example we give the data on the density of vacancies $c = n/N$ at the melting point for gold, silver, and copper, and the relevant values of E and $\Delta S'/k$ (Table 5.1).

Table 5.1.

	$c \times 10^4$	E [eV]	$\Delta S'/k$
Au	2.7 ± 0.6	0.98	1.22
Ag	1.7 ± 0.5	1.04	0.9
Cu	2.0 ± 0.5	1.07	0.5

It is clear that the simple equation (5.3) already yields a completely reliable estimate of the equilibrium density of defects.

Similar calculations can be carried out in more complicated cases. Let us consider, e. g., an ionic crystal. During the formation of point defects the condition of electric neutrality of the crystal as a whole must be fulfilled. Therefore defects are born in pairs (Fig. 5.1) consisting either of a vacancy and an interstitial ion (Frenkel defect) or of two unlike vacancies (Schottky defect), or else of the antipode of the Schottky defect, i. e., two unlike interstitial ions. For the equilibrium concentration of these pairs, the following formula, similar to (5.3), is applicable:

$$n = \sqrt{N_1 N_2} \exp(-E/2kT), \qquad (5.4)$$

where E is the pair formation energy, and N_1 and N_2 are the number of points

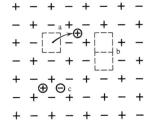

Fig. 5.1. Point defects in a NaCl-type crystal. (a) Frenkel defect; (b) Schottky defect; (c) antipode of the Schottky defect. The vacancies are denoted by squares, and interstitial ions by circles

at which the first and second partners of the pair can reside. Since the energy of an interstitial ion greatly exceeds that of a vacancy, the equilibrium concentration of point defects in ionic crystals usually depends on that of Schottky defects.

The formation of point defects is accompanied by appreciable displacements of the atoms surrounding the defect. The atoms around a vacancy mainly shift towards the vacant point. An interstitial atom, on the contrary, crowds out the surrounding atoms. As a result, an interstitial atom usually even increases, rather than decreases, the crystal volume, while a vacancy increases it by less then one atomic volume. Thus, for instance, in aluminum an interstitial atom increases the crystal volume by 0.5 Ω (where Ω is the atomic volume) and a vacancy, by 0.8 Ω. The x-ray density corresponding to the mean interatomic distance changes then as if a dilatation center with a volume of $+1.5\ \Omega$ were implanted into the crystal in the case of an interstitial atom, and $-0.2\ \Omega$ in the case of a vacancy (i.e., vacancies increase the x-ray density of the crystal). A Frenkel pair, which consists of an interstitial atom and a vacancy, reduces the x-ray and dilatometric density of the crystal in the same way: the volume change is $+1.3\ \Omega$. The difference between the dilatometric and x-ray density corresponds to an increase in the crystal volume by exactly Ω for a vacancy and a decrease by exactly Ω for an interstitial atom. Therefore, a comparison of the dilatometric density of a crystal with its x-ray density makes it possible, in principle, to determine the difference between the number of interstitial atoms and vacancies, which is equal to the difference between the number of atoms and lattice sites. The elastic field arising around point defects as dilatation centers and elastic dipoles exerts the determining influence on the interaction of point defects among themselves and with other lattice defects, particularly with dislocations [1.7].

Knowing the interatomic-interaction law, one can calculate with the aid of a computer the atomic structure of a point defect and pinpoint the site to which each atom moved on appearance of the defect. In a macroscopic description of the displacements of the atoms around the defect one assigns the vector field of displacements $u(x)$ which depends continuously on coordinate x. Corresponding the change in interatomic distances is the distortion field

$$u_{ij} \equiv u_{j,i} \equiv \partial u_j / \partial x_i. \tag{5.5}$$

The symmetric part of the distortion tensor gives the deformations

$$\varepsilon_{ij}(x) = \frac{1}{2}(u_{ij} + u_{ji}), \tag{5.6}$$

and the antisymmetric part, the lattice rotations

$$\omega_i = \frac{1}{2} e_{ijk} u_{jk}. \tag{5.7}$$

Corresponding to the interatomic interaction forces is a stress field σ_{ij}, which can be found from deformations ε_{ij} with the aid of Hooke's law

$$\sigma_{ij} = c_{ijkl}\varepsilon_{kl} = c_{ijkl}u_{kl}, \tag{5.8}$$

where c_{ijkl} is the elastic-moduli tensor. The following notation is adopted in $(5.5-8)$: summation is carried out over repeated indices; subscripts placed after a comma mean differentiation with respect to the corresponding coordinates according to the type (5.5); tensor e_{ijk} is the antisymmetric unit tensor, and its nonzero components are equal to $+1$ or -1, depending on whether subscripts i, j, k form an even or odd permutation of numbers 1, 2, and 3.

The set of equations $(5.5-8)$ is an ordinary set of equations of the *theory of elasticity* of anisotropic bodies (see [1.6]). It is inapplicable for the description of distortions in the core region of a point defect. It is not correct, for instance, to assume that vector u for an interstitial atom and for a vacancy corresponds to the distance to the crystal boundary, since the theory of elasticity presupposes that vectors u are small compared to the interatomic distances. Hereafter, in describing the elastic field of lattice defects we have to resort to the *internal-stress theory* which introduces the concept of proper-distortion tensor u_{ij} and proper-deformation tensor $\varepsilon_{ij}^0 = \frac{1}{2}(u_{ij}^0 + u_{ji}^0)$, which describe the defect structure in macroscopic terms. Equations $(5.6-8)$ are now preserved for elastic-distortion tensor u_{ij} and elastic-deformation tensor ε_{ij}, except that the sum of tensors u_{ij} and u_{ij}^0 corresponds to the gradients of the distortion field, i.e.,

$$u_{j,i} = u_{ij} + u_{ij}^0. \tag{5.9}$$

In a macroscopic description, point defects are characterized by local proper deformation $\varepsilon_{ij}^0(x)$ concentrated in a volume of atomic order. For a dilatation center with extra volume δV

$$\int (dx)\varepsilon_{ij}^0(x) = \delta_{ij}(\delta V/3). \tag{5.10}$$

Here, (dx) is the volume element, δ_{ij} is the symmetric unit tensor ($\delta_{ij} = 1$ for $i = j$ and $\delta_{ij} = 0$ for $i \neq j$).

To calculate the elastic field of a point defect we use the Fourier transformation. It can be shown that the Fourier components of stress tensor σ_{ij} are proportional to those of the tensor of proper deformations ε_{ij}^0

$$\tilde{\sigma}_{ij}(k) = -c_{ijkl}^*(n) \cdot \tilde{\varepsilon}_{kl}^0(k). \tag{5.11}$$

Here, $c_{ijkl}^*(n)$ is the so-called plane tensor of the elastic moduli related to the ordinary tensor of the elastic moduli c_{ijkl} by the expression

$$c_{ijkl}^*(n) = c_{ijkl} - c_{ijmn}n_m \Lambda_{np}^{-1} n_q c_{pqkl}, \tag{5.12}$$

where Λ_{np}^{-1} is a tensor reciprocal to tensor $\Lambda_{np} = n_m c_{mnpq} n_q$, $n = k/|k|$ is a unit vector along the direction of wave vector k. Convolution $c_{ijkl}^*(n)$ with vector n with respect to any index is equal to zero, which ensures the fulfilment of the equilibrium conditions for stresses $n_j \tilde{\sigma}_{ij}(k) = 0$. Substituting the Fourier image of the proper deformation for the dilatation center into (5.11) and performing the reciprocal Fourier transformation, we obtain for the stress field

$$\sigma_{ij}(r) = -\frac{1}{(2\pi)^3} \int (dk) e^{ikr} c_{ijkl}^*(n) \tilde{\varepsilon}_{kl}^0(k). \tag{5.13}$$

At large distances r from the dilatation center the stresses decrease as r^{-3}, and

$$\lim_{r \to \infty} r^3 \sigma_{ij}(r) = \frac{\delta V}{24\pi^2} \oint dn \lim_{\varepsilon \to 0} \frac{d^2}{d\varepsilon^2} c_{ijll}^* \left(n + \varepsilon \frac{r}{r}\right) \tag{5.14}$$

where, in distinction to (5.13), the variable n is orthogonal to r, and $\tilde{\varepsilon}_{kl}^0(0)$ is taken to be equal to $\delta_{kl}(\delta V/3)$ in accordance with (5.10). In the elastic-isotropic case $c_{ijll}^* = 2G(1 + v)(1 - v)^{-1}(\delta_{ij} - n_i n_j)$, where G is the shear modulus, and v is the Poisson coefficient. Hence, in a spherical coordinate system (r, θ, φ) it follows for stresses outside the dilatation center that

$$\sigma_{rr} = -\frac{(1 + v)G\delta V}{3\pi(1 - v)r^3} \qquad \sigma_{\theta\theta} = \sigma_{\varphi\varphi} = -\frac{1}{2}\sigma_{rr}. \tag{5.15}$$

The energy of interaction of the dilatation center with an external field is determined, to a first approximation, by the value of pV_0, where $p = -\sigma_{ii}/3$ is the pressure at the site of the defect. As a result, vacancies, as dilatation centers with a negative extra volume, are drawn into compressed regions, while interstitial atoms, as centers of dilatation with a positive extra volume, tend to shift into negative-pressure regions. In the elastic – isotropic approximation the dilatation centers do not interact because from (5.15) it follows that $p = -(\sigma_{rr} + \sigma_{\theta\theta} + \sigma_{\varphi\varphi})/3 = 0$. Consideration of the elastic anisotropy gives, for the interaction energy, nonzero corrections decreasing with the distance as r^{-3}.

In analyzing the balance of the fluxes of vacancies and interstitial atoms towards dislocations under irradiation one has to take into account stress-nonlinear effects of interaction of point defects with the external field and to allow for changes in the rigidity of bonds near defects, which bring about elastic polarization of the defect in the stress field. Polarization of vacancies with decreased rigidity helps them to be drawn into high-stress regions. Polarization of rigid interstitial atoms produces the opposite effect. The energy of interaction of the dilatation centers, with due regard for the elastic polarization of the defects in the elastic-isotropic approximation, is not equal to zero and decreases with the distance as r^{-6}. In some cases (for instance, for pores) this polarization interaction is considerable.

The interaction of point defects gives rise to various complexes. The formation and decomposition reactions of complexes can be considered by analogy to the general rules of reactions in gas mixtures or dilute solutions [5.1]. The simplest example is the formation reaction of bivacancies in metals $V + V \rightleftarrows V_2$. According to the mass action law the concentrations of vacancies $[V]$ and bivacancies $[V_2]$ are related by the condition

$$[V_2] = K[V]^2. \tag{5.16}$$

Reaction constant $K = \alpha \exp(U/kT)$, where α is the number of possible orientations of bivacancies in a crystal, which is equal to half the coordination number (i.e., the number of nearest neighbors for a given lattice site) and U is the binding energy of the vacancies in a bivacancy. This binding energy is usually appreciably less than the energy of formation of a single vacancy. Therefore, the concentration of bivacancies does not exceed that of single vacancies.

A further joining of vacancies may give either a three-dimensional formation (cavern, pore) or a plane (of atomic thickness − "pancake"), depending on the type of the crystal and the environmental conditions (Fig. 5.2a). The pancake may become unstable and collapses on attainment of a certain critical size. A defect of quite a different type − *a prysmatic dislocation loop* − arises (Fig. 5.2b). This lattice defect will be discussed in more detail in Sect. 5.3. Here, we shall only note that the T-shaped symbols in Fig. 5.2b indicate the breaking off of the atomic plane. Hereafter we use these symbols for schematic description of the dislocation structure of a crystal.

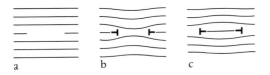

Fig. 5.2a−c. Formation of prismatic dislocation loops. (a) flat vacancy pancake; (b) vacancy loop; (c) interstitial loop

Passing on to interstitial atoms it should be mentioned that this term is not quite suitable. Interstitial atoms are far from being always settled in interstices characteristic of the lattice of a given crystal. In face-centered-cubic metals, for example, an interstitial atom is not incorporated into tetrahedral (1/4, 1/4, 1/4) or octahedral (1/2, 1/2, 1/2) interstices, but displaces some atom from its lattice site (see Fig. 5.3), forming a pair (*dumbbell*) with it oriented along one of the ⟨100⟩ directions. In bcc metals, interstitial dumbbells also arise, but in the ⟨110⟩ direction. Both in fcc and bcc metals the interstitial atoms can form configurations of another type, so-called *crowdions*. An example of a crowdion in a bcc metal is presented in Fig. 5.4. A single extra atom is located over a length of several interatomic distances in the ⟨110⟩

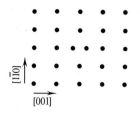

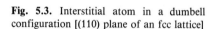

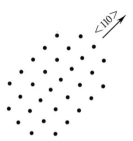

Fig. 5.3. Interstitial atom in a dumbell configuration [(110) plane of an fcc lattice]

Fig. 5.4. Crowdion along the $\langle 110 \rangle$ axis. The $\{001\}$ plane of an fcc lattice is shown

direction. Plane clusters of interstitial atoms (Fig. 5.2) form dislocation loops similar to those formed upon the collapse of plane clusters of vacancies (but of opposite sign).

In crystals consisting of atoms of several sorts, the atoms occupying lattice sites of a foreign sublattice may serve as point defects. An increase in the concentration of such defects corresponds to the initial stage of disordering.

5.2.2 Role of Impurities, Electrons, and Holes

Impurity (foreign) atoms, both incorporated into interstices and located at lattice sites, are themselves point defects of the lattice and affect the concentration of the intrinsic lattice defects considerably, sometimes increasing their concentration by several orders of magnitude. Thus, in pure crystals of NaCl the concentration of vacancies V_{Na} and V_{Cl} is determined from the mass action law for the reaction of the formation of Schottky defects

$$[V_{Na}][V_{Cl}] = A \exp(-U/kT). \tag{5.17}$$

From the condition of electric neutrality

$$[V_{Na}] = [V_{Cl}] = \sqrt{A} \exp(-U/2kT). \tag{5.18}$$

Comparing the above expression with (5.3) and considering that in a NaCl crystal the number of lattice sites for Na and for Cl is the same, i.e., $N_1 = N_2 = N$, we find that in our case $A = 1$.

In a NaCl crystal with a $CaCl_2$ impurity the Ca^{2+} ions replace Na^+ and must be additionally neutralized by cation vacancies V_{Na} (the formation of interstitial Cl^- ions is energetically less favourable). The condition of electric neutrality takes the form

$$[V_{Na}] = [V_{Cl}] + [Ca^{2+}],$$

and from (5.6) there follows for the concentration of cation vacancies

$$[V_{Na}] = \tfrac{1}{2}[Ca^{2+}] + \sqrt{\tfrac{1}{4}[Ca^{2+}] + A \exp(-U/kT)}. \tag{5.19}$$

As the temperature decreases, the concentration of $[V_{Na}]$ approaches that of the impurity, while the concentration of $[V_{Cl}]$ becomes lower than in pure crystals of NaCl. The effect of the impurity on the vacancy concentration becomes negligible only in the case of high temperatures and slightly doped crystals, when the concentration of Schottky defects exceeds that of the impurity. For heavily doped crystals, up to the melting point, the concentration of cation vacancies is equal to that of a divalent impurity, and anion vacancies are practically absent.

The change in concentration of anion and cation vacancies on doping sharply affects the electrical and mechanical properties of crystals. The ionic conductivity usually depends on one of the types of vacancies. Therefore the investigation into the temperature behaviour of the ionic conductivity of crystals helps to find the vacancy concentration and to determine, with the aid of equations of the type (5.19), the concentration of impurities of other valencies. During plastic deformation, dislocations sweep out charged vacancies and set up considerable volume and surface charges in the crystal (*A. V. Stepanov's effect*). These charges, in turn, exert the determining effect on deformation and fracture.

The neutralization of charged point defects can be achieved not only with the aid of other lattice defects, but also by using the corresponding disturbances in the electron structure of the crystal, predominantly electrons in the conduction band or holes in the valence band. In the former case, a charged point defect serves as an acceptor (the capture of an electron produces a hole in the valence band capable of migrating through the crystal), and in the latter, as a donor (an electron is added in the conduction band). In semiconductors, the value and type of conductivity can be controlled in this way.

In ionic crystals, complexes of point defects of electrons and holes form different color centers. The simplest example is an F center, which is an anion vacancy compensated by an electron smeared over all the surrounding cations. Schemes of the structure of some other color centers in ionic crystals will be discussed in [Ref. 1.7, Chap. 15].

5.2.3 Effect of External Influences

As a rule, point defects can occupy several equivalent configurations in the unit cell of a crystal. Under external influences defects gradually assume positions which have become energetically more favorable (for instance, the dumbbells of the interstitial atoms turn to lie along the extension axis). An induced anisotropy arises, which does not disappear at once upon termination of the external influence. The crystal "memorizes" the direction of the external influence for a certain time, responding with delay to each influence by the appropriate redistribution of the point defects. This is called the effect of *orientational* or *directed* ordering. It is also observed in internal friction (*Snoek effect*), elastic and magnetic aftereffect, stabilization of magnetic and ferroelectric domains (defects capable of reorienting help the domain to memorize the direction of polarization), and rubberlike elasticity of polysynthetic twins and martensitic alloys (on heating, a crystal spontaneously resumes the shape lost as a result of previous deformation).

External influences may lead not only to redistribution of point defects, but also to the appearance of new defects in concentrations greatly exceeding the thermodynamic equilibrium value. The nonequilibrium concentration of point defects can be attained (frozen) in the course of rapid quenching from a high temperature.

During the growth of crystals from the melt, nonequilibrium defects have no time to diffuse to the crystal surface and either settle on dislocations or precipitate (coagulate) in the crystal bulk, forming various clusters. In large dislocation-free silicon ingots, clusters of interstitial atoms are the main source of deterioration in crystals and devices based on them. The thermal stability and yield stress of such crystals directly depend on the size of prysmatic dislocation loops arising on the basis of interstitial clusters.

If a change in crystal temperature is accompanied by the decomposition of the solid solution, the impurity atoms leaving the lattice sites strongly increase the vacancy concentration (and those leaving interstices, the concentration of the interstitial atoms). The concentration of the vacancies (or the interstitial atoms) then increases until the decomposition of the solid solution ceases or until the point defects begin to coagulate. In a wide range of ionic crystals, the heating results in such a rapid increase in the number of cation Frenkel pairs that this process can be interpreted as a phase transformation with a shift of the cations to interstitial positions. The number of allowed interstitial positions usually greatly exceeds the number of ions in the unit cell, the cations are distributed statistically among these positions and move through the crystal very easily, thus ensuring a very high (10^{-2} Ω^{-1} cm^{-1} and higher) ionic conductivity. Both in the phase transformation and in the passage of the ion current, cooperative phenomena play a substantial role; the cations "help" each other to move to interstitial positions in the former case, and to migrate from one interstice to another in the latter. Typical examples of this class of crystals, which are referred to as *superionic* (or solid electrolytes) are zirconium oxide, silver iodide, silver chalcogenides, copper halides, sulphates of univalent metals, polyaluminates, and various compounds based on them.

In the course of plastic deformation, extra vacancies and interstitial atoms are produced by the intersection of dislocations, decomposition of dislocation dipoles, and nonconservative motion of dislocations [Ref. 1.8, Chap. 12]. Preferred formation of both vacancies (for instance, in torsion of plastic crystals) and of interstitial atoms (on uniaxial deformation at high stresses and moderate temperatures) may take place, depending on the deformation conditions. In the former case the activation energy of the process of plastic deformation may be reduced to the energy of vacancy migration, and in the latter it may be increased to the energy necessary for the formation of interstitial atoms.

A high concentration of point defects of various types can be achieved by irradiating a crystal with fast particles, as well as with x and gamma rays. In materials used in critical parts of nuclear reactors, up to $10^{16} - 10^{17}$ new Frenkel pairs per second appear in each cubic centimeter, i.e., each atom is knocked out of its lattice site more than once a week, on the average. Further, the interstitial atoms and vacancies recombine, settle on dislocations, and also coagulate with the formation of new dislocation loops and pores. Mass formation of pores results in *radiation swelling* of the material. The volume of the construction elements then changes by tens of per cent. The main cause of swelling is a small (about one per cent) difference in the interstitial and vacancy fluxes towards dislocations. Since interstitial atoms are stronger dilatation centers than vacancies, they are preferentially segregated on dislocations, whereas the vacancies are preferentially absorbed by pores. The swelling reduces if the interstitial atoms are captured by impurity atoms and have time to recombine with the vacancies and also if the pores form a regular lattice (usually an fcc lattice in fcc metals or a bcc lattice in bcc metals). When some anisotropic crystals, especially fissile ones, are irradiated, the crystals change their shape considerably (up to 1000%) without an appreciable change in their volume. This phenomenon, which is called the *radiation growth* of crystals, is due to the diffusion transport of atoms from planes with one orientation to planes with another. The dissolution and growth of atomic planes is achieved through the precipitation of vacancies and interstitial atoms, respectively, on dislocations. The selection of the direction of radiation growth depends on the anisotropy of 1) the elastic fields of differently oriented dislocations, 2) the elastic field of the cascades of radiation defects due to fast particles, and 3) the interaction of point defects with dislocations. The chemical composition of the material changes under irradiation because of nuclear reactions; transmutant atoms appear, which can be regarded as new point defects. By such *radiation doping* it is possible to control the amount of the donors and acceptors in semiconductors. Among other examples of defects of typically radiation origin, mention should be made of the *thermal peak* (the region of local superheating in which the atoms experience high-amplitude vibrations for some time) and the *displacement peak* (a small region with a completely disordered crystal lattice). These peaks may be the source of Frenkel defects, crowdions, pores, and dislocation loops.

5.3 Dislocations

Dislocations in crystals are specific linear defects disturbing the regular alternation of the atomic planes. In distinction to point defects, which disturb the short-range order, dislocations disturb the long-range order in a crystal, distorting its entire structure. A crystal with a regular lattice can be depicted as a family of parallel atomic planes (Fig. 5.5a). If one of the planes breaks off

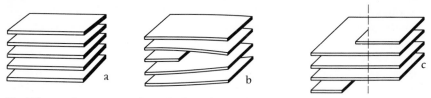

Fig. 5.5a – c. Arrangement of atomic planes in the perfect crystal (**a**), and in the crystal with an edge dislocation (**b**) and with a screw dislocation (**c**)

inside the crystal (Fig. 5.5b), its edge forms a linear defect, which is called an *edge dislocation*. Examples of edge dislocations arising on collapse of a vacancy pancake or on incorporation of a layer of interstitial atoms are given in Fig. 5.2. Figure 5.5c illustrates another simple type of dislocation, the *screw dislocation*. Here, none of the atomic planes terminates inside the crystal, but the planes themselves are only approximately parallel and merge with each other so that the crystal actually consists of a single helically bent atomic plane. On each circuit around the dislocation axis this "plane" ascends (or descends) by one screw pitch equal to the interplanar distance. The axis of the helical stair forms the dislocation line.

5.3.1 Burgers Circuit and Vector

The principal geometric characteristic of a dislocation is the so-called *Burgers vector*. To find the Burgers vector, one must compose, from translation vectors, a circuit so that it would close in the ideal lattice. Then this (Burgers) circuit, constructed around the dislocation line, is broken (Fig. 5.6). The translation vector drawn to close both ends of the circuit is called the Burgers vector of the dislocation. It can be ascertained that the Burgers vector is independent of the concrete choice of the Burgers circuit (it suffices to take into account that all the circuits which do not enclose the dislocation remain closed). In the case of an edge dislocation (Fig. 5.6a) the Burgers vector is per-

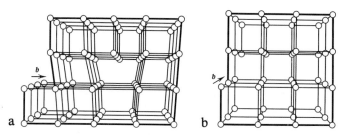

Fig. 5.6a, b. Scheme for determining the Burgers vector of dislocations. (**a**) Edge dislocation; (**b**) screw dislocation

pendicular to the dislocation line, and its length is equal to the extra inter-planar distance corresponding to the broken-off plane. In the case of a screw dislocation (Fig. 5.6b) the Burgers vector is parallel to the dislocation and equal in length to the screw pitch. At other values of the angle between the dis-location line and the Burgers vector, dislocations of mixed orientation are obtained. The choice of direction of the Burgers vector is conventional and depends on the choice of the direction of the dislocation and the direction of the scanning of the Burgers circuit. The circuit is usually assumed to be drawn in a clockwise direction (looking along the conventional direction of the dis-location), so that the dislocations in Fig. 5.6 should be assumed directed towards the reader. The screw dislocations shown in Figs. 5.6b can be called left-handed; the atomic planes around them turn as a left-handed screw and the Burgers vectors are antiparallel to the direction of the dislocations. Around right-handed dislocations, the atomic planes turn as a right-handed screw, while the Burgers vectors are parallel to the direction of the disloca-tions.

Generally, a dislocation is an arbitrary space curve around which the Burgers vector remains constant (and equal to some translation vector of the lattice), although the orientation of the dislocation may change. Thus, dis-location loop *ABCDEA* in Fig. 5.7 consists of an edge dislocation (more precisely, a segment with an edge orientation) *AB*, a right-handed dislocation *BC*, a mixed ("45-degree") dislocation *CD*, a left-handed dislocation *DE*, and a helicoidal dislocation *EA*.

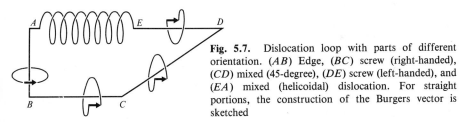

Fig. 5.7. Dislocation loop with parts of different orientation. (*AB*) Edge, (*BC*) screw (right-handed), (*CD*) mixed (45-degree), (*DE*) screw (left-handed), and (*EA*) mixed (helicoidal) dislocation. For straight portions, the construction of the Burgers vector is sketched

The condition for the conservation of the Burgers vector along the dislocation means that the dislocation may nor terminate or originate inside the crystal (particularly on an inclusion) and must either close on itself forming a loop, or emerge at the free surface, or branch out to other disloca-tions. In the last case the sum of the Burgers vectors of the dislocations after the branching must be equal to the Burgers vector of the initial dislocation. This can be ascertained by constructing a Burgers circuit around all the dislocations after the branching point (Fig. 5.8a). By analogy with the Kirchhoff theorem for branching lines of electric currents, the result can be formulated as follows: if we assume that all the dislocations are directed towards the branching point (node), the sum of their Burgers vectors must be

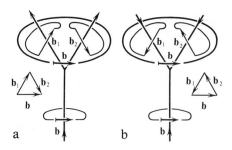

Fig. 5.8a, b. Dislocation reactions in a triple node. (a) Node as a point of dislocation branching; (b) node as a merger point of three dislocations

equal to zero (Fig. 5.8 b). The construction of Burgers circuits shows that the atoms of a crystal with a dislocation cannot be brought into one-to-one correspondence with the atoms of the perfect crystal so that the neighboring atoms of the ideal lattice also remain neighbors in the crystal with the dislocation. Only one point in the initial perfect crystal corresponds to the first and last atoms in the circuit constructed in the real crystal. If, however, we attempt to describe the structure of the real crystal by assigning vectors u which define the displacement of a given atom from a lattice point, we inevitably find that on travelling around the dislocation a misfit of the magnitude of the Burgers vector b is accumulated. Assuming the displacement field $u(r)$ to be continuous, we discover that the dislocations are branching lines of this field. In tracing any circuit around the dislocation, the displacement vector acquires an increment of magnitude of the Burgers vector

$$\oint du = -b . \tag{5.20}$$

In distinction to (5.5), tensor $u_{ij} = u_{j,i}$ is no longer eddy free and corresponds to the elastic-distortion tensor, which satisfies the following condition in accordance with (5.20):

$$\oint u_{ij} dx_i = -b_j , \tag{5.21}$$

i.e., the circulation of the elastic-distortion tensor around dislocations yields the Burgers vector. To eliminate the ambiguity in the description of the displacement field of the dislocation, one can construct any surface S bounded by the dislocation line and require that on this surface the displacement vector experiences a jump by the value of the Burgers vector. This condition is equivalent to introducing the proper distortion

$$u_{ij}^0 = b_j dS_i \tag{5.22}$$

localized on each elements dS of surface S, or the proper deformation

$$\varepsilon_{ij}^0 = (b_i dS_j + b_j dS_i)/2 . \tag{5.23}$$

The formation of a dislocation can now be represented as the result of the following operations:

a) in the perfect crystal, a cut on surface S is made;
b) the sides of the cut are displaced by Burgers vector b, and extra atoms are removed or additional atoms introduced in order to close up the cut sides, if necessary.

The displacement of the cut sides by the Burgers vector leads to discontinuity in the Burgers circuits and circulation of the elastic-distortion tensor (5.21).

Relations (5.20, 21) show that the role played by dislocations in the theory of elasticity is similar to the role of vortex lines in hydrodynamics or flux lines in magnitostatics: the field becomes complicated, and linear singularities arise, characterized by the curl corresponding to the velocity vector or the magnetic vector. The velocity and magnetic potentials become multivalued functions of the coordinates which branch round the vortex and flux lines.

In the case of dislocations the role of the potential is played by displacement vector u, which is found to be an ambiguous function of the coordinates and changes, according to (5.20), by the value of the Burgers vector on each tracing of a circuit drawn around the dislocation line. Determining the circulation of tensor u_{ij} according to (5.21), the dislocations serve as eddy lines of the elastic-distortion field. A peculiarity of dislocations in crystals is the discreteness of the set of possible values of the Burgers vector; from the very method for constructing the Burgers circuit it follows that the Burgers vector is equal to one of the translation vectors. In this respect dislocations in crystals are similar to quantized eddies in superfluid helium or quantized currents in type II superconductors, for which the circulation of velocity or magnetic field must be a multiple of Planck's constant.

5.3.2 Elastic Field of Straight Dislocation

From relation (5.20) it follows that the field of displacements around a straight dislocation contains the branching term

$$u_i = (b_i/2\pi)\theta, \tag{5.24}$$

which increases linearly with azimuth θ and is independent of the distance to the dislocation axis. For a screw dislocation in an elastic-isotropic medium, (5.24) completely describes the elastic dislocation field.

In a cylindrical system of coordinates (r, θ, z) with the z axis running along the dislocation, only the component

$$u_{\theta z} = b/2\pi r \tag{5.25}$$

is different from zero in the distortion tensor. Accordingly, in the stress tensor, only two components

$$\sigma_{z\theta} = \sigma_{\theta z} = Gb/2\pi r, \tag{5.26}$$

where G is the shear modulus, are different from zero. A similar calculation for an edge dislocation yields

$$
\begin{aligned}
\sigma_{rr} &= \sigma_{\theta\theta} = -Gb\sin\theta/2\pi(1-v)r, \\
\sigma_{r\theta} &= \sigma_{\theta r} = Gb\cos\theta/2\pi(1-v)r
\end{aligned}
\tag{5.27}
$$

where azimuth θ is measured from the Burgers vector in the direction of the extra half plane. As follows from (5.26, 27), the stresses increase approaching the dislocation axis as r^{-1} and reach values of the order of 10^{-1} G at a distance of the order of the lattice spacing. To calculate the stress field of a straight dislocation in an anisotropic medium we expand the field of the proper deformations of the dislocation into a two-dimensional Fourier series and use (5.11). In a cylindrical system of coordinates (r, θ, z), whose axis coincides with the dislocation line, we introduce unit vectors n and $m = \partial n/\partial\theta$ corresponding to axes r and θ and express proper distortion (5.22) as

$$
u_{ij}^0(x) = b_j m_i \delta(m_k x_k)(n_l x_l)/2 |n_l x_l|.
\tag{5.28}
$$

Relation (5.28) implies that the cut has been made along a plane with the normal m, and the cut sides are displaced by half the Burgers vector; the positive (with respect to the azimuth) cut side is displaced by $b/2$ along the half plane lying on the side of the positive direction of unit vector n, and the remaining part of the cut by $b/2$ in the opposite direction. For wave vector $k = kn$, the Fourier component of field (5.28) has the form

$$
\tilde{u}_{ij}^0(k) = -\frac{1}{(2\pi)^2}\frac{b_j m_i}{ik}.
\tag{5.29}
$$

Because of the arbitrary orientation of the cut, it can be assumed that all the Fourier components $u_{ij}^0(x)$ have the form (5.29). Considering the symmetry of tensor c_{ijkl}^* as regards the interchange of subscripts k and l, we substitute into (5.11) the proper distortion u_{lk}^0 for the proper deformation ε_{kl}^0. For the Fourier components of the stress tensor we obtain

$$
\tilde{\sigma}_{ij}(k) = \frac{1}{(2\pi)^2}\frac{c_{ijkl}^*(n)b_k m_l}{ik}.
\tag{5.30}
$$

The inverse Fourier transformation yields

$$
\sigma_{ij}(r) = \frac{b_k}{(2\pi)^2}\int_0^\infty dk\int_0^{2\pi} d\theta\, m_l c_{ijkl}^*(n)\sin(krn)
\tag{5.31}
$$

$$
= \frac{b_k}{(2\pi)^2 r}\sum_{n=0}^\infty (-1)^n\int_0^{2\pi} d\theta\, m_l c_{ijkl}^*(n)\cos(2n+1)\theta
\tag{5.32a}
$$

$$= \frac{b_k}{(2\pi)^2} \int_0^\infty d\theta \frac{m_l c_{ijkl}^*(n)}{(nr)} . \tag{5.32b}$$

The form (5.32a) is obtained by expanding $\sin(krn)$ into a series with respect to the Bessel functions and by term-by-term integration of this series. The form (5.32b) corresponds to the conventional replacement

$$\int_0^\infty \sin(krn)dk = (rn)^{-1}$$

which holds true for a sufficiently smooth integrand.

In the elastic-isotropic case

$$c_{ijkl}^*(n) = G\left(\frac{2v}{1-v}\delta_{ij} + \delta_{ik}\delta_{jl} + \delta_{il}\delta_{jk}\right) \tag{5.33}$$

where, with one of the axes chosen parallel to unit vector n, subscripts i, j, k, l run through only two values corresponding to unit vectors orthogonal to unit vector n. As a result, convolution $m_l c_{ijkl}^*$ retains only the following three components of the plane tensor of elastic moduli:

$$c_{zmzm}^* = G; \qquad c_{mmmm}^* = 2(1-v)^{-1}G; \qquad c_{zzmm}^* = 2v(1-v)^{-1}G .$$

Substituting these equations into (5.32) we can obtain the familiar equations (5.26, 27) for a screw and an edge dislocation in an elastic-isotropic medium.

Since the stresses decrease with distance from the dislocation as r^{-1}, the energy of the elastic dislocation field enclosed in a hollow cylinder with an outer radius R and an inner radius r_0 depends logarithmically on R and r_0,

$$E = E_0 \ln(R/r_0) . \tag{5.34}$$

Orientation factor E_0, which depends on the Burgers vector, the dislocation orientation, and the elastic properties of the crystal can be calculated with the aid of (5.32).

For the components of the stress tensor acting on plane P, which passes through the dislocation line, (5.32) must be multiplied by normal p_j. Projecting unit vector p onto the directions of unit vectors n and m and taking into account that the convolution of tensor $c_{ijkl}^*(n)$ with vector n with respect to any subscript gives zero and that $(nr) = (mp)(qr)$ if radius vector r and unit vector q, orthogonal to the dislocation, lie in plane P, we obtain from (5.32b)

$$p_j\sigma_{ij}(r) = (qr)^{-1}b_j B_{ij} \tag{5.35}$$

where

$$B_{ij} = \frac{1}{(2\pi)^2} \oint d\theta\, m_k c^*_{iklj} m_l .$$ (5.36)

Since the dislocation energy is determined by the work of stresses (5.32) on proper deformations (5.23) arising on the formation of a dislocation,

$$E_0 = b_i p_j \sigma_{ij}(q)/2 ,$$ (5.37)

whence it follows, with due regard for (5.35), that

$$E_0 = b_i B_{ij} b_j /2 .$$ (5.38)

The calculation of the interaction energy of two parallel dislocations is quite similar to that for the proper dislocation energy. Suppose the former has a Burgers vector $b^{(I)}$, and the latter, a Burgers vector $b^{(II)}$. To calculate the stress field $\sigma^I_{ij}(x)$ of the first dislocation we use (5.35), replacing b_j by $b_j^{(I)}$. The orientation factor E_{12} of the interaction energy of the dislocations is numerically equal to the interaction force of dislocations situated at a unit distance and can be found by the following expression, similar to (5.37):

$$E_{12} = b_i^{(II)} p_j \sigma_{ij}^{(I)}(q) ,$$ (5.39)

whence it follows, with an allowance for (5.35), that

$$E_{12} = E_{21} = b_i^{(I)} B_{ij} b_j^{(II)}.$$ (5.40)

By summing the stresses caused by individual dislocations we can construct the stress field induced by dislocation ensembles. Thus, for a series of parallel dislocations lying in a common plane P, the stress tensor components acting on this plane are given, according to (5.35), by the sum

$$p_i \sigma_{ij}(x) = \sum_n \frac{b_i^{(n)} B_{ij}}{x - x_n} ,$$ (5.41)

where $x = (qr)$ is the distance in this plane in the direction perpendicular to dislocations, and $b^{(N)}$ and $x^{(N)}$ are the Burgers vector and the coordinate of the Nth dislocation, respectively. Expression (5.40) serves as the basis for the theory of dislocation pile-ups, thin twins, plane cracks, and other stress sources which consist of rows of parallel dislocations or can be modelled by such rows (see Chap. 12). If the parallel dislocations are closely spaced, their distribution can be described macroscopically by assigning, not the coordinates of the individual dislocations, but their distribution density $\rho(x)$ on the given portion of plane $P = (x, y, 0)$. Then sum (5.41) turns into an integral, and the dislocation equilibrium equation into an integral equation. The dislocation density $\rho(x)$ makes it possible to directly judge the difference between the macroscopic deformations and stresses on the two sides of plane P.

Denoting the difference by brackets, we have for elastic distortions, by (5.21),

$$[u_{ij}] = \rho(x)q_i b_j \qquad (i,j = 1,2) . \tag{5.42}$$

Hence, for the stress difference

$$[\sigma_{ij}] = \rho(x)c^*_{ijkl}(p)q_k b_l , \qquad (i,j = 1,2,3) . \tag{5.43}$$

Equation (5.43) describes, in particular, the stress bands framing the misfit grain boundaries, slip bands, twin boundaries, growth bands, and other similar crystal defects.

Quite unexpectedly, (5.43) proves useful in calculating the interaction force of nonparallel (crossing) dislocations. Let us consider two crossing straight dislocations of Burgers vectors b^I and b^{II} along unit vectors $\tau^{(I)}$ and $\tau^{(II)}$, which make a certain angle θ with each other and are parallel to plane P. To calculate the interaction force of these dislocations, we take into account the translation symmetry of the problem and note that the addition to the first dislocation of another N-1 dislocations of the same type and disposed parallel to the first one in the common plane with normal P increases the sought-for force F, acting on the second dislocation, exactly by a factor of N. Passing to the continuous distribution of dislocations of the first type with density ρ, we ascertain that a unit length of the second dislocation will be acted upon by a force $f = \rho F \sin \theta$. On the other hand, according to (5.43), the second dislocation lies, in this limiting case, in the uniform stress field

$$\sigma_{ij}^{(I)} = [\sigma_{ij}^{(I)}]/2 = \rho c^*_{ijkl}q^I_k b^{(I)}_l/2 \tag{5.44}$$

and experiences the following force directed along unit vector p:

$$f = b_i^{(II)}q_j^{II}\sigma_{ij}^I . \tag{5.45}$$

A comparison of the expressions obtained for f yields the formula

$$F = (2\sin\theta)^{-1}b_i^{(I)}q_j^{(I)}c^*_{ijkl}(p)q_k^{(II)}b_l^{(I)} \tag{5.46}$$

which transforms, in the elastic-isotropic case (5.33) into the Kroupa equation

$$F = (2\sin\theta)^{-1}G\left\{\frac{2v}{1-v}(q^{(I)}b^{(II)})(q^{(II)}b^{(II)}) + (q^{(I)}b^{(II)})(q^{(II)}b^{(I)})\right.$$

$$\left. + (q^{(I)}q^{(II)})[(b^{(I)}b^{(II)}) - (b^{(I)}p)(b^{(II)}p)]\right\}, \tag{5.47}$$

derived by integrating the interaction force along the dislocation. The plus sign in (5.46, 47) corresponds to the repulsion, and the minus sign, to the attraction of dislocations. The magnitude of the force is independent of the distance between the dislocations, which could be predicted before the calculations of similar considerations. Repulsion of crossing dislocations plays the determining role in the effects of elastic interaction of gliding dislocations with so-called forest dislocations, which cross the slip plane at different angles and interfere with plastic deformation. In the case of attraction, the elastic interaction stimulates the drawing of dislocations closer together, the formation of dislocation nodes, and the progress of dislocation reactions.

According to (5.34, 38) the elastic energy per unit length of dislocation is proportional to the product of the elastic modulus by the square of the Burgers vector and depends logarithmically on the inner and outer truncation radii r_0 and R. The inner radius corresponds to a region of the order of several interatomic distances, which is called the dislocation core, where the lattice distortions are large and cannot be described in terms of the theory of elasticity. The energy density of the dislocation core may reach a value of the same order as the latent heat of fusion of the crystal; therefore, sometimes dislocations with a hollow core are formed. The dislocation energy is made up of the energy of inelastic distortions in the dislocation core and the energy of elastic deformations around the dislocation. As a rule, the basic contribution comes from the second term, which can be estimated by assuming the value of R in (5.34) to be of the order of the crystal radius, and the value of r_0, of the order of the radius of the dislocation core (it is usually taken that $r_0 \sim b$). The dislocation energy is about $0.5 \, Gb^2$ per centimeter of dislocation length, which corresponds to one or several electronvolts per interatomic distance.

5.3.3 Dislocation Reactions

As dislocation energy is proportional to b^2, one can formulate a simple rule for evaluating the energetic advantage of the formation of different types of dislocations. Suppose, for instance, that a dislocation has a Burgers vector $(b_1 + b_2)$, where b_1 and b_2 make an acute angle, i.e., $(b_1 b_2) > 0$. Then $(b_1 + b_2)^2 > b_1^2 + b_2^2$, and (neglecting the effect of the elastic anisotropy on the dislocation energy) the total energy of two dislocations with Burgers vectors b_1 and b_2 turns out to be lower than the energy of the initial dislocation. In particular, dislocations with a large Burgers vector find it energetically favorable to split into several unit dislocations with Burgers vectors equal to one of the minimum translation vectors. Although crystal growth is sometimes accompanied by the appearance of giant dislocations with a Burgers vector equal to several tens and hundreds of lattice spacings, the main bulk of dislocations in crystals consists of dislocations of unit strength, and the specific dislocation structure of a given crystal strongly depend on the set of minimum translation vectors. In crystals of the type, CsI, for instance, the minimum vectors of translation $a \langle 100 \rangle$ are arranged along the edges of the unit cube. It is easy to see that in these crystals dislocations cannot form triple nodes of the type depicted in Fig. 5.8: dislocation reactions (branching and dislocation fusions) can proceed only with the participation of at least four dislocations according to a scheme of the type $b_1 + b_2 \rightleftarrows b_3 + b_4$. In face-centered cubic crystals and in crystals with a diamond lattice the minimum vectors of translation $a/2 \langle 110 \rangle$ form the tetrahedron shown in Fig. 5.9 (the Thompson tetrahedron). The edges of each face of this tetrahedron correspond to a reaction of the type

$$(a/2)[1\bar{1}0] + a/2[011] \rightleftarrows (a/2)[101]$$

(for instance, $AB + BC = AC$ for face δ), which ensures the possibility of the appearance of a dislocation net with triple points, and also of two dislocations merging into one. The energetic advantage of such a reaction follows from the calculation of the sum of the squares of the Burgers vectors: before the reaction $b_1^2 + b_2^2 = a^2/2 + a^2/2 = a^2$, and after it $b_3^2 = a^2/2$, which corresponds to halving the energy.

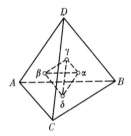

Fig. 5.9. Thomson tetrahedron. The tetrahedron edges correspond to the minimum translation vectors $a/2 \langle 110 \rangle$. Points α, β, γ and δ lie at the centers of the tetrahedron faces

By taking into account the effect of the elastic anisotropy on the dislocation energy one can carry out a more comprehensive analysis of the dislocation structure of the crystal. By calculating the orientation factor for different directions of the dislocation it is possible to construct the indicating surface characterizing the energetic advantage of differently oriented dislocations. As in the analysis of the anisotropy of the surface energy with the aid of the Wulff diagram [Ref. 5,1, Chap. 9], it is convenient here to construct the indicating surface for the value E_0^{-1}. Then the surface points farthest from the center define the most favorable orientations of the dislocation in an infinite crystal. The concave parts of the surface correspond to energetically unstable orientations. The dislocations disposed along these directions acquire a zigzag shape, and the orientation of the straight segments making up the zigzags corresponds to the ends of the concave parts (more precisely, to the points lying on the common tangent to indicating surface E_0^{-1}). It can also be shown that the equilibrium orientation of dislocations intersecting the surface of the growing crystal corresponds to those points of the surface E_0^{-1} which are determined by the normal parallel to that of growth surface, and during the crystal growth the angular dislocations can only arise on faces whose orientation corresponds to the concave parts of the surface E_0^{-1}.

The construction of indicating surfaces corresponding to different components of tensor B_{ij} (5.36) enables one to investigate the energetic advantage of such dislocation configurations which simultaneously involve dislocations with different Burgers vectors. Thus, two dislocation rays with Burgers vectors $b^{(1)}$ and $b^{(2)}$ having their origin in a common dislocation node do not affect one another if, and only if, their orientations correspond to points with a common tangent on the indicating surface $[b_i^{(1)} B_{ij} b_j^{(2)}]^{-1}$.

5.3.4 Polygonal Dislocations

The above statements concerning equilibrium configurations of dislocation nodes, angular dislocations, and dislocations intersecting the free surface of the crystal are based on investigations of the relevant three-dimensional elastic fields. At the vertex of an angular dislocation, at the node, or at the point of emergence of the dislocation at the free surface a specific feature of an elastic field arises: along the rays issuing from the pole, the stresses and deformations are inversely proportional to the distance. Accordingly, the forces F of interaction and self-action of the dislocations also vary as r^{-1}, and the moments of these forces with respect to the pole

$$m = r \times F \tag{5.48}$$

are independent of the distance and are determined exclusively by the orientation of the dislocation ray. This permits us to dispense with unwieldy calculation of the intricate three-dimensional field and to investigate directly the orientational dependence of the interaction and self-action moments of the dislocation rays. The most effective method for solving such problems, based on taking into account the similarity and symmetry of elastic fields induced by differently oriented dislocations, is called the straight-dislocation technique. Regarding dislocations of any configuration we only have to know that the small element of the dislocation line, when viewed from point M at angle $d\varphi$, makes the following contribution to the component u, under investigation, of the stress tensor or the distortion tensor

$$du(M) = \Phi\left(\frac{r}{r}\right)\frac{d\varphi}{r} \tag{5.49}$$

where r is the radius vector of the dislocation element, which is measured from observation point M. Orientation factor $\Phi(r/r)$ depends on the crystal anisotropy and can be described by an arbitrary antisymmetric directional function, i.e., $\Phi(r/r) = -\Phi(-r/r)$. Two important conclusions follow from (5.49)

a) the dislocation segments lying along the beams issuing from point M make no contribution to $u(M)$; and

b) the summation of the contribution of dislocations lying in one plane P and possessing an identical Burgers vector b amounts to the summation of reciprocal distances r^{-1}, and all the dislocations $r_{(i)}^{-1}(\varphi)$, where i is the dislocation number and φ is the azimuth in a polar system of coordinates with center at the observation point, can be replaced by a single dislocation whose shape is given by

$$r(\varphi) = [\sum_i r_{(i)}^{-1}(\varphi)]^{-1}. \tag{5.50}$$

When summing the reciprocal distances, it is necessary to ascribe to function $r_{(i)}^{-1}(\varphi)$ a sign depending on that of the dislocation, assuming, for instance, the elements directed towards increasing azimuth φ to be positive.

Considering that the equation of a straight line in polar coordinates has the form

$$r^{-1} = \lambda^{-1} \cos(\varphi - \varphi_0) \tag{5.51}$$

and that the summation of expressions of the type (5.51) again yields an expression of the same type, we conclude that the equivalent source of stresses for polygonal dislocations is a polygonal dislocation. According to (5.50) the form of this dislocation is defined by the following simple rules:

Rule 1. All the nonintersecting segments remain unmoved.

Rule 2. The point of intersection of like (with respect to their direction) dislocations approaches the observation point by half the distance.

Rule 3. The point of intersection of unlike dislocations recedes to infinity.

Using these rules, it is possible to express the elastic field of polygonal dislocations lying in a common plane P via the field of straight dislocations lying in the same plane. (Observation point M must then also lie in plane P). Let us consider, for instance, an antisymmetric dislocation cross consisting of two semiinfinite rays issuing from point O at azimuths α_1 and $\pi + \alpha_1$, and two rays entering point O at azimuths α_2 and $\pi + \alpha_2$. For observation point M determined by radius vector $r = OM$ with azimuth φ the elastic field of the antisymmetric cross is equivalent to the field of a single straight dislocation A with azimuth φ located at a distance $d = r \sin(\varphi - \alpha_1) \sin(\varphi - \alpha_2) \times$ cosec $(\alpha_1 - \alpha_2)$. If we add to the antisymmetric cross two straight dislocations B and C with azimuths α_1 and α_2 passing through pole O, the cross will turn into an angular dislocation with a doubled Burgers vector. The elastic field of this dislocation at observation point M will coincide with the sum of the fields of three straight dislocations A, B, and C with a single Burgers vector. With the aid of (5.35) the obtained result for the field of the angular dislocation can be written as follows:

$$p_j \sigma_{ij}(r, \varphi) = \frac{b_j}{2r} \left[\frac{B_{ij}(\alpha_1)}{\sin(\varphi - \alpha_1)} + \frac{B_{ij}(\alpha_2)}{\sin(\varphi - \alpha_2)} \right.$$
$$\left. + \frac{B_{ij}(\varphi) \sin(\alpha_2 - \alpha_1)}{\sin(\varphi - \alpha_1) \sin(\varphi - \alpha_2)} \right], \tag{5.52}$$

where $B_{ij}(\alpha)$ stands for tensor (5.36) of a dislocation lying in plane P at azimuth α.

A plane dislocation node consisting of n rays issuing from the common pole O at azimuths α_k which possess Burgers vectors $b^{(k)}$ can be composed of n angular dislocations with Burgers vectors $b^{(k)}$. Performing the appropriate summation of expressions of the type (5.52) with due regard for the vanishing of the sum of all the vectors $b^{(k)}$, we obtain

$$p_j \sigma_{ij}(r, \varphi) = \frac{1}{2r} \sum_{k=1}^{n} b_j^{(k)} \frac{B_{ij}(\alpha_k) + B_{ij}(\varphi) \cos(\varphi - \alpha_k)}{\sin(\varphi - \alpha_k)}. \tag{5.53}$$

In a similar way, a polygonal loop can be composed from angular dislocations. If the loop apices are specified by radius vectors $r^{(k)}$ with azimuths φ_k, the summing of expressions (5.52) gives for stresses at the origin

$$p_i \sigma_{ij} = \frac{b_j}{2} \sum_{k=1}^{n} \frac{B_{ij}(\varphi_k)}{r^{(k)}} \frac{\sin(\alpha_k^+ - \alpha_k^-)}{\sin(\varphi_k - \alpha_k^+) \sin(\varphi_k - \alpha_k^-)} \tag{5.54}$$

where α_k^+ and α_k^- are the azimuths of the polygon sides which adjoin the kth vertex (transition from α_k^- to α_k^+ is performed in the direction of increasing azimuth).

The calculation of the interaction and self-action forces of the dislocation requires a limiting transition $\varphi \to \alpha_k$ for convolutions of the type $b_i \sigma_{ij} p_j$. Then in the formulae the expressions of the type (5.38, 40) appear as well as the derivatives with respect to dislocation direction. For the moment acting on the ith ray of the dislocation bundle from the side of the jth ray we get from (5.52)

$$M^{(i)(j)} = \frac{1}{2} \left[E_{(i)(j)}(\alpha_j) \operatorname{cosec}(\alpha_i - \alpha_j) + E_{(i)(j)}(\alpha_i) \operatorname{ctg}(\alpha_i - \alpha_j) \right.$$
$$\left. - \frac{\partial}{\partial \alpha} E_{(i)(j)}(\alpha_i) \right] \tag{5.55}$$

where $E_{(i)(j)}$ is the orientation factor (5.40) for the interaction of parallel dislocations with Burgers vectors $b^{(i)}$ and $b^{(j)}$. Moment (5.55) reduces to zero if the following condition is met:

$$E_{(i)(j)}(\alpha_j) = \sin(\alpha_i - \alpha_j) \frac{\partial}{\partial \alpha_i} \left[\operatorname{cosec}(\alpha_i - \alpha_j) E_{(i)(j)}(\alpha_i) \right]. \tag{5.56}$$

The tangent to polar diagram $E_{(i)(j)}^{-1}(\alpha)$, constructed at the point with azimuths α_i, intersects, subject to condition (5.56), the diagram at the point with azimuth α_j. Hence it is clear that the ith and jth rays do not interact only when a common tangent to diagram $E_{(i)(j)}^{-1}$ passes through the points with azimuth α_i and α_j. In the particular case $n = 2$, the so-called first Lothe theorem on the moment of interaction of the arms of an angular dislocations follows from (5.55): the arm with azimuth α_1 is acted upon by the moment

$$M_{12} = E_0(\alpha_2)\operatorname{cosec}(\alpha_1 - \alpha_2) - E_0(\alpha_1)\operatorname{ctg}(\alpha_1 - \alpha_2) + \frac{\partial}{\partial\alpha}E_0(\alpha_1). \quad (5.57)$$

Since the equilibrium of the two arms of an angular dislocation requires the fulfillment of condition $M_{12} = M_{21} = 0$, corner points on the dislocation arise if, and only if, there is a well on the indicating surface $E_0^{-1}(\alpha)$. The equilibrium orientation of the arms of an angular dislocation corresponds to the azimuths of the points lying on the common tangent to diagram $E_0^{-1}(\alpha)$. The sectors of forbidden dislocation orientations lie between these points in diagram $E_0^{-1}(\alpha)$. Dislocations whose average azimuth occurs in such a sector acquire a zigzag shape, so that the orientation of the arms of the angular dislocations corresponds to the boundaries of the sectors of the forbidden orientations. For a large number of crystals of different symmetry, the orientational factors $E_0(\alpha)$ have been calculated both analytically and numerically, the sectors of forbidden orientations have been determined, and good agreement of theory with experimental data has been obtained on zigzag and polygonal dislocations with stable corner points.

For a dislocation parallel to the crystal surface the stress relaxation on this surface gives rise to a force pushing the dislocation out of the crystal. The force is found to be exactly equal to that of interaction of the given dislocation with the "image dislocation" obtained by mirror mapping of the dislocation from the surface. Using (5.38), we obtain for the image force per unit length of the dislocation

$$F = (2d)^{-1}b_i B_{ij}b_j = d^{-1}E_0 \qquad (5.58)$$

where d is the distance to the surface. The analog of force (5.58) is the interaction force of the dislocation with a grain, phase, or twin boundary. For a dislocation spaced distance d from the boundary of an anisotropic bicrystal,

$$F = d^{-1}(E_0 - E_s) \qquad (5.59)$$

where E_0 is orientation factor (5.38) for a dislocation in an unbounded medium with the same elastic moduli as in a semiinfinite medium containing the dislocation, and E_s is the orientation factor of an identical dislocation placed on the interface. At $E_s = 0$, (5.58) follows from (5.59) as a particular case. These interaction forces of dislocations with interfaces manifest themselves most prominently in whiskers and thin films, where they may reach values of several per cent of Gb and lead to effectively pushing the dislocations out of the crystal or, conversely, pulling them into the crystal.

If a dislocation intersects the free surface of a crystal, the stress relaxation on the surface gives rise to an orienting moment

$$M = \cos \alpha \frac{\partial}{\partial \alpha} [E_0(\alpha) \sec \alpha] \tag{5.60}$$

where azimuth α is measured from the normal to the surface (second Lothe theorem). If the self-energy of the dislocation is independent of the direction, moment (5.60) orients the dislocation perpendicularly to the surface. Equilibrium orientations of dislocations are generally found by drawing tangent planes to indicating surface $E_0^{-1}(\alpha)$. Indeed, the vanishing of moment (5.60) corresponds to the minimum of linear energy $E_0(\alpha) \cdot \sec \alpha$ per crystal layer of unit thickness, or the maximum of quantity $E_0^{-1}(\alpha) \cdot \cos \alpha$ corresponding to the maximum distance from the center of polar diagram $E_0^{-1}(\alpha)$ to a plane parallel to the crystal surface. The reliability of the theoretical prediction of the equilibrium orientations of dislocation rays arising during the synthesis of crystals was checked most thoroughly in practice by comparing data on the orientation of dislocations as obtained with the aid of x-ray diffraction topography with the calculation results. Discrepancy between theory and experiment was observed only when surface $E_0^{-1}(\alpha)$ had flattened areas and, hence, taking into account the elastic part of the dislocation energy did not ensure the finding of energetically favorable orientations. Usually, deviations from theory can be explained by the effect of the orientation on the energy of the dislocation core.

Inspection of dislocations in nonplastic crystals (quartz, calcite, fluorite, topaz, diamond, etc.) shows that nucleation of dislocations during crystal growth usually occurs on small inclusions, and that the dislocations acquire a V shape. The apex of the V accommodates the inclusion, the sides of the V form dislocation rays terminating on the growth surface. It is easy to see that the configuration described requires that (5.60) should have two solutions $M(\alpha) = 0$ corresponding to different azimuths and, consequently, diagram $E_0^{-1}(\alpha)$ must have two points with tangents of the same orientation. Since this is impossible if surface $E_0^{-1}(\alpha)$ is convex, it can be concluded that V-shaped dislocations arise only when indicating surface $E_0^{-1}(\alpha)$ contains concave portions, and they form at those orientations of the growth surface which correspond to the sectors of forbidden orientations of dislocations. In the case of quartz and calcite this prediction of anisotropic dislocation theory has been confirmed by experiment.

5.3.5 Curved Dislocations

The diagram technique for polygonal dislocations used in the preceding section can be generalized for the case of curved dislocations. Although (5.54) can be used directly for numerical calculations of the elastic field of a plane dislocation loop, the construction of the field of an elementary source (5.49) is of definite interest. We proceed from expression (5.52) for the angular-dislocation field and take the limit at $\varphi \to \alpha_2$. Putting $\alpha_1 = \alpha$ and $\alpha_2 = \varphi$, we have

$$p_j \sigma_{ij}(r, \varphi) = - (b_j/2d) \left\{ B_{ij}(\alpha) + B_{ij}(\varphi) \cos (\varphi - \alpha) \right.$$
$$\left. - \left[\frac{\partial}{\partial \varphi} B_{ij}(\varphi) \right] \sin (\varphi - \alpha) \right\} \tag{5.61}$$

where $d = r \sin (\alpha - \varphi)$ is the distance from the observation point to a arm with azimuth $\alpha = \alpha_1$. Field (5.61) can be interpreted as the field of a dislocation arm with azimuth α, since the second arm of the angular dislocation, which is directed along the radius vector of the observation point, does not, by (5.49), make any contribution to the elastic field at the observation point. Elementary source (5.49) corresponds to the difference of two proximate angular dislocations so arranged that their vertices lie on a common arm with azimuth α, while the second arms, when extended, pass through the observation point. The field of this source is found by differentiating (5.61) with respect to azimuth φ, which yields

$$p_j \frac{\partial}{\partial \varphi} \sigma_{ij}(r, \varphi) = - (b_j/2r) \left[B_{ij}(\varphi) + \frac{\partial^2}{\partial \varphi^2} B_{ij}(\varphi) \right]. \tag{5.62}$$

Comparing (5.62) and (5.49) and taking into account that $d\varphi$ in (5.49) corresponds to the change in the azimuth of a point lying on the dislocation, rather than in the azimuth of the observation point, we obtain for the orientation factor of the elementary source

$$\Phi(\varphi) = (b_j/2) \left[B_{ij}(\varphi) + \frac{\partial^2}{\partial \varphi^2} B_{ij}(\varphi) \right]. \tag{5.63}$$

Hence, for the field of a plane closed loop considered in the plane of the loop, the following Brown formula follows:

$$p_j \sigma_{ij}(x) = (b_j/2) \oint \frac{d\varphi}{r} \left[B_{ij}(\varphi) + \frac{\partial^2}{\partial \varphi^2} B_{ij}(\varphi) \right] \tag{5.64}$$

where $r = |x - x'|$ is the distance from the observation point x to point x' running through the loop contour, and φ is the azimuth of the radius vector $r = x - x'$. In particular, for the self-action force follows

$$F(x) = b_i \sigma_{ij} p_j = \oint \frac{d\varphi}{r} \left[E_0(\varphi) + \frac{\partial^2}{\partial \varphi^2} E_0(\varphi) \right] \tag{5.65}$$

where point x is located on the loop contour. It is significant that the elastic field of the loop (5.64) is built up of the fields of dislocations disposed, not along the tangents to the loop, but along the secants passing through the observation point.

The results obtained can be generalized for the three-dimensional case by passing from the consideration of stresses (5.35) acting on the plane containing the dislocation line to the study of all the components of the elastic field of the dislocation and, accordingly, from the derivatives with respect to the azimuth in a fixed plane to those with respect to directions along arbitrarily oriented vectors. In place of (5.61) we have for the dislocation ray

$$u(x) = \frac{1}{2} [u(x, \tau) - \tau_\alpha \frac{\partial}{\partial x_\alpha} u(\tau, x)] . \tag{5.66}$$

Here, $u(x)$ is any component of the stress, deformation, or distortion tensor; radius vector x is measured from the apex of the angular dislocation to the observation point; $u(x, \tau)$ is the field of the dislocation passing through the origin in the direction of vector τ. From (5.66) it follows for an arbitrary (not necessarily plane) polygonal loop

$$u(x) = \frac{1}{2} \sum_{k=1}^{n} (x_\alpha^{(k+1)} - x_\alpha^{(k)}) \frac{\partial}{\partial x_\alpha} [u(x^{(k+1)} - x^{(k)}, x - x^{(k+1)})$$
$$- u(x^{(k+1)} - x^{(k)}, x - x^{(k)})] . \tag{5.67}$$

The limiting transition to a curved space loop gives

$$u(x) = \frac{1}{2} \oint \tau_\alpha \tau_\beta \frac{\partial^2}{\partial x_\alpha \partial x_\beta} u(\tau, x' - x) ds , \tag{5.68}$$

where $ds(x')$ is an element of the dislocation loop disposed along unit vector $\tau(x')$. Accordingly, the orientation factor of the elementary source has the following form instead of (5.63):

$$\Phi(r) = \frac{1}{2} \tau_\alpha \tau_\beta \frac{\partial^2}{\partial x_\alpha \partial x_\beta} u(\tau, r) . \tag{5.69}$$

Assuming $u \equiv \sigma_{ij}$, we obtain the following general formula for the stress field of an arbitrary dislocation loop from (5.68)

$$\sigma_{ij}(x) = \frac{1}{2} \oint \tau_\alpha \tau_\beta \frac{\partial^2}{\partial x_\alpha \partial x_\beta} \sigma_{ij}(\tau, x' - x) ds \qquad (5.70)$$

which is called the Indenbom – Orlov equation.

5.4 Stacking Faults and Partial Dislocations

The ordinary (perfect) dislocations with a Burgers vector equal to that of lattice translation, which were discussed in Sect. 5.3, are purely linear defects. The atomic structure of the crystal around them in any local area corresponds to an elastically distorted structure of the perfect crystal. Quite a different type of dislocation is formed by the boundaries of stacking faults terminating inside the crystal. Imagine the structure of a monatomic close-packed crystal in the form of the successive layers of spheres (Fig. 5.10). Each next layer can be stacked on the substrate, where the spheres occupy positions A, in two ways B and C. Layer B can be covered by A or C, and layer C by A or B. The face-centered-cubic structure corresponds to the sequence $...ABCABC...$, and the hexagonal close-packed structure to $...ABABAB....$.

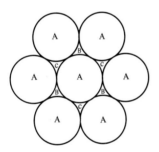

Fig. 5.10. Close packing of atoms in cubic and hexagonal crystals

By reversing the normal alternation of atomic planes we obtain a mirror-symmetric configuration, i.e., a twin. Thus, sequence $...ABC{\downarrow}BAC...$ corresponds to the twin boundary along (111) in the face-centered-cubic lattice (the arrow indicates the position of the boundary). In the case of a twin interlayer the sequence of stacking at the second interlayer boundary is reversed again.

In the limiting case of a single-layer twin of the type $...ABC{\downarrow}B{\uparrow}CAB...$ we have only one irregularly stacked layer, which is an intrinsic-type stacking fault. Such a defect appears either when one layer is withdrawn from a face-centered cubic crystal as in Fig. 5.2ab (i.e. the atoms in this layer are replaced by vacancies, and then the formed slit closes) or when a layer and all layers above it are shifted to the adjacent position (for instance, if layer A lying on layer C is moved to position B). A double-layer twin of the type $...ABC{\downarrow}BA{\uparrow}BCA...$ forms a stacking fault of the extrinsic type and is obtained by forcing an extra plane into the crystal (assembling interstitial atoms in the same plane as in Fig. 5.2c).

The edges of stacking faults which break off inside the crystal form linear defects, which are called partial dislocations. For a partial dislocation the misfit of the ends of the Burgers circuit corresponds to a displacement of atomic layers on contraction (or expansion) of the area occupied by the stacking fault. Therefore the Burgers vector of a partial dislocation constitutes only a fraction of the minimum translation vector.

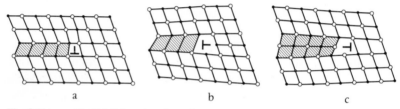

a   b   c

Fig. 5.11a–c. Partial dislocations in an fcc crystal. (a) Shockley dislocation, (b) Frank negative dislocation, (c) Frank positive dislocation. Stacking faults are hatched

Figure 5.11 illustrates diagrammatically the structure of partial dislocations in a face-centered-cubic lattice. The edge of an intrinsic-type stacking fault forms a dislocation either with a Burgers vector $a/2 \langle 112 \rangle$ lying in the plane of the fault (Shockley dislocation, Fig. 5.11a) or a dislocation with a Burgers vector $a/3 \langle 111 \rangle$ perpendicular to the fault plane (negative Frank dislocation, Fig. 5.11b). The edge of an extrinsic-type stacking fault always forms a positive Frank dislocation (Fig. 5.11c). On a Thompson tetrahedron (Fig. 5.9), the Burgers vectors of partial dislocations correspond to segments joining the vertices of the tetrahedron to the centers of the adjacent (for Shockley dislocations) or opposite (for Frank dislocations) faces. It can be clearly seen that the length of all these vectors is less than that of the minimum translation vectors corresponding to the edges of the Thompson tetrahedron.

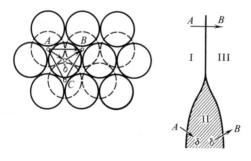

Fig. 5.12. Splitting of a total dislocation into partial Shockley dislocations. The stacking fault is hatched. Burgers vectors are denoted by arrows. The corresponding translation vectors are given in the left-hand side on the atomic scheme

In crystals permitting the formation of stacking faults, perfect dislocations lying in close-packed planes can split into partial dislocations bounding ribbons of stacking faults. For instance, a dislocation with a Burgers vector AB lying in the δ plane of the Thompson tetrahedron can split into two Shockley dislocations

$$AB = A\delta + \delta B$$

(see Fig. 5.12). A similar dislocation lying in the β plane can split into a Shockley and a Frank dislocation

$$AB = A\beta + \beta B.$$

In both cases the intermediate stacking faults linking partial dislocations are of the intrinsic type. When two dislocations with Burgers vectors $A\gamma + \gamma B$ and $B\delta + \delta C$ split in slip planes γ and δ, respectively, meet according to the scheme of Fig. 5.12, the partial dislocations with Burgers vectors γB and $B\delta$ merge together according to the reaction

$$\gamma B + B\delta = \gamma\delta$$

to form an energetically favorable "stair-rod dislocation" of the type $(a/6)$ $\langle 110 \rangle$. This results in a characteristic two-face defect called the Lomer-Cottrell barrier (see scheme of Fig. 5.13).

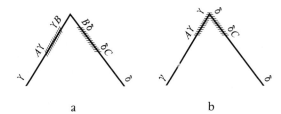

Fig. 5.13 a, b. Formation of the Lomer-Cottrell barrier. Split dislocations in intersecting splitting planes γ and δ (**a**). "Dislocation top" (Lomer-Cottrell barrier) (**b**). Stacking faults are hatched

In face-centered metals and crystals with a diamond lattice, *stacking-fault tetrahedra* occasionally arise from such deffects. Suppose, for instance, the vacancies are condensed into a flat layer lying in the δ plane and bounded by segments parallel to AB, BC, and CA. When the vacancy layer is closed according to the scheme of Fig. 5.2b, a dislocation loop with Burgers vector δD (Frank dislocation) arises, which encloses an intrinsic stacking fault. Further on, the dislocations can dissociate according to the reactions

along AB	$\delta D \rightarrow \delta\gamma + \gamma D$
along BC	$\delta D \rightarrow \delta\alpha + \alpha D$
along CA	$\delta D \rightarrow \delta\beta + \beta D.$

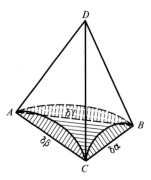

Fig. 5.14. Formation of a tetrahedron from stacking faults

Partial Shockley dislocations with Burgers vectors γD, αD, and βD can move in the γ, α, and β planes, respectively (see Fig. 5.14), and, meeting along the tetrahedron edges parallel to AD, BD, and CD, react according to schemes of the type of Fig. 5.13. As a result, a tetrahedron with edges from partial edge dislocations of the stair-rod type arises from the intrinsic stacking faults (the dislocation along AB has Burgers vector $\delta\gamma$, the characteristic of the other dislocations being obtained by circular permutation of the indices).

The specific features of stacking faults and partial dislocations in a hexagonal close-packed structure are analyzed similarly to the above-discussed case of a face-centered-cubic lattice. Here we have only one system of close-packed planes (parallel to the base) in which stacking faults can arise. Accordingly, no sessile dislocations or tetrahedra from stacking faults can form in hexagonal close-packed metals. Stacking faults correspond to inter-layers with a face-centered-cubic structure; a single-layer stacking fault is described by a sequence of stacking of planes of the type $...ABAB|CBCB...$, a double-layer by $...ABAB|C|ACA...$, a triple-layer by $...ABAB|CA|BA...$, etc. (violations of the sequence of stacking of the planes are indicated by separation lines). In face-centered-cubic metals, the collapse of single-layer vacancy "pancakes" shown in Fig. 5.2a gives a stacking fault surrounded by a Frank partial dislocation with a Burgers vector orthogonal to the plane of the loop, which means that the prismatic loop arising on collapse (Fig. 5.2b) corresponds to a partial dislocation of edge orientation. In hexagonal close-packed metals such a collapse must be accompanied by a displacement so as to ensure tight closing of the atomic planes. As a result, a loop with a partial dislocation is formed whose Burgers vector is tilted relative to the basal plane. The collapse of a double-layer vacancy pancake naturally produces an edge prismatic loop, without a stacking fault.

Multilayer close-packed structures permit a great diversity of stacking faults. Atoms in one layer or a group of layers, atoms of one or several sorts, etc., may be stacked irregularly. Perfect dislocations in such structures may split, not only into two partial (half-) dislocations, but also into four (quarter-) dislocations or even six partial dislocations.

Even in comparatively simple structures, dislocations splitting simultaneously in several equivalent or nonequivalent planes are sometimes observed. In particular, screw dislocations in bcc metals, when splitting, form intricate stars from ribbons of stacking faults, which are rearranged under external effects. If a screw dislocation with a Burgers vector $(a/2)$ [111] lying in slip plane $(1\bar{1}0)$ splits according to the reaction

$$\frac{1}{2}[111] = \frac{1}{8}[110] + \frac{1}{4}[112] + \frac{1}{8}[110],$$

three partial dislocations arise which lie in the same slip plane, the dislocation with a Burgers vector $(a/4)$ [112] being situated in the center. The Burgers vectors of all three dislocations lie in the slip plane, therefore the split dislocation is a glissile one. If the initial screw dislocation splits according to the reaction

$$\frac{1}{2}[111] = \frac{1}{8}[110] + \frac{1}{8}[101] + \frac{1}{8}[011] + \frac{1}{4}[111],$$

a star arises consisting of three ribbons of stacking faults which adjoin the central dislocation with Burgers vector $(a/4)$ [111]. Only one of the ribbons lies in slip plane $(1\bar{1}0)$, thus the split dislocation is a sessile one. The transition from the glissile configuration to the sessile one under the effects of stresses and thermal fluctuations determines the mobility of the dislocations and the plastic properties of the crystal.

The magnitude of the dislocation splitting depends on the stacking-fault energy γ and can be found from the equation

$$d = \gamma^{-1}E_{12}, \tag{5.71}$$

where E_{12} is the orientation factor of the energy of interaction of partial dislocations bounding the stacking fault. The stacking-fault energy usually varies from $10^{-3}\,Gb$ to $10^{-1}\,Gb$, i.e., from several ergs per cm^2 to hundreds of ergs per cm^2. In the latter case the surface tension of the stacking fault draws the partial dislocations together so strongly that the distance between them is of the same order as the size of the dislocation core, and one should rather speak of the splitting of the dislocation core than of the splitting of the perfect dislocation into two partial ones. The stacking-fault energy changes appreciably on crystal alloying. In fcc metals, the introduction of impurities of other valences may reduce γ by one order or more. As the point of phase transition from the cubic to the hexagonal close-packed phase is approached, the stacking-fault energy tends to zero, and the splitting of dislocations sharply increases.

A phenomenon similar to dislocation splitting arises in crystals of ordering alloys. Let us consider the simple case of ordering illustrated in Fig. 5.15.

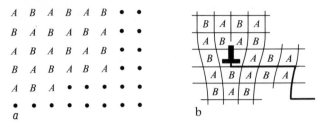

Fig. 5.15a, b. Effect of dislocations on the superstructure due to ordering: **(a)** Ordering in the perfect crystal, **(b)** break-off of the antiphase boundary on a dislocation

Atoms of two sorts A and B form two face-centered lattices inserted in one another (as in NaCl). When atoms A and B are interchanged, an antiphase domain will appear, which is a structure differing from Fig. 5.15a by a shift by the vector of translation $a \langle 100 \rangle$ of the unified cubic lattice. At the boundary of the antiphase domains identical atoms settle opposite each other. In the absence of dislocations the boundaries of antiphase domains form closed surfaces, but on each dislocation with a Burgers vector $a \langle 100 \rangle$ the antiphase boundary breaks off (Fig. 5.15b). Conversely, each such dislocation serves as a source of a new antiphase boundary and disturbs the superstructure. Thus, as a result of ordering (or any other phase transformation with a change in the number of atoms in the unit cell) an ordinary perfect dislocation may turn into a partial one, whose Burgers vector is no longer the translation vector of the crystal in the new phase, and the role of the stacking fault will be played by the antiphase boundary. An analog of split dislocations here is "superdislocations", i.e., dislocation pairs joined by the antiphase boundary and possessing a summary Burgers vector equal to the translation vector of the crystal in the new phase. As an example, Fig. 5.16 shows schematically the dislocations in an ordered alloy of the type AuCu$_3$. An ordinary dislocation of the type $(a/2) \langle 110 \rangle$ disturbs the superstructure and is associated with the termination of the boundary of the antiphase domains. Only a superdislocation with Burgers vector $a \langle 110 \rangle$, consisting of a pair of ordinary dislocations, can serve as a complete dislocation for the superstructure. Dislocations $(a/2) \langle 110 \rangle$ making up the superdislocation are partial dislocations for the superstructure and are related by the stacking fault, i.e., the antiphase boundary.

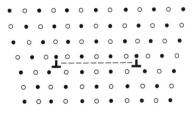

Fig. 5.16. Superdislocation in an ordered alloy of the AuCu$_3$ type

Consideration of dislocation splitting is just the first step towards getting a detailed account of the atomic structure of the dislocation core, which cannot be rendered by assigning a Burgers vector. Many properties of dislocations, such as their ability to move in a crystal, to absorb and emit point defects, etc. [Ref. 5.1, Chap. 12], are due to various atomic-scale kinks and jogs on the dislocation line. Even atomically smooth dislocations with identical Burgers vectors may have different nuclear structures and, thus, also different properties if the extra plane can be broken off along crystallographically nonequivalent sections. By way of example we consider the diamond lattice (Fig. 5.17). Networks {111} are here arranged in pairs, (aa'), (bb'), (cc'), etc. In each pair the atomic planes are spaced at 1/3 the interplanar distance between the pairs and are linked by a tripled number of bonds. If we form an edge dislocation by inserting an extra plane from above (Fig. 5.17), the termination of this plane at the upper or lower network of the pair (for instance, at level b' or b) will yield edge dislocations with a different structure of the core and with different properties. In the second case, for instance, when the dislocation moves in slip plane {111}, three times fewer bonds have to be broken than in the first case, i.e., the second type of dislocation will be more mobile. The number of broken bonds determining the efficiency of the influence of the dislocation on the electrical properties of the crystal will also be different.

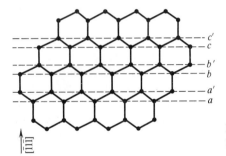

Fig. 5.17. Structure of the diamond lattice in projection onto the $(1\bar{1}0)$ plane

In the case of equidistant atomic planes, different dislocations with an identical Burgers vector can also be formed. Examples are the edge dislocations with Burgers vectors $(a/2)$ ⟨110⟩ lying in the ⟨100⟩ planes of the diamond lattice. Figure 5.18 shows the mutual arrangement of the successive networks {100}. It can be clearly seen that the direction of the slight shift for neighboring networks differs by 90°, because a fourfold screw axis, and not a simple one, runs along the cube edge. Accordingly, breaking off the extra plane on different storeys gives edge dislocations with different gliding abilities.

In multicomponent crystals, dislocations may differ in the type of atoms filling the edge of the extra half plane. In crystals of InSb, for instance, both In and Sb dislocations, which differ drastically in their mechanical, chemical,

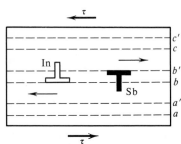

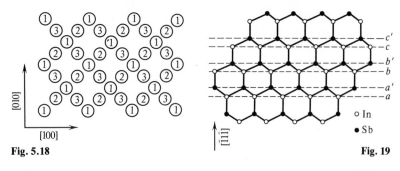

Fig. 5.18

Fig. 19

Fig. 5.18. Structure of the diamond lattice in projection onto the {001} plane. Nodes of networks located at different heights are labelled 1, 2, and 3

Fig. 5.19. Atomic structure of InSb in projection onto the (1̄10) plane. In contrast to Fig. 5.17 the adjacent networks a and a', b and b', c and c' are filled with atoms of different sorts

◁ **Fig. 5.20.** Movement of In and Sb dislocations

and electrical properties, can be distinguished on this basis. In Fig. 5.19 the structure of InSb is shown in the same projection as the diamond structure in Fig. 5.17. If we restrict ourselves to easily gliding dislocations, then in In dislocations extra planes must be inserted from above, and in Sb dislocations, from below. In a homogeneous field of stress τ, In and Sb dislocations lying in parallel slip planes will move in different directions (Fig. 5.20). On plastic bending of a crystalline beam (Fig. 5.21) a crystal of InSb will contain predominantly either In or Sb dislocations, depending on the sign of bending (during bending the extra planes must naturally enter from the convex side of the beam). As a result, the mechanical, electrical, and chemical properties of the crystal will depend on the sign of bending.

Fig. 5.21. Preferential formation of In and Sb dislocations on plastic bending of an InSb beam

5.5 Continuum Description of Dislocations

5.5.1 Dislocation-Density Tensor

In a macroscopic examination of a crystal with dislocations, (5.20) must be applied to circuits enclosing a large number of dislocations; by b is meant the total Burgers vector of all the dislocations crossing the area bounded by this circuit. Using Stokes's theorem, the integral relation (5.21) is rewritten in the differential form

$$\beta_{ij} = - e_{ikl} \frac{\partial}{\partial x_k} u_{lj} . \tag{5.72}$$

Here, β_{ij} is the dislocation-density tensor describing the macroscopic distribution of dislocations. The components of the ith line of this tensor are numerically equal to those of the summary Burgers vector of all the dislocations crossing a unit area perpendicular to the ith axis. The operation assigned by (5.72) and familiar to us from (5.21) is called the circulation of the tensor, in this case of the elastic-distortion tensor. Recall that elastic-distortion tensor u_{ij} is composed of the symmetric tensor of (macroscopic) elastic deformations ε_{ij} (5.6) and the antisymmetric tensor of elastic rotations ω_{ij}, which is equivalent to the axial vector of rotations ω (5.7)

$$u_{ij} = \varepsilon_{ij} + \omega_{ij} = \varepsilon_{ij} + e_{ijk} \omega_k . \tag{5.73}$$

The diagonal components of tensor u_{ij} correspond to elastic extensions (contractions) of the lattice along the coordinate axes, and the nondiagonal components describe elastic shears along the coordinate planes (the first subscript stands for the plane, and the second for the direction of the shear). If u_{ij} has been found from experimental (for instance, x-ray) data, the macroscopic dislocation density can be calculated from (5.72).

Using relation (5.73), we can separate out in (5.72) the dependence of the dislocation density on the elastic deformations and rotations of the lattice

$$\beta_{ij} = - e_{ikl} \frac{\partial}{\partial x_k} \varepsilon_{lj} + \frac{\partial \omega_i}{\partial x_j} - \delta_{ij} \frac{\partial \omega_k}{\partial x_k} . \tag{5.74}$$

Of special interest is the case of an unstressed crystal ($\varepsilon_{ij} = 0$), when (5.74) relates the distribution of dislocations with the lattice-curvature tensor $\kappa_{ij} = \partial \omega_i / \partial x_j$

$$\beta_{ij} = \kappa_{ij} - \delta_{ij} \kappa_{kk} . \tag{5.75}$$

As an example, Fig. 5.22 shows a crystal plate cut out perpendicularly to the y

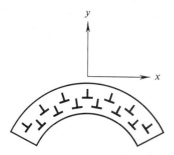

y

x

Fig. 5.22. Bending of a crystal plate containing parallel edge dislocations

axis and bent around the z axis. In the curvature tensor, one component $\kappa = \kappa_{zx}$ is not equal to zero. According to (5.75), only the corresponding component $\beta_{zx} = \kappa_{zx}$ in the dislocation-density tensor is different from zero, which can be interpreted as a uniform distribution in the crystal of edge dislocations with a Burgers vector along the x axis situated parallel to the bending axis with a density

$$N = \kappa/b = 1/bR\,, \tag{5.76}$$

where R is the curvature radius of the bent plate. Equation (5.76) can be derived from simple geometric considerations by counting the number of extra planes with interplane distance b, which must be introduced from the convex side of the plate being bent in order to obtain its curvature radius R.

In principle, integration of (5.72) enables one to solve the reverse problem and construct the field of stresses and rotations of the lattice for a given distribution of dislocations. Here, (5.72) must be supplemented by the equation of equilibrium of stresses σ in the bulk of the body

$$\frac{\partial \sigma_{ij}}{\partial x_i} = 0 \tag{5.77}$$

and on its surface

$$\sigma_{ij} n_j = 0\,, \tag{5.78}$$

where n is a normal to the free surface of the body.

5.5.2 Example: A Dislocation Row

We restrict ourselves to the one-dimensional problem of dislocations arranged with a uniform density in some plane P with a normal n, i.e.,

$$\beta_{ij}(r) = \beta_{ij}^0 \delta(r_k n_k)\,, \tag{5.79}$$

where β_{ij}^0 are constants characterizing the orientation and Burgers vectors of dislocations lying in plane P, and radius vector r is reckoned from any point of plane P. In our case, the stresses throughout the crystal volume satisfy (5.78), while (5.72) takes the form

$$\beta_{ij}^0 = -\, e_{ikl} n_k [u_{lj}]\,,\tag{5.80}$$

where the brackets denote the difference between the values of the given quantity on the different sides of plane P. The solution of the sets of equations (5.78, 80) with due regard for generalized Hooke's law, is

$$[\sigma_{ij}] = c_{ijkl}^* e_{kmn} n_m \beta_{nl}^0\,.\tag{5.81}$$

To ascertain this, it will suffice to multiply both sides of (5.80) from the left vectorially by normal n and calculate the convolution of the expression obtained with the plane tensor of elastic moduli (5.12) introduced in Sect. 5.2, taking into account the equilibrium equation (5.77).

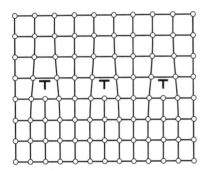

Fig. 5.23. Horizontal row of edge dislocations

By analogy with (5.81), we can obtain expression (5.11) for the Fourier components of the stress tensor, which was used in Sects. 5.2, 3. Directly from (5.81) follows (5.32) for the stress field of a straight dislocation. Equation (5.81) changes to (5.43) subject to the condition $p(x) = $ const. Let us consider, by way of example, a horizontal row of edge dislocations parallel to the z axis and lying in common plane $y = 0$ (Fig. 5.23). In this case the dislocation-density tensor has one nonzero component $\beta_{zx} = b/h$, where h is the distance between the dislocations. From (5.81) we obtain for jumps of macroscopic stresses

$$[\sigma_{ij}] = c_{ijxx}^*(b/h)\,.\tag{5.82}$$

The same result follows from (5.43) at $p = h^{-1}$ and with the above-indicated choice of orientations of the Burgers vector and dislocations. Other similar examples will be considered in Sect. 5.6.

5.5.3 Scalar Dislocation Density

A macroscopic description of the dislocation structure of a crystal with the aid of the dislocation-density tensor is insufficient in many cases. Thus, in uniformly deformed crystals the macroscopic stresses and the macroscopic lattice curvature are nonexistent and, hence, the dislocation-density tensor reduces to zero in accordance with (5.72), although the crystal may contain a large number of dislocations. In such cases the dislocation structure of the crystal is best characterized by a scalar quantity, the total length of the dislocation lines in a unit volume of the specimen. This quantity is called the *scalar dislocation density*, or simply the dislocation density. In practice, it is usually measured by counting the average number of dislocation lines crossing a unit area within the specimen. Averaging of the results obtained over areas of different orientation yields an estimate of the average dislocation density. In many cases such an estimate is an important characteristic of the crystal quality (for instance, in the semiconductor technology). A more detailed analysis of the dislocation   structure of a crystal requires isolation of the density of dislocation of different types, which is achieved by using modern diffraction methods of investigating the real structures of crystals (Sect. 5.8).

5.6 Subgrain Boundaries (Mosaic Structures) in Crystals

5.6.1 Examples of Subgrain Boundaries: A Tilt Boundary and a Twist Boundary

The presence in crystals of misoriented regions (subgrains) tilted at small angles to each other was noted even in the early investigations on crystal morphology. Soon after x-ray diffraction by crystals was discovered, it was established, that inside visible grains the crystal does not have an ideal structure. Contrary to theory (i. e., dynamic theory, see [Ref. 1.6, Chap. 4]), the diffracted beams propagated in the angular range of several minutes, rather than several seconds, and had an intensity exceeding the calculated value by two orders. The researchers had to assume the presence in the crystal of a mosaic of small (about 1 μm across) slightly misoriented subgrains, which had not manifested themselves in morphological investigations, but which affected the coherence of the diffracted waves. The origin of mosaics in crystals was not established until methods for investigating the dislocation structure of crystals were developed. It turned out that subgrains are comparatively dislocation-free regions, and subgrain boundaries consist of dislocation networks. Dislocations inside subgrains and at their boundaries proved to be elementary sources of the summary field of rotations of the crystal lattice, which characterizes the mosaic pattern of the crystal. The interpretation of the structure of dislocation networks along the subgrain boundaries is a decisive step towards general interpretation of the dislocation structure of the crystal and an analysis of the mechanism of its formation.

Figure 5.24 demonstrates the simplest dislocation model of a symmetric boundary of two single-crystal subgrains, typical of polyganization subgrains, kink boundaries, and strain and accommodation bands [Ref. 5.1, Chap. 12].

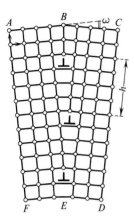

Fig. 5.24. Vertical row of edge dislocations

Through faces AB and BC more atomic planes enter the subgrains than leave them through faces DE and EF. All the extra planes terminate inside the bicrystal in the only disturbed region, i.e., at the subgrain boundary. The edge of each broken-off plane forms an edge dislocation, so that the entire subgrain boundary is represented as a "vertical" row of edge dislocations. As can be seen from Fig. 5.24, the angle of misorientation of subgrains ω is determined by the ratio of Burgers vector b to distance h between the dislocations at the boundary

$$\omega = b/h . \qquad (5.83)$$

The dislocation structure of a subgrain boundary can generally be judged from relation (5.75). There it is necessary to switch from lattice curvature $\partial \omega_i / \partial x_j$ to the local bending of the curvature at the subgrain boundary $\omega_i n_j$ (ω is the misorientation, and n is a normal to the boundary) and from the bulk to the surface distribution of dislocations,

$$\beta_{ij} = \omega_i n_j - \delta_{ij} \omega_k n_k . \qquad (5.84)$$

If the misorientation axis of the subgrains is parallel to their boundary, then $\omega_k n_k = 0$, and for the dislocation density if follows

$$\beta_{ij} = \omega_i n_j . \qquad (5.85)$$

Such a boundary is called a "tilt boundary." It may be represented by a row of dislocations parallel to the rotation axis which have a Burgers vector directed along a normal to the boundary. This row of dislocations should be distributed along a boundary with a linear density $\rho = b^{-1}\omega$, i.e., at distances $h = b/\beta = b/\omega$ from each other. This model is illustrated in Fig. 5.24.

Suppose now the misorientation axis of the subgrains is perpendicular to their boundary ("twist boundary"). Assuming $\omega_i = \omega n_i$, we obtain from (5.84)

$$\beta_{ij} = \omega(n_i n_j - \delta_{ij}) . \tag{5.86}$$

If we choose coordinate axes 1 and 2 in the boundary plane, then $n_1 = n_2 = 0$, and only two diagonal components,

$$\beta_{11} = \beta_{22} = -\omega , \tag{5.87}$$

will be nonzero in the dislocation-density tensor, which can be interpreted as a square network of screw dislocations arranged at the boundary at distances $h = b/\omega$ from each other.

5.6.2 The Dislocation Structure of the Subgrain Boundary in General

The above-mentioned schemes of structure with tilt and twist boundaries are just the simplest examples. In each particular case the structure of dislocation networks realizing the surface density of dislocations, which is given by equations of the type (5.84 – 86) should be refined depending on the arrangement of the boundary and the misorientation axis relative to the Burgers vectors of dislocations typical of the given crystal. Suppose, for instance, that a normal to the tilt boundary in a face-centered cubic crystal does not coincide with any one if the close-packed directions of the type $\langle 110 \rangle$. Then this boundary cannot be made up of single dislocations of the same type, since these dislocations must have a Burgers vector equal to one of the minimum translation vectors $(a/2) \langle 110 \rangle$, and (5.85) requires that the average Burgers vector of the dislocations be perpendicular to the boundary. If the normal to the boundary lies in the $\{100\}$ or $\{111\}$ planes, the tilt boundary can be constructed from two families of parallel edge dislocations. In the general case of the tilt boundary it is necessary to use three families of dislocations with noncoplanar Burgers vectors, along which the vector of the normal to the boundary could be resolved.

Such three families permit constructing a dislocation model, not only of an arbitrarily oriented tilt boundary, but also of any boundary of the general form. To prove this, we construct three vectors $\tilde{b}^{(1)}$, $\tilde{b}^{(2)}$, and $\tilde{b}^{(3)}$ reciprocal to the above-mentioned three noncoplanar Burgers vectors $b^{(1)}$, $b^{(2)}$, and $b^{(3)}$, so that the following orthonormalization condition is fulfilled:

$$b^{(i)}\tilde{b}^{(j)} = \delta_{ij}.$$ (5.88)

Vectors $\tilde{b}^{(j)}$ are actually unit vectors of the lattice, which is reciprocal to the one constructed on vectors $b^{(i)}$, and are found according to the familiar rule

$$\tilde{b}_i^{(1)} = \frac{e_{ijk}b_j^{(2)}b_k^{(3)}}{e_{lmn}b_l^{(1)}b_m^{(2)}b_n^{(3)}}$$ (5.89)

(relations for $\tilde{b}^{(2)}$ and $\tilde{b}^{(3)}$ are obtained from (5.89) by circular permutation of superscripts 1, 2, and 3). Scalar multiplication of the dislocation-density tensor from the right by vectors $\tilde{b}^{(n)}$ yields vectors $V_i^{(n)} = \beta_{ij}\tilde{b}_j^{(n)}$, with the aid of which the dislocation-density tensor can be written as

$$\beta_{ij} = V_i^{(1)}b_j^{(1)} + V_i^{(2)}b_j^{(2)} + V_i^{(3)}b_j^{(3)},$$ (5.90)

which is equivalent to isolating three bundles of dislocations with Burgers vectors $b^{(1)}$, $b^{(2)}$, and $b^{(3)}$; the density and direction of these bundles are given by vectors $V^{(1)}$, $V^{(2)}$, and $V^{(3)}$, respectively.

For tilt boundary (5.85)

$$V_i^{(n)} = \omega_i(n_j\tilde{b}_j^{(n)}),$$ (5.91)

i.e., all the dislocations are arranged along the rotation axis with densities proportional to misorientation angle ω and to the projection of vectors $\tilde{b}^{(n)}$ onto the normal to the boundary. For twist boundary (5.86)

$$V_i^{(n)} = -\omega[\tilde{b}_i^{(n)} - n_i(\tilde{b}_j^{(n)}n_j)],$$ (5.92)

i.e., the dislocations are arranged along the projections of vectors $\tilde{b}^{(n)}$ onto the interface with a density proportional to this projection and the misorientation angle.

According to (5.90), for the three families of dislocations with non-coplanar Burgers vectors the dislocation structure of an arbitrary boundary is interpreted unambiguously. As a rule, however, the crystal structure permits a more diversified set of possible Burgers vectors. In this case analysis of the structure of subgrain boundaries requires additional consideration. If, for instance, it is known that the subgrain boundaries are formed as the result of plastic deformation by gliding [Ref. 1.7, Chap. 12], additional conditions arise, which limit the orientation of the dislocations in the subgrain boundaries (the dislocations must lie in glide planes). We shall also note the limitations associated with the condition concerning the energy minimum of the dislocations making up the boundary. As a first (and rather crude) approximation, this condition can be replaced by that for the minimum of the summary length of the dislocation lines making up the subgrain boundary. As

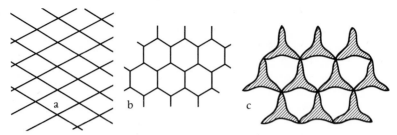

Fig. 5.25 a – c. Formation of a hexagonal dislocation network. (**a**) Initial rhombic network of two families of screw dislocation, (**b**) splitting of quaternary nodes into tertiary nodes, (**c**) splitting of tertiary nodes (stacking faults are hatched)

a result, quaternary nodes in dislocation networks are, as a rule, energetically unfavorable and dissociate into pairs of triple nodes joined by dislocations with a Burgers vector defined according to the general rule of Fig. 5.8.

Let us consider, for instance, the twist boundary in the close-packed plane (111) of a face-centered-cubic crystal. By (5.86), such a boundary can be composed of two families of screw dislocations forming a rhombic network of the type shown in Fig. 5.25a. There are, all in all, three Burgers vectors of the type $(a/2) \langle 110 \rangle$ lying in the boundary plane: $(a/2) [1\bar{1}0]$, $(a/2) [10\bar{1}]$, and $(a/2) [0\bar{1}1]$, i.e., AB, BC, and CA in the notation of Fig. 5.9. Accordingly, three different versions of a rhombic network consisting of two dislocation families are possible. But it is energetically more advantageous to use all three dislocation families simultaneously, split the quaternary nodes of the type depicted in Fig. 5.26a into triple nodes according to the scheme of Fig. 5.26b, and form a hexagonal network of screw dislocations (Fig. 5.25b) with nodes corresponding to a dislocation reaction of the type $AB + BC = AC$.

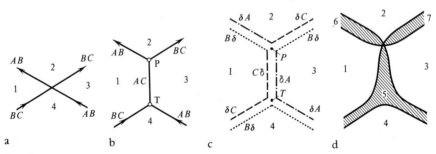

Fig. 5.26a – d. Splitting of nodes in Fig. 5.25. (**a**) Quaternary node, (**b**) splitting of a quaternary node into two tertiary nodes, (**c**) splitting of perfect dislocations into partial dislocations, (**d**) result of reaction between partial dislocations. (*1 – 4*) Numbers of the boundary unit cells adjoining a quaternary node (**a**) or tertiary nodes (**b**). Relative shifts of the crystal parts separated by the plane of the drawing in adjacent regions 1 and 2, 2 and 3, 3 and 4, 4 and 1 (**a**); 1 and 2, 2 and 3, 3 and 4, 3 and 1, 4 and 1(**b**) differ by the Burgers vector of the perfect dislocation separating them. The corresponding relative shifts in regions 5, 6, and 7 (stacking fault ribbons) differ from the shifts in the adjacent regions 1 – 4 by the Burgers vectors of the partial dislocations separating them

If the perfect dislocations making up the subgrain boundary dissociate into partial ones to form intermediate ribbons of stacking faults (Fig. 5.26 b), additional reactions between partial dislocations near the network nodes are possible. As a result, the stacking fault ribbons either broaden and merge together (extended nodes) or constrict (constricted nodes). Figure 5.26 d illustrates such reactions between partial Shockley dislocations in the scheme of Fig. 5.26 b and shows that extended and constricted nodes in the hexagonal scheme of Fig. 5.25 b must arise alternately (Fig. 5.25 c).

To any irregular and nonuniform dislocation network at the subgrain boundary can be correlated a two-dimensional net in a regular lattice constructed from Burgers vectors of the dislocations forming the network. To each polygon in the dislocation network is correlated a lattice site, to each node, a polygon constructed from Burgers vectors of the dislocations forming the given node, and to each dislocation forming a segment of the dislocation network, the Burgers vector of this dislocation. In Fig. 5.27 such a construction is carried out for the dislocation nodes and meshes of the dislocation network 1, 2, 3, 4 (5, 6, 7) shown in Fig. 5.26. In the analysis of various irregularities in the structure of dislocation networks the technique for reconstruction of the network of the Burgers vectors, illustrated in Fig. 5.27, often serves as an effective means of investigating the mechanism of subgrain formation.

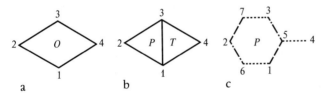

Fig. 5.27 a – c. Reactions between the Burgers vectors for the scheme of Fig. 5.26. (a) Quaternary node O, (b) splitting of node O into two tertiary nodes P and T, (c) rearrangement of tertiary nodes on splitting of the dislocations into partial ones

Two-dimensional dislocation networks can form subgrain boundaries with different degrees of misorientation within the range from several seconds to several degrees. In very slightly misoriented subgrains (cells), the distance between dislocations is of the same order as the subgrain size, the dislocation network transforms from a plane to a three-dimensional form, and the individual contribution from each dislocation to the lattice-distortion field must be taken into account. For strongly misoriented subgrains (fragments), the above equations must be refined allowing for the finiteness of the misorientation angle. In place of (5.83), for example, we obtain

$$\omega = 2 \arcsin (b/2h) . \tag{5.93}$$

With very large (several tens of degrees) subgrain misorientation angles, the dislocations in the boundaries draw so close together that analysis of the boundary structure requires taking the atomic structure of the dislocation core into consideration. The use of some analogy between grain and twin boundaries (see Sect. 5.7) proves more effective here.

5.6.3 Subgrain Boundary Energy

The subgrain boundary energy is made up of the elastic energy of the stresses around the dislocations forming the boundary. The stresses caused by individual dislocations are so superimposed on each other that the total stress rapidly relaxes at distances of the order of spacing h between the dislocations in the boundary. As a result, in the dislocation energy equation (5.34) the crystal size must be replaced by h, which reduces the dislocation energy to

$$E = E_0 \ln(h/r_0) \tag{5.94}$$

and explains the energetic advantage of *polygonization*, i.e., uniting the dislocations initially scattered throughout the crystal volume into plane walls and networks dividing the crystal into unstressed blocks. The surface energy of the block (subgrain) boundaries γ is found by summing the energy of the dislocations making up the boundary. The total length of the dislocation lines is of the order h^{-1} cm per cm^2 of the boundary and the surface energy of the subgrain boundary is

$$\gamma = E_0 h^{-1} \ln(h/r_0) = E_0 \omega b^{-1}(b/\omega r_0), \tag{5.95}$$

where $E_0 b^{-1}$ usually reaches values of the order of 10^3 erg/cm^2.

The calculated absolute values of the subboundary energy and of its functional dependence on the misorientation angle are in good agreement with experiment. A more comprehensive theory, taking into account the specifics of the contribution from different dislocation families making up the boundary, explains the anisotropy of the surface energy (the dependence of γ on the boundary orientation). Allowance for the dislocation structure of the subboundaries also explains the accelerated migration of point defects along the boundaries, the segregation of impurities at the boundaries, the effect of the boundaries on the electrical properties of semiconductors and ionic crystals, the mobility of boundaries under external influences on the crystal, and also the different effects of interaction of individual dislocations with subboundaries.

5.6.4 Incoherent Boundaries

So far we have been discussing coherent subboundaries which were constructed from dislocations in such a way that [in accordance with (5.84)]

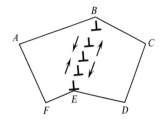

Fig. 5.28. Formation of stresses on deviation of the vertical wall of edge dislocations

the superposition of the fields of individual dislocations did not yield macroscopic stresses. If the relation between the orientations of the dislocation lines and the Burgers vectors, required by (5.84), is not fulfilled, an incoherent boundary arises, which causes macroscopic stresses, i.e., stresses arising from the interaction of the blocks separated by this boundary. Suppose, for instance, the boundary shown in Fig. 5.24 deviates from the symmetric position by turning through a small angle φ in a clockwise direction (Fig. 5.28). Of the ω/b planes terminating on a unit segment of the boundary, $b^{-1} (\omega/2 + \varphi)$ planes will pass through segment AB, and $b^{-1} (\omega/2 - \varphi)$, through segment BC. As a result the left-hand block will be contracted, and the right-hand one extended; the difference between the deformations along the boundary will be

$$[\varepsilon] = \omega \varphi. \tag{5.96}$$

The general case of incoherent subgrain boundary can be obtained from (5.74) by passing on from the volume to surface distribution of dislocations and replacing the gradients of lattice deformations and rotations by the differences of their values on the two sides of the boundary

$$\beta_{ij}^0 = - e_{ikl} n_k [\varepsilon_{lj}] + \omega_i n_j - \delta_{ij} \omega_k n_k. \tag{5.97}$$

The expression obtained differs from (5.80) only by the division of the distortion jump into a symmetric part (deformation jump $[\varepsilon_{lj}]$) and an antisymmetric part associated with misorientation of the blocks ω. The stress jump at the interface is again determined by (5.81), and therefore it is possible to eliminate the strain jump from expression (5.97) by expressing it via the stress jump with the aid of the generalized Hooke's law. Some simple relationships for incoherent block boundaries, however, follow immediately from (5.97). Calculating the trace (sum of the diagonal elements) of tensor equality (5.97), we obtain

$$\omega_i n_i = - \tfrac{1}{2} \beta_{ii}, \tag{5.98}$$

i.e., the rotation about an axis perpendicular to the boundary is determined by the average density of screw dislocations, as in the case of a coherent boundary (5.87). Performing scalar multiplication of (5.97) from the right by the vector of the normal to the boundary, we obtain for rotations about axes lying in the boundary plane

$$\omega_i - n_i(\omega_k n_k) = \beta_{ij} n_j + e_{ikl} n_k [\varepsilon_{ij}] n_j \,, \tag{5.99}$$

i.e., the contribution to this rotation is made by the boundary components of the Burgers vectors perpendicular to the boundary and by the jumps of tangent deformations. For boundaries parallel to the symmetry plane or perpendicular to the twofold (four- or sixfold) axis there are no such jumps because of the continuity of the tangent stresses acting on the subgrain boundary. In this particular case (5.99) agrees with the expression (5.85) found previously.

A typical example of incoherent boundaries are so-called "irrational twins" (Brilliantov – Obreimov bands), which outwardly resemble slip bands and, therefore, have for a long time puzzled investigators probing into the mechanism of plastic deformations in ionic crystals (see [Ref. 1.7, Chap. 12]). In crystals of the type NaCl these "twins" can indeed be easily confused with slip bands (in both cases the boundaries are arranged in {110} planes, the strain jump and the internal stresses are similar). The slip bands, however, are characterized by condition $\beta_{ij} n_j = 0$ (the Burgers vectors of all dislocations lie in the slip band) and, by (5.99), there are no lattice rotations about axes parallel to the slip band [{110} planes in NaCl are planes of symmetry; for them, from the continuity of tangential stresses the continuity of tangential strains follows, and the second term in the right-hand side of (5.99) is equal to zero]. "Irrational twins", on the contrary, form incoherent tilt boundaries (the summary Burgers vector of dislocations is nearly perpendicular to the boundary) and cause stresses because of the deviation of the boundary from the correct position, as in the scheme of Fig. 5.28. The irrational-twins mystery was resolved when components parallel and perpendicular to the boundary were isolated from the edge components of the dislocations forming the boundary. According to (5.43, 81), the first components caused stress bands framing the boundary, and, to (5.99), the other components caused a rotation of the lattice about an axis lying in the boundary plane. As in (5.96), the ratio of the deformation jump to the misorientation angle depended on the angle of the deviation of the boundary from the symmetric orientation corresponding to the {110} plane.

In analyzing stresses associated with subgrain boundaries one must specially consider the case of boundaries consisting, not of plane, but of three-dimensional dislocation networks. It is possible to single out in the structure of such networks dislocation dipoles and loops determining the difference in the strain of the boundary area with respect to adjacent subgrain and corresponding to the boundary – subgrain-interaction stresses localized in the boundary area. These stresses are sometimes called oriented microstresses. The subgrain-interaction stresses (so-called second-kind stresses) cause broadening of the lines of the x-ray spectrum, while oriented microstresses result in a general shift of the lines to the side corresponding to the stresses in the subgrain volume.

5.7 Twins

A twin is a crystal with consistently mutually misoriented regions (twin components) whose atomic structure is related geometrically by some symmetry operation (twinning operation). Twinning operations include reflection in a plane (reflection twins), rotation about a definite crystallographic axis (axial twins), reflection at a point (inversion twins), translation for a part of the lattice spacing (translation twins), and combinations of these. Each twin component can occupy in the crystal one continuous region or several scattered regions simultaneously, forming a so-called polysynthetic twin. In a narrow sense of the word a twin must possess two components related by a single second-order operation (reflection in a plane, rotation through 180°, inversion, translation by half the lattice spacing), so that a repetition of the operation restores the structure of the initial component. There are, however, also multicomponent twins ("triplets", "quadruplets", "sixtuplets", etc.), for which twin operations are rotations about three-, four- and six-fold axes, translations by a fractional part of the lattice spacing, and also combinations of several crystallographic operations.

Twins may appear during the growth of crystals and their intergrowing, during recrystallization, phase transformations, and also under mechanical, thermal, electrical, magnetic, etc. effects on single and polycrystals. Twin components may differ in optical, mechanical, electrical, magnetic, and other properties if the anisotropy of the relevant characteristics is not invariant to twinning operations.

5.7.1 Twinning Operations

A complete set of operations relating the components of a given twin depends not only on the mutual orientation of the components, but also on their own symmetry. Two symmetry operations of a twin are equivalent if they differ by an operation inherent in a given crystal. Therefore a reflection twin may simultaneously be an axial twin, etc.

Suppose component I is transformed to component II by twinning operation f; then the product

$$f_i = f g_i. \tag{5.100}$$

(where g_i — an operation belonging to symmetry group G_I of component I) again gives a twinning operation transforming structure I to structure II. Here, the different operations g_i and g_j correspond to the two operations f_i and f_j, and for the finite groups the number of twinning operations is equal to the order of group G_I. Generally, if we neglect the multiplication twinning of translations, the number of twinning operations will be equal to the order of the point symmetry group of the twin components.

Operations g'_j constituting symmetry group G_{II} of the second component of the twin are found by the rule

$$g'_k = f g_k f^{-1} \qquad\qquad (5.101)$$

where any one of operations (5.100) can be used as twinning operation f. If the twinning operation commutes with all the operations of group G_I, i.e., $f g_i = g_i f$, it follows from (5.101) that symmetry groups G_I and G_{II} coincide, $g'_j = g_j$. If, besides, a repetition of the twinning operation restores the structure of the initial component, i.e., if $f = f^{-1}$, an addition to group G_I of twinning operations (5.100) forms a supergroup

$$G = G_I + f G_I, \qquad\qquad (5.102)$$

for which G_I serves as a subgroup of index 2, and the set of twinning operations serves a conjugate class. An addition to group G_I of operations $R f_i$, where R is the operator of the sign change, forms the antisymmetry group

$$G' = G_I + R f G_I, \qquad\qquad (5.103)$$

describing the change in the structure of the components on application of the twinning operation. Operation R may be taken to mean, in particular, an operation of changing the sign of the deviation of the atomic coordinates from their average for components I and II.

Fig. 5.29a, b. Twins in β-quartz. (a) Dauphine twins, (b) Brasilian twins. One of the twin components is shaded

By way of example we consider Dauphine twins in β-quartz (Fig. 5.29a). The symmetry group of β-quartz $D_3^4 - P3_121$ (or $D_3^6 - P3_221$) includes rotations about a threefold screw axis $\bar{C}_3$ and three twofold axes u_2 perpendicular to it. Twinning operations for Dauphine twins are rotations about a sixfold axis C_6 and twofold axes C_2 parallel to axis $\bar{C}_3$, and also rotations about each of the three twofold axes perpendicular to it and positioned between axes u_2. Superposition of all operations g_i and f_i forms supergroup $D_6^4 - P6_222$ (or $D_6^5 - P6_422$), which is the symmetry group of α-quartz. Operations g_i and $R f_i$ form the antisymmetry group $P6'_4 22$ (or $P6'_2 22$).

Figure 5.29b illustrates another type of twins in β-quartz, the so-called Brazilian twins, arising on the intergrowth of left-hand and right-hand quartz along prism plane $(11\bar{2}0)$, which serves as a reflection plane. In addition to a reflection in this plane, twinning operations include a reflection in prism planes $(1\bar{2}10)$ and $(\bar{2}110)$, and an inversion and a rotation about sixfold mirror-rotation axis S_6. The space symmetry groups of component (D_3^4 and D_3^6) are not identical here, but the point symmetry groups (classes) of the component do coincide, which permits describing the point symmetry of the Brazilian twins in terms of antisymmetry classes. Supplementing the point symmetry group of β-quartz with twinning operations according to (5.102), we obtain supergroup $D_{3d} = \bar{6} : 2$, and — for twinning operations as antisymmetry operations according to (5.103) — the sought-for antisymmetry group $\bar{6}' : 2$.

In the case of multicomponent twins (triplets, quadruplets), whose components I, II, III ... have a common symmetry group G_I and are coupled pairwise by cyclic operations $f_{I,II}, f_{I,III}, f_{II,III}$, etc., supplementing group G_I with all possible twinning operations (5.100) yields the supergroup

$$G = G_I + f_{I,II}G_I + f_{I,III}G_I + \ldots, \tag{5.104}$$

for which G_I is a subgroup of index 3, 4, etc., and the twinning operations form 2, 3, etc., conjugate classes, respectively.

Generally, multicomponent twins with identical symmetry components can be considered with the aid of representation theory. A deviation of the structure of the components from the average is described by representations of the supergroup (5.102, 104). One-dimensional real representations can be interpreted in terms of antisymmetry, and one-dimensional complex representations, in the language of color symmetry, etc. An analysis of the structure and properties of twins based on representation theory is especially fruitful in the case of twins formed on phase transformations, when supergroup G has a simple physical meaning and describes the structure of a more symmetric phase, and the twinning operations are symmetry operations vanishing on phase transformation. In particular, if the transformation is accompanied by an increase in the number of atoms in the unit cell of the crystal (formation of a superstructure), the resulting twins (antiphase domains) are translation twins.

As an example we consider twins (domains) in ferroelectric $Cd_2(MoO_4)_3$. In the paraphase the crystal structure is described by space group D_{2d}^3. On ferroelectric transformation the volume of the unit cell doubles, and axis S_4 and two twofold axes u_2 perpendicular to axis S_4 are lost. As a result, a "quadruplet" arises, all the components of which have symmetry C_{2v}^8. Expansion of (5.104) has the form

$$D_{2d}^3 = C_{2v}^8 + S_4 C_{2v}^8 + S_4^2 C_{2v}^8 + S_4^3 C_{2v}^8. \tag{5.105}$$

Twinning operations $S_4 C_{2v}^8$ and $S_4^3 C_{2v}^8$ relate domains of opposite polarity (positive and negative ferroelectric domain). Twinning operations $S_4^2 C_{2v}^8$ relate antiphase domains of the same polarity. These operations include translations by the length of vectors equal to the paraphase translation vectors lost on transformation.

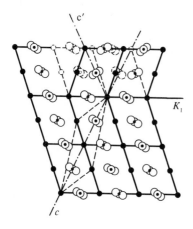

Fig. 5.30. Calcite twinning

5.7.2 Twinning with a Change in Crystal Shape

In twinning, the unit cell often changes its shape; this, in turn, changes the external shape of the part of the crystal which has flipped to the twin orientation. Figure 5.30 illustrates one of the classical examples of twinning with a change in shape, the twinning of calcite $CaCO_3$. Here, the twinning operation is a reflection in the (110) plane (twinning plane). The flipping of the lattice to the symmetric position is accompanied by a macroscopic shear strain along the twinning plane in twinning direction [001], the shear angle being 34°22′. The shape of the unit cells changes identically in all the sublattices, but the motion of the atoms cannot be thought of as simple shear; a mutual displacement of the sublattices occurs, which transfers the atoms into a configuration symmetric to the initial (for instance, mirror symmetric in the case of reflection twins). In calcite, the mutual displacement of the sublattices results in a rotation of groups of three oxygen atoms through 52°30′ about the axes passing through the carbon atoms parallel to [1$\bar{1}$0].

Macroscopically, the motion of atoms during twinning with a change in shape can be represented as the result of a uniform shear without a change in crystal volume

$$u_{ij}^0 = s p_i t_j . \tag{5.106}$$

Here, p is a normal to the twinning plane, t is the unit vector of the twinning direction lying in this plane, and s is the value of the twinning shear (see Fig. 5.31). Angle $\varphi = \arctan s$ is called the twinning angle. Each point at a distance d from the twinning plane moves by a distance sd along the twinning direction upon a twinning shear. The twinning shear transforms a sphere of unit radius into an ellipsoid (twinning ellipsoid).

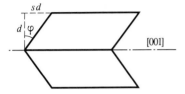

Fig. 5.31. Twin shear in twinning with a change in shape

Owing to distortion (5.106), in the field of external stress σ_{ij} the potential energy of the twin components differs by

$$\Delta W = \sigma_{ij} u_{ij}^0 = \sigma_{ij} \varepsilon_{ij}^0. \tag{5.107}$$

The twinning process is thus affected only by the tangential stress acting on the twinning plane in the twinning direction. The value ΔW determines the thermodynamic force acting on the twin boundary. The change in the sign of the stress changes the sign of ΔW and reverses the force. Accordingly, twinning is replaced by *detwinning*, which displaces from the crystal the twin component for which $\Delta W < 0$. If domains differ not only in twinning shear, but also in electric or magnetic polarization, expression (5.103) must be supplemented by the product of polarization times the corresponding electric or magnetic field.

Even for a small twinning shear the physical properties of a twin component may differ significantly. Twinning with a change in shape is inevitably accompanied, in particular, by a rotation of the optic indicatrix. This rotation is usually more prominent than the shear. For Rochelle salt, for instance, a rotation of the indicatrix exceeds the twinning shear 30 − fold, and therefore the twin components (domains) differ drastically in their extinction positions (Fig. 5.32).

Circular sections of the twinning ellipsoid determine two possible orientations of the twin boundary which do not cause macroscopic stresses. With an intermediate orientation of the twin boundary the contact of the twin components is equivalent to the disposition on the twin boundary of the dislocations whose macroscopic density at a small twinning shear can be calculated by the formula

$$\beta_{ij} = e_{ikl} n_k u_{lj}^0 = s \, [n \times p]_i t_j, \tag{5.108}$$

where n is the normal to the twin boundary.

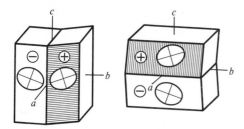

Fig. 5.32. Rotation of optic indicatrices in twinning with a change in shape (the twins are domains in Rochelle salt)

The macroscopic relation (5.108) corresponds to a model in which the dislocations are disposed at the twin boundary parallel to trace $[n \times p]$ of the twinning plane and possess Burgers vectors along the twinning direction t. The thermodynamic force (5.107) acting along the normal to the boundary can be converted to the force acting on these dislocations in a direction tangent to the boundary (and perpendicular to the dislocation). From (5.108) it is possible to calculate, with the aid of (5.81), the stresses caused by the twin boundary

$$[\sigma_{ij}] = -c^*_{ijkl}u^0_{kl}. \tag{5.109}$$

The actual structure of the twin boundary depends on the concrete type of crystal. In the case of ferromagnetic domains − twin components differing in spontaneous magnetization − there is, at the boundary, a broad (several hundred interatomic distances) transition band along which the direction of magnetization changes continuously from one domain to the next. In ferroelectrics, the boundaries of twin (domains) have a broad transition band only near the Curie point.

More often, however, the interaction forces of atomic layers with different orientations are weak as compared with the orientation forces due to the energetic disadvantage of the intermediate (nontwin) configurations. The transition band is also practically absent, i.e., the twin boundary is sharp, even on the atomic scale. Here, a twin with a boundary parallel to the twinning plane must have an equilibrium configuration without any stress concentration. The energy localized in such a twin amounts to the surface energy proportional to the area of the boundary. If the twin boundary deviates from the twinning plane, the interface must have a stepped shape, i.e., microscopically, the boundary must consist of parts parallel to the crystallographic twinning plane, and the boundary must deviate in a jump by the thickness of one atomic layer (Fig. 5.33). To each step corresponds a *"twinning dislocation"* with a Burgers vector of value

$$b = sa, \tag{5.110}$$

where s is the twinning shear and a is the distance between the atomic planes parallel to the twinning plane. It is easy to see that the scheme of Fig. 5.33 corresponds precisely to the macroscopic expression (5.108).

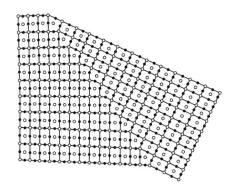

Fig. 5.33. Atomic steps (twinning dislocations) at the boundary of a wedgelike twin interlayer

The concept of the fitting of twin components with the aid of twinning dislocations permits broad generalizations for the case of grain boundaries and different phase boundaries separating crystals with different orientations or different structures (different lattice spacings). Here, too, parts with a good coherence are revealed which are separated by transition steps resembling twinning dislocations and are called grain boundary and epitaxial dislocations, respectively. The density of these dislocations and the macroscopic stresses are defined by equations similar to (5.108, 109). Generally, ordinary perfect dislocations are also present at interfaces. Their interaction with boundary dislocations may weaken the macroscopic stresses caused by the boundary and exert a substantial effect on the course of twinning, phase transformation, or the movement of grain boundaries.

The orientation and shape of twin, domain, and phase boundaries mainly depends on the stresses (5.109) caused by these boundaries. Since the convolution of tensor $c_{ijkl}^{*}(n)$ with vector n in respect to any index gives zero, no stresses should arise when the boundary is parallel to the plane preserved on distortion of u_{kl}^{0}. Such an invariant plane by no means exists for any arbitrary distortion. During twinning, there are two invariant planes corresponding to circular sections of the twinning ellipsoid. In the general case of distortion with an invariant plane, tensor u_{kl}^{0} can be represented in the form (5.106), where p is a normal to the invariant plane, and t is a unit vector making an arbitrary angle with unit vector p.

In practice, twins usually have the shape of parallel-side interlayers and flat wedges parallel to the circular sections of the twinning ellipsoid. In the case of distortion with an invariant plane, phase precipitations also have a platelike shape, and the platelets orient themselves parallel to the invariant plane. If the distortion of u_{kl}^{0} does not have an invariant plane, phase precipitations form at first thin platelets, whose orientation corresponds to the minimum of the elastic energy

$$W = \frac{1}{2} c_{ijkl}^{*} u_{ij}^{0} u_{kl}^{0}$$

(5.111)

caused by stresses (5.109) in a unit volume of the platelet. With a further development of the phase transformation, thicker platelets arise, which are either divided into layers of different orientation so that the average distortion receives an invariant plane parallel to the platelet surface, or lose the coherent bond with the matrix because of the epitaxial dislocations (misfit dislocations) forming on the platelet surface and smoothing out the difference in the spacings of the lattice of the contacting phases. In a plastic matrix, stresses (5.109) can also be eliminated by plastic deformation with emergence of perfect dislocations at the boundary, which serves as the source of stresses, and by formation of dislocation walls and networks compensating these stresses. Usually, the boundaries of nuclei in the early stages of growth of a new phase contain only epitaxial dislocations. In later stages, they are joined by perfect dislocations, the result being an appreciable decrease in stress and a qualitative change in the kinetics of the phase transition.

The mutual arrangement of twins and phase segregations also obeys the law of the elastic-energy minimum. The sides of twin platelets and those of the new phase induce elastic fields similar to the fields of dislocation loops with an effective Burgers vector. The association of platelets of different orientation is often due to the condition of mutual compensation of these vectors. Consistent allowance for elastic fields induced by phase boundaries helped to formulate the elastic domain theory, which explained many details of very complex heterophase structures arising on martensitic and diffusion phase transformations. Generalization of this theory for the case of domain structures consisting of domains inducing not only elastic, but also electric or magnetic fields is an urgent task for the theory of imperfect crystals.

5.7.3 Twinning Without a Change in Shape

When there is no twinning shear and the twinning ellipsoid degenerates into a sphere, the macroscopic shape of the crystal does not change in twinning, and a twin boundary of any orientation does not cause macroscopic stresses. The boundaries of such twins usually have smooth rounded outlines. A crystallographic habit arises only in the case of a very pronounced anisotropy of the boundary energy. In twinning without a change in shape, the optic indicatrices of the components coincide.

The structure of the components in twinning without a change in shape differs only in the mutual displacement of sublattices, with the shape of the unit cells preserved. Characteristic examples are illustrated in Fig. 5.34. In Fig. 5.34a, a small displacement of two identical sublattices leads to the formation of a translation twin; identical cells are shifted by half the lattice spacing. The components of a translation twin have the same physical properties. This does not mean, however, that such a twin can only arise on phase transformation. It is readily apparent from Fig. 5.16 that the movement of ordinary dislocations in the superlattice of an ordered alloy always forms a

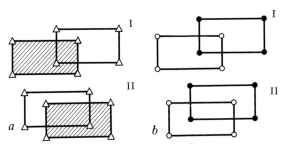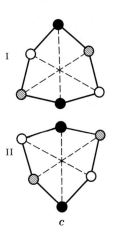

Fig. 5.34a–c. Atomic structure of twin components in twinning without a change in shape. **(a)** Antiphase domains (translation twins), **(b)** inversion twins of the type of 180° domains in triglycine-sulphate, **(c)** axial twins of the type of Dauphine twins in quartz. (*I, II*) Twin components

translation twin, whose components (antiphase domains) are related by translation by the length of the Burgers vector.

Figure 5.34b demonstrates a twin arising on mutual displacement of two different sublattices. In twinning, the polar direction reverses, so that the twin components may be 180° ferroelectric domains with opposite electrical polarizations. The twinning operations include inversion. Therefore, not only the optic indicatrices, but all the elastic constants of the components coincide. Examples are 180° domains in triglycinesulphate.

In Fig. 5.34c, three hexagonal sublattices merging into each other on rotation about a threefold screw axis perpendicular to the drawing shift by equal distances along twofold axes, so that the component "rotates" through 180° or 60° in the plane of the drawing. This axial twin is absolutely similar to the Dauphine twins of β-quartz (Fig. 5.29a). The optic indicatrices of the components coincide; there is no twin shift, but one can distinguish elastic constants s_{14} corresponding to shears in the basal plane (the plane of the drawing) caused by tangential stresses acting on this plane. As a result, in distinction to the preceding case, a twin can be obtained (and destroyed) by applying definite mechanical stresses. The elastic energy of the Dauphine twins of β-quartz in a given external stress field differs by

$$\Delta W = -2s_{14}[\sigma_{23}(\sigma_{11} - \sigma_{22}) + 2\sigma_{12}\sigma_{13}] . \tag{5.112}$$

Twinning in quartz is thus affected only by tangential stresses acting on the basal plane and on the planes parallel to the optic axis. The value of ΔW determines the thermodynamic force acting on the boundary of the Dauphine twins and causing its displacement. In contrast to twins with a change in shape (5.107), a change in the sign of the stresses here does not affect the direction of boundary displacement. To change the sign of ΔW it is necessary to change the type of the stressed state. With a uniaxial stressed state

$$\Delta W = -2s_{14}\sigma^2 \sin^3 \theta \cos \theta \cos 3\varphi , \tag{5.113}$$

where the polar angle θ and azimuth φ of the loading axis are measured from the z and y axes of the crystallophysic coordinate system, respectively. The symmetry of ΔW is described by antisymmetry group $6':2$, as would be expected.

Note that in a given stress field, those components of twins without a change in shape are energetically favorable which show the minimum rigidity and, hence, have the maximum elastic energy. This seemingly strange statement follows from the fact that the complete potential energy of twins in an external field, calculated with an allowance for the work of the given forces on displacements of the specimen surface is precisely equal to the elastic energy taken with the opposite sign. Accordingly, the component of the Dauphine twin for which $\Delta W > 0$ proves energetically favorable. For quartz, $s_{14} < 0$, and consequently mechanical twinning is also favorable for uniaxial loading along the directions satisfying the condition $\cos \theta \sin 3\varphi > 0$.

General analysis of the physical properties of twin components requires a comparison of their twinning operations with the anisotropy ot their corresponding characteristics. If the twinning operations coincide with the symmetry elements of the optic indicatrix, no twin shear occurs, and the components do not differ in their refractive indices, but may differ in the sign of optical activity (Brazilian twins, domains in triglycinesulphate, etc.). If the twinning operations (for instance, inversion) coincide with the symmetry elements of the elastic-constant tensor, then the elastic properties of the components are identical, etc.

5.8 Direct Observation of Lattice Defects

One of the most important achievements of modern crystallography is the development of a number of methods for studying the arrangement and principal characteristics of dislocations, stacking faults, and even point defects of the lattice. Some of these methods found immediately industrial application in the quality control of single and polycrystals.

5.8.1 Ionic Microscopy

Atomic resolution was first achieved with the aid of an ionic projector, which gives on a fluorescent screen an image of the proper anode in the form of a very fine needle. The projector is filled with hydrogen or helium at very low pressure. The gas atoms are ionized near the anode needle point, where the electrostatic field may reach 10^7 V/cm, and move further in the radial direction, creating on the screen image of the inhomogeneities of the electric field which are present at the anode surface. If the anode is cooled to the temperature of liquid nitrogen or helium, resolution of the atomic structure of the needle point is achieved [Ref. 1.6, Fig. 1.19]. Such images reveal subgrain boundaries, individual dislocations, and even individual point defects (vacancies, interstitials and impurity atoms).

5.8.2 Electron Microscopy

Direct resolution of the atomic structure of a crystal is also provided by transmission electron microscopes. They enable the observation of atomic rows and planes disposed along the direction of transmission [Ref. 1.6, Fig. 4.107] and the location of sites at which the mutual arrangement of the atomic planes indicates the presence of dislocations, where atomic planes break off (edge dislocations) or bend, shifting to the next storey (screw dislocations). As an example, Fig. 5.35 shows a image of a single-crystal silicon film with a single dislocation. The spacing between the atomic planes is 3.138 Å. Some electron microscope observations have shown a sharp diffraction contrast from clusters of point defects and individual heavy atoms.

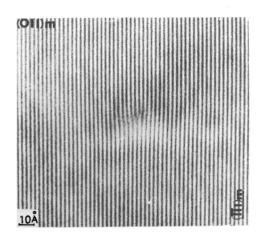

Fig. 5.35. Direct electron micrograph of a unit dislocation in silicon [5.2]

When using low-resolution microscopes it is possible to investigate the real structure of crystals by the *moiré method* proposed by A. V. Shubnikov. If a beam passes through crystal platelets with different lattice spacings or orientations placed on each other, the screen shows periodically arranged moiré bands corresponding to sites with increased summary density. The distance between the moiré bands is equal to a/ε, if platelets with a relative difference between the lattice spacings $\varepsilon = \Delta a/a$ are used, and a/ω if the platelets are turned through an angle ω (in both cases a is the lattice spacing). The distortions of the moiré bands help to judge the distortions in the crystals. In particular, under certain conditions dislocations cause "dislocations" (band terminations) in the moiré patterns. The moiré method proved to be most convenient in investigating the real structure of epitaxial films.

The method most widely used for investigating lattice defects in thin $(10^{-5} - 10^{-4}$ cm) crystal platelets is diffraction electron microscopy. Owing to the small wavelength of the electrons (0.04 Å for a microscope with a working voltage of 100 kV) the lattice defect pattern can be studied in the column (beam) approximation. According to the kinematic theory of diffraction the amplitude of diffracted wave E_1 varies along the beam as

$$dE_1/dz = F \exp(2\pi i s z), \qquad (5.114)$$

where F is the structure factor, and parameter s is equal to the z component of vector s, which characterizes the deviation from the condition of regular Wulff – Bragg reflection (see [Ref. 1.6, Fig. 4.26]). The amplitude of the passing wave is taken to be unity. In the perfect crystal $s = $ const, and in the phase plane, relation (5.114) corresponds to the motion of the terminus of the vector of complex amplitude E_1 along a circle of radius $R = F/2\pi s$ (Fig. 5.36). By integrating expression (5.114) or by direct inspection of Fig. 5.36 we ascertain that the intensity of the diffracted wave on emerging from a crystal of thickness t

$$I = |E_1|^2 = F^2 \frac{\sin^2 \pi t s}{(\pi s)^2}. \qquad (5.115)$$

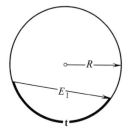

Fig. 5.36. Trajectory of the terminus of the complex amplitude vector on the phase plane for the perfect crystal

As the crystal thickness increases, I changes periodically from zero to $(F/\pi s)^2$ with a period $t_k = s^{-1}$. If the specimen is bent or is not uniform throughout its thickness, *extinction bands* of equal slope ($s = $ const) or equal thickness ($t = $ const) appear on its image. An allowance for the interaction of the incident and the diffracted wave required by the dynamic theory of diffraction leads to a formula similar to (5.115), the only difference being that s is replaced by

$$s^* = \sqrt{s^2 + t_0^{-2}}, \qquad (5.116)$$

where t_0 is the extinction length for a given reflection. As a result, when s diminishes, the period of intensity oscillations $t^* = (s^*)^{-1}$ tends to t_0.

If the scattering centers are shifted by vector u, the diffracted-wave phase will change by an angle $\alpha = 2\pi g u$, where g is the reciprocal-lattice vector corresponding to a given reflection. As a result, in place of (5.114) we have

$$dE_1/dz = F\exp(2\pi isz + 2\pi igu),\qquad(5.117)$$

and for crystals distorted by lattice defects ($u \neq$ const) the phase diagrams are distorted. Thus, in a face-centered cubic crystal an intrinsic stacking fault shifts the crystal parts that it separates by $u = (a/6)\langle112\rangle$. For reflection (hkl), vector $g = (1/a)[hkl]$ and the beam crossing a stacking fault with $u = (a/6)[112]$ acquires a phase difference $\alpha = \pi/3\,(h + k + 2l)$, i.e., either $\alpha = 0$ and the stacking fault does not give a diffraction contrast, or $\alpha = 2\pi/3$ as on Fig. 5.37, and the trajectory of the terminus of the amplitude vector on the phase diagram experiences a kink by an angle of 120° and shifts to a different circle on the same radius [1]. The diffracted-wave amplitude varies from 0 to $F(1 + \sqrt{3}/2)/\pi s$ depending on the beam path length before and after the stacking fault (t_1 and t_2, respectively, in Fig. 5.37). As a result the tilted stacking faults induce a series of extinction bands parallel to the trace of the wave front.

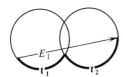

Fig. 5.37. Trajectory of the terminus of the complex amplitude vector on the phase plane for a intrinsic-type stacking fault in an fcc crystal

Smooth changes in displacement vector near dislocations cause smooth distortions in phase diagrams of the type shown in Fig. 5.38. As is seen from Fig. 5.5, the atomic planes perpendicular to an edge dislocation or parallel to a screw dislocation are not curved. On reflection from these planes $gu = 0$; dislocations do not give a diffraction contrast. As a first approximation, one may take into account only the branching term (5.24) of the displacement field around a dislocation. Then for a dislocation perpendicular to the beam and situated at a distance x from it

$$\alpha = 2\pi gu \approx gb\,\mathrm{arctg}\,(z/x).\qquad(5.118)$$

The intensity of the dislocation contrast is, accordingly, determined, as a first approximation, by scalar product gb. For reflections with $gb = 0$ the diffraction image of dislocations disappears, and this indicates the direction of the Burgers vector of the dislocation. Finer details of the diffraction pattern indicate the sign and magnitude of the Burgers vector as well as the inclination of the dislocation relative to the wave-front plane.

---

1 The values $\alpha = \pi$ and $\alpha = +\pi/6$ are not possible since a face-centered-cubic lattice does not give reflections with an odd sum $h + k + l$.

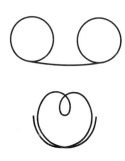

Fig. 5.38. Distortion of phase diagrams on smooth change in displacement vector near a dislocation

The width of the diffraction pattern of a dislocation for reflections with gb values of the order of unity is about $t_0/3$ (from 70 to 300 Å for different crystals). As a result, the electron microscope makes it possible to investigate the dislocation structure of even severely deformed metals, resolve the dislocation structure of subgrain boundaries misoriented by angles of about one degree, observe the dissociation of dislocations and dislocation nodes, etc.

To narrow down the dislocation image and improve the resolution, diffraction electron microscopy resorts to various techniques for reducing extinction length t_0. An example is the so-called weak-beam method, where the specimen is set in a position corresponding to a strong reflection $n(\bar{h}\,\bar{k}\,\bar{l})$, but the dislocation image is observed in a weak beam (hkl), where the image width is only of the order of $10-20$ Å.

In high-voltage microscopes $(1-3$ MeV) the image is, as a rule, formed by series of beams of the type $n(hkl)$ and is as narrow as in the weak-beam method. A still more important advantage of high-voltage microscopes is the possibility of transmission through crystal platelets of comparatively large (up to several micrometers) thickness, which eliminates a number of difficulties resulting from the disturbance of the real structure of crystals in the course of the preparation of specimens for electron-microscopic investigation.

A rigorous analysis of the diffraction image of a real crystal requires the use of dynamic, rather than kinematic, diffraction theory. In this theory the wave field of the electrons inside the crystal is resolved along the Bloch waves

$$\psi_k = \exp(\mathrm{i}kr) \sum_h A_h \exp(\mathrm{i}k_h r) \tag{5.119}$$

where k_h are the reciprocal-lattice vectors. The Bloch waves satisfy the Schroedinger equation for the potential field induced in the crystal by ions and electrons and propagate freely in the crystal. With a given electron energy the resolved values of the wave vectors obey some dispersion equation $F(k) = 0$, which in the k space corresponds to a multibranched dispersion surface. Since a dispersion surface can be interpreted as an isoenergetic surface in the reciprocal-lattice space, an analysis of its shape does not differ in the least from an usual analysis, for instance that of the shape of a Fermi surface. In

Sect. 4.3), the two-beam contrast occuring most frequently[2]. The most diverse techniques are employed here depending on the quality of the crystal, its size, and the absorption. The best resolution is obtained by using monochromatic radiation from narrow (point or line) sources. Figure 5.39 illustrates the basic methods of x-ray diffraction topography which reveal, not only the mosaic structure of the crystal, but also individual dislocations. Figure 5.39a schematizes photography of the surface of a crystal in a monochromatized beam reflected at the Bragg angle (Berg – Barrett method). The photoplate is placed near the surface irradiated. Due to extinction, only a thin surface layer (1 μm for aluminium and 50 μm for lithium fluoride) takes part in the reflection. If this layer contains dislocations, the primary extinction is weakened near them, and the diffracted wave is enhanced. Subgrains are revealed by the diffraction contrast.

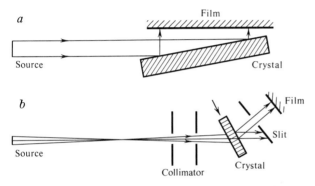

Fig. 5.39. Basic methods of x-ray topography. **(a)** Berg – Barrett method, **(b)** Lang method

Figure 5.39b illustrates the scheme of photography of a weakly absorbing crystal in the passing beam (Lang method). A crystal brought into a precisely reflecting position is scanned under the incident beam, and the photoplate is moved synchronously with it. Such an x-ray diffraction topograph pictures the distribution of the dislocations in the bulk of large crystals with a surface up to several square centimeters. By taking two photographs with the use of reflections (hkl) and $(\bar{h}\bar{k}\bar{l})$, one obtains a stereocouple, which helps to judge the spatial arrangement of dislocations.

When photographing strongly absorbing crystals in the transmitted beam, use is made of anomalous passage of x-rays (Borrmann effect). As with electrons, the wave field propagating along the reflecting planes is composed of two waves. In the s-type wave the intensity maxima coincide with the lattice points, and in the p-type wave they are disposed interstitially. As a result, the

---

[2] Dynamical scattering in an ideal crystal in the case of a simultaneous appearance of three beams may be used for direct experimental determination of their phases [5.3], see also [Ref. 1.6 (Sect. 4.7.6)].

s-type wave experiences considerable photoelectric absorption and damps out rapidly, whereas the p-type wave passes through the crystal almost unhindered and produces a shadow pattern of all the defects encountered en route on a photoplate placed behind the specimen. The contrast of the image then mainly depends on the effect of the interbranch scattering, rather than on the change in the level of photoelectric absorption of x-rays in the distorted regions. Dislocations with $(gb) = 1$ interfere most effectively with the propagation of p-type waves and produce the deepest shadows on x-ray diffraction topographs. With a special position of dislocations, when the dislocation lines are parallel to the diffraction vector, the image contrast is similar to the electron-microscopic dislocation contrast.

The theory of the x-ray image of lattice defects is generally constructed similar to the theory of the electron-microscopic image. One can, as a rule, restrict oneself to the two-wave approximation, but it is necessary to take into account, not only the dynamic effects of the transmitted and the reflected wave, but also the difference in the directions of their propagation. Local perturbations of the x-ray wave field propagate in all the directions between those of the wave vectors of the transmitted and the reflected wave, which results in the smearing out of the images of the lattice defects in the direction of the diffraction vector.

The rigorous solution of the image theory problems requires the consideration of space-inhomogeneous problems corresponding to exitation on dispersion surfaces, no longer of individual points, but of extended regions. In place of individual Bloch waves, wave packets have to be considered as well as the wave equations found for the amplitudes of these packets from the Maxwell equation. The two-wave approximation yields Takagi equations relating the amplitudes of the transmitted E_0 and the diffracted E_1 wave

$$\left(\frac{\partial}{\partial z} + \operatorname{tg}\theta\,\frac{\partial}{\partial x}\right)E_0 = \frac{i\pi}{t_0}\exp[2\pi i(gu)]E_1$$

$$\left(\frac{\partial}{\partial z} - \operatorname{tg}\theta\,\frac{\partial}{\partial x}\right)E_1 = \frac{i\pi}{t_0}\exp[-2\pi i(gu)]E_0 . \tag{5.123}$$

Here, t_0 is the extinction length, θ is the Bragg angle, the z axis is perpendicular to the crystal surface, and the x axis is parallel to the reflection vector. For simplicity, the symmetric Laue case is considered, where the reflecting planes are perpendicular to the crystal surface.

In slightly distorted parts of the crystal, the wave field can be decomposed into Bloch waves corresponding to the solution of set (5.123) for a coordinate-linear field $u = u(z,x)$. The adaptation of the Bloch waves to the atomic structure of the distorted crystal results in bending of the trajectories of the Bloch waves. Introducing the notation $\tan\theta = c$ and $dx/dz = v$ for the tangent of the Bragg angle and for the angular coefficient of the trajectories

$y = y(z)$ of the Bloch waves, we get equations

$$\pm \frac{d}{dz}\left(\frac{m_0 v}{\sqrt{1 - (v/c)^2}}\right) = f \tag{5.124}$$

similar to those for the trajectory of relativistic particles in the external field. Here, signs $\pm$ correspond to different branches of the dispersion surface (s- and p-type waves behave as particles and antiparticles and deflect in different directions in the external field). The "rest mass" m_0 is determined by the splitting of the dispersion surface at the Brillouin zone boundary

$$m_0 = \frac{\pi}{t_0}\,\text{ctg}\,\theta. \tag{5.125}$$

"External force" f depends on the curvature of reflecting planes or of the density of their distribution

$$f = \pi\left(\frac{\partial^2}{\partial z^2} - \text{tg}^2\,\theta\frac{\partial^2}{\partial x^2}\right)(gu). \tag{5.126}$$

To calculate the sectional image of a lattice defect in the beam approximation, one has to construct the beam trajectories (i.e., the trajectories of the Bloch waves) by (5.124) and find, for each point of the exit surface, two trajectories (for an s- and a p-type wave) arriving at this point from the illuminated point on the specimen surface. The initial angular coefficients of the trajectories then have to be varied. Further, the Bloch wave phases are calculated along the trajectories, and the phase shift is determined at the point where the waves meet and interfere. In the axial section of the topograph, for small (gu), the result differs but little from the electron-microscopic (5.122). For dislocations in a special position the limiting value of integral (5.122), with the scattering plane tending to the dislocation, equals $\Gamma_0 = \pi(gb)$, which corresponds to $(gb)/2$ additional extinction lines on each side of the dislocation. Investigation into the shape of the extinction contours makes it possible to determine the Burgers vector of dislocation with high accuracy. For surface films, growth bands, and subgrain and domain boundaries, a beam approximation (5.124) also helps to reveal the important details of the x-ray pattern. Calculation of the trajectories then reveals the effects of total internal reflection, the effect of beam focusing for different types of waves, and the waveguide effects. The first-named effect leads to long-range surges of the field from strongly distorted regions, the second, to an increase in image brightness and in the high sensitivity of the x-ray diffraction topographic detection of band- and column-type distortions, while the third effect explains the experimentally observed channeling of the radiation along the dislocations and stacking faults.

In the region of strong distortions, beam approximation (5.124) is inapplicable because of strong interbranch scattering. Investigation into the wave field requires here the solution of Tagaki equations (5.123).

For the simplest stacking fault it suffices to consider the generation of a new wave field as a result of the interbranch scattering of Bloch waves. As with electron microscopy, on crossing a stacking fault the diffracted wave acquires an additional phase shift $\alpha = 2\pi(gu)$. This shift is equivalent to a change in the amplitude of the Bloch waves. If the stacking fault plane lies between the directions of the transmitted and the diffracted waves, the stacking fault serves as a mirror reflecting the wave field, and as a waveguide. If the stacking fault plane intersects the direction of the transmitted and the diffracted waves, the stacking fault may serve as a diffraction lens creating a bright image of the source (slit) at a point symmetric across the fault plane. The short-range dislocation field can also serve as a diffraction lens for the Bloch waves and produce a bright spot on sectional topographs. At high (gb) values this spot reveals an internal structure in the form of a bundle of (gb)/2 interference bands corresponding to spatially modulated interbranch scattering.

To increase the sensitivity of x-ray diffraction topographic investigations of distortions, wide use is made of various double-crystal devices using the interference of wave fields scattered by different crystals or different parts of one and the same crystal. The sensitivity of these methods to deformations and rotations of the lattice is of the order of the ratio between the wavelength of x-radiation and the beam width. The effect of diffraction focusing of the Bloch waves enables one to measure rotation angles of 10^{-3} angular second and to clearly detect lattice rotations from a single dislocation at distances of about 1 mm. The opportunities offered by the x-ray moiré technique can be evaluated as follows. If the difference in the structure of two adjacent crystal blocks is described by displacement vector u, then the additional phase shift arising at block boundaries is characterized by an angle $\alpha = 2\pi gu$, as in a stacking fault. With uniform deformation, u depends linearly on the coordinates, and phase factor $\exp(i\alpha)$ undergoes periodic changes, restoring its value each time the product gu varies by unity, i.e., u changes by an interplanar distance $d = g^{-1}$ corresponding to the given reflection. As a result, oscillations in image intensity (*x-ray moiré*) arise and make it possible to investigate small (of the order of d/D, where D is the distance between the moiré bands) deformations and rotations of the lattice. At $d = 3$ Å and $D = 3$ mm, the moiré corresponds to a change in lattice parameter $\Delta d/d = 10^{-7}$ or to a misorientation $\omega = 0.02''$.

Various effects influence the formation of an x-ray diffraction pattern, depending on the geometry of the lattice defects and the photography conditions, namely: diffraction smearing-out of beams by local distortions and diffraction focusing of beams by stacking faults and dislocations, geometric reflection of x-rays by two-dimensional defects and complete internal reflection with field channeling along the defect, etc. Therefore, image analysis often involves considerable difficulties.

Fig. 5.40 a, b. Mosaic structure. **(a)** x-ray topogram of an MgO crystal, × 15 (Courtesy of V. F. Miuskov), **(b)** electron micrograph of a Ge crystal, × 20,000 (Courtesy of N. D. Zakharov)

The width of an x-ray dislocation image usually equals tens of micrometers, which exceeds by three orders the width of the electron-microscopic dislocation image. Therefore, resolution of the dislocation structure of crystals by means of x-ray techniques is possible only for comparatively perfect crystals with a dislocation density not higher than 10^5 cm^2, and resolution of dislocations in subgrain boundaries is possible only for very slightly (not more than by several seconds) misoriented subgrains.

By way of example, Fig. 5.40a shows an x-ray diffraction topograph of the mosaic structure of MgO obtained by the Lang method. The dislocation

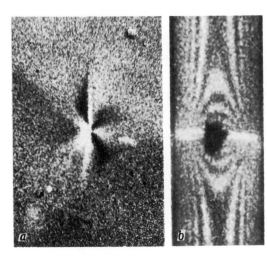

Fig. 5.41 a, b. X-ray topogram of a dislocation in silicon (Courtesy of E. V. Suvorov), when viewing (**a**) along the dislocation, (**b**) normal to the dislocation

structure of boundaries in a mosaic of $200-500$ µm subgrains misoriented by an angle of several seconds has been resolved. Larger subgrains misoriented by angles of about one minute clearly differ in contrast and have sharp boundaries. Subgrains turned by large angles drop out of the image altogether and seem to be white. For comparison, Fig. 5.40 b shows an electron micrograph of subgrains in a single crystal of silicon. Here, the dislocation structure of subgrain boundaries misoriented by tens of minutes is resolved. Figure 5.41 a presents a picture of dislocation in silicon obtained by the Borrmann method. The distortion field around the dislocation can clearly be seen, the areas with stresses of opposite signs differ qualitatively. For further comparison, Fig. 5.41 b presents a sectional topograph of a dislocation in a silicon crystal. Here, the distortion field can be characterized quantitatively by the pattern of interference (extinction) bands.

5.8.4 Photoelasticity Method

The stress field around isolated dislocations parallel to the direction of observation is revealed still more clearly by the use of the photoelasticity method based on the effect of stresses on the light-wave velocity. In an initially isotropic medium the stresses cause birefringence, whose intensity is proportional to the difference between the principal stresses in the wave-front plane. Proportionality factor C is called the photoelastic constant of the material. In crystals, changes in the coefficients of the optic indicatrix are expressed linearly in terms of the components of the stress tensor with the aid of photoelastic coefficients forming a fourth-rank tensor (see [Ref. 1.7]). Figure 5.42 gives photos of stresses around edge dislocations in silicon obtained with the aid of a polarization infrared microscope equipped with an

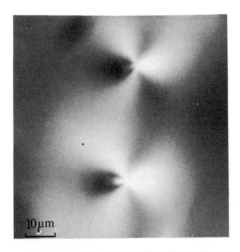

Fig. 5.42. Stress rosettes around edge dislocations in silicon (Courtesy of V. I. Nikitenko)

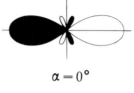

$$\alpha = 0°$$

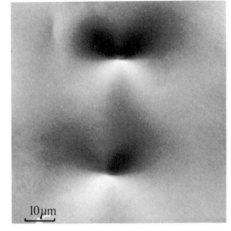

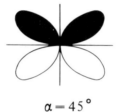

$$\alpha = 45°$$

electron-optical image converter, which permits observation of the image in a visible light. For comparison, the same figure presents calculated bire-fringence rosettes. When the rays are directed along the dislocations, the difference between the principal stresses in the plane of the wave front is, by (5.27), $2\sigma_{r\theta} = Gb/\pi(1 - v) \cdot \cos\theta/r$, whence the birefringence

$$\Delta = n_1 - n_2 = C\frac{Gb}{\pi(1 - v)}\frac{\cos\theta}{r}. \tag{5.127}$$

When a crystal is observed in crossed nicols, the intensity of the propagated light is defined by the difference between the diffractive indices for oscillations along planes disposed at an angle of 45° to the polarizer and analyzer axes. If the axes of the crossed polarizer and analyzer make an angle

φ with the Burgers vector, the observed birefringence is weakened by a factor of $\sin 2(\theta - \varphi)$. As a result, the shape of the intensity rosettes is given by the equation

$$r = \text{const} \cos \theta \sin 2(\theta - \varphi) \,. \tag{5.128}$$

For $\varphi = 0$ we obtain the rosette of Fig. 5.42a, and for $\varphi = 45°$, that of Fig. 5.42b. Determination of the sign and value of the birefringence near the dislocations gives the sign, direction, and magnitude of the Burgers vector.

The photoelasticity method also helps to investigate the elastic field of dislocation ensembles, growth bands and pyramids, slip bands, twins, incoherent subboundaries, as well as to obtain information on spontaneous deformation of individual domains in a polydomain specimen, and sometimes on the structure of the domain boundaries.

5.8.5 Selective Etching Method

During the dissolution of a crystal, the dissolution nuclei are formed more readily in regions with higher energy. Since high internal stresses around dislocations cause local increases in crystal energy, the ends of the dislocation lines on the crystal surface can sometimes be revealed by the action of various etchants on the crystal. An efficient choice of an etchant and etching conditions produces pronounced etch pits at the dislocation ends whose shape helps to judge the inclination of the dislocation to the crystal surface. Repeated etching with repolishing and continuous etching in polishing solutions permits tracing the position of dislocations in the specimen bulk. Repeated etching without repolishing and continuous etching under load are helpful in studying

Fig. 5.43. Jumpwise movement of dislocations in a NaCl crystal (Courtesy of V. N. Rozhansky, A. S. Stepanova)

the kinetics of dislocation movement. As an example, Fig. 5.43 gives a photo illustrating the jumpwise movement of dislocations in a NaCl crystal. After a dislocation leaves the etch pit, the latter ceases to deepen and, spreading out, acquires a flat bottom. The size and shape of the etch pit therefore indicate the moment when the dislocation arrived at the given point and the time it has been there.

The possibility of the formation of clearly defined dislocation pits on etching low-index (close-packed) faces is ensured by the anisotropy of the crystal dissolution rates. Detection of dislocations on an arbitrarily oriented surface usually requires preliminary deposition of some impurity on the dislocations, which facilitates the formation of etching nuclei. Such decoration (coloring) of dislocations in transparent crystals enables one to observe directly the three-dimensional arrangement of dislocations.

The selective etching method is also used to detect various clusters, precipitates, domain and twin boundaries, fast-particle tracks, and other similar defects.

5.8.6 Investigation of the Crystal Surface

The points of emergence of dislocations at a crystal face can be detected by investigating the atomic structure of the face. By constructing a Burgers circuit on the crystal surface as in Fig. 5.5, it is easy to ascertain that the end of the dislocation (not necessarily of a screw one) corresponds to the end of a step, whose height is equal to the normal component of the Burgers vector. By observing steps on crystal faces with the aid of an optical, electron, or ion microscope directly or after decoration with foreign particles one can judge the density and the arrangement of the dislocations in the crystal. Decoration of the crystal surface permits direct observation of the dislocation ends, precipitates, and fine details of domain structure. The possibilities for decorating individual point defects are being investigated. The effect of dislocations on the microrelief of crystal faces and the importance of this effect for crystal growth and dissolution (evaporation) are studied in detail in the dislocation theory of crystal growth [5.1].

Bibliography

Anselm, A.I.: *Vvedeniye v teoriyu poluprovodnikov* (Introduction to Semiconductor Theory) (Fizmatgiz, Moscow 1962) [German transl.: *Einführung in die Halbleitertheorie* (Akademie, Berlin 1964)]

Bata, L. (ed.): *Advances in Liquid Crystal Research and Applications*, v. 1,2 (Pergamon, Oxford; Akademiai Kiadò, Budapest 1981)

Becker, P. (ed.): *Electron and Magnetization Densities in Molecules and Crystals* (Plenum, New York 1980)

Belov, N.V.: *Kristallokhimiya silikatov s krupnymi kationami* (Izd-vo Akad. Nauk SSSR, Moscow 1961) [English transl.: *Crystal Chemistry of Large Cation Silicates* (Consultants Bureau, New York 1965)]

Belov, N.V.: *Ocherki po strukturnoi mineralogii* (Essays on Structural Mineralogy) (Nedra, Moscow 1976)

Belov, N.V.: *Struktura ionnykh kristallov i metallicheskikh faz* (The Structure of Ionic Crystals and Metallic Phases) (Izd-vo Akad. Nauk SSSR, Moscow 1947)

Belyaev, L.M. (ed.): *Ruby and Sapphire* (Natl. Bureau of Standards, Washington 1980)

Bersuker, I.B.: *Electronnoye stroyenie i svoistva koordinatsionnykh soedinenii* (Electron Structure and Properties of Coordination Compounds) (Khimiya, Leningrad 1976)

Blundell, T.L. , Jonson, L.N.: *Protein Crystallography* (Academic, London 1976)

Boky, G.B.: *Kristallokhimiya* (Crystal Chemistry) (Nauka, Moscow 1971)

Born, E., Paul, G.: *Röntgenbeugung am Realkristall* (Thiemig, Munich 1979)

Born, M., Huang, Kun.: *Dynamical Theory of Crystal Lattices* (Clarendon, Oxford 1954)

Bragg, L., Claringbull, G.F.: *Crystal Structures of Minerals* (Bell, London 1965)

Bueren, H.G. van: *Imperfections in Crystals* (North-Holland, Amsterdam 1960)

Chebotin, V.N., Perfiliev, M.V.: *Elektrokhimia tvërdykh elektrolitov* (Electrochemistry of Solid Electrolytes) (Khimiya, Moscow 1978)

Chistyakov, I.G.: "Ordering and Structure of Liquid Crystals", in *Advances in Liquid Crystals,* ed. by G.H. Brown (Academic, New York 1975)

Cottrell, A.H.: *Theory of Crystal Dislocations* (Blackie, London 1964)

Coulson, C.A.: *Valence* (Clarendon, Oxford 1952)

Crawford, J.H., Slifkin, L.M. (eds.): *Point Defects in Solids* (Plenum, London 1972)

Damask, A.C., Dienes, G.J.: *Point Defects in Metals* (Gordon and Breach, London 1963)

Dashevsky, V.G.: *Konformatsiya organicheskikh molekul* (Conformation of Organic Molecules) (Khimiya, Moscow 1974)

Dederichs, P.H., Zeller, R.: "Dynamical Properties of Point Defects in Metals", in Point Defects in Metals II, *Springer Tracts in Modern Physics*, Vol. 87, Springer, Berlin, Heidelberg, New York 1980)

Dickerson, R.E., Geis I.: *The Structure and Action of Proteins* (Harper and Row, New York 1969)

Donnay, J.D.H.: *Crystal Data,* 2nd ed. (Washington, American Crystallographic Association 1963)

Donohue, J.: *The Structures of the Elements* (Wiley, New York 1974)

Evans, R.C.: *An Introduction to Crystal Chemistry*, 2nd ed. (Cambridge University Press, Cambridge 1964)

Farge, Y., Fontana, M. P.: *Electronic and Vibrational Properties of Point Defects in Ionic Crystals* (North-Holland, Amsterdam 1979)

Fermi, E.: *Moleküle und Kristalle* (Barth, Leipzig 1938)

Fersman, A.E.: *Geokhimiya* (Geochemistry), Vols. 1–4 (Khimteorizdat, Moscow 1933–1939)

Fridkin, V.M.: *Photoferroelectrics*, Springer Series in Solid-State Sciences, Vol. 9 (Springer, Berlin, Heidelberg, New York 1979)

Friedel, J.: *Dislocations* (Pergamon, Oxford 1964)

Geil, P.H.: *Polymer Single Crystals* (Wiley Interscience, New York 1963)

Gilman, J.J.: *Micromechanics of Flow in Solids* (McGraw-Hill, New York 1969)

Hagenmuller, P., Gool, W. van (eds.): *Solid Electrolytes. General Principles, Characterization, Materials, Applications* (Academic, London 1978)

Hirth, J.P., Lothe, J.: *Theory of Dislocations* (McGraw-Hill, New York 1968)

Indenbom, V.L., Jan, R.V., Kratochvil, J., Kröner, E, E., Kroupa, F., Ludwig, W., Orlov, A.N., Saada, G., Seeger, A., Sestàk, B.: *Theory of Crystal Defects* (Academia, Prague 1966)

Indenbom, V.L., Chukhovsky, F.N.: Usp. Fiz. Nauk *107*, 229 (1972) [English transl.: The image problem in X-ray optics. Sov. Phys. Uspekhi *15*, 298 (1972)]

Indenbom, V.L.: Ein Beitrag zur Entstehung von Spannungen und Versetzungen beim Kristallwachstum. Krist. Tech. *14*, 493 (1979)

Itogi Nauki. Ser. Khimich. (Advances in Science. Chemistry), Vols 1–7. (VINITI, Moscow 1966–1970)

Itogi Nauki i Tekhniki. Ser. Kristallokhimiya. (Advances in Science and Technology. Crystal Chemistry), Vols. 8–13 (VINITI, Moscow 1972–1979)

Iveronova, V.I., Katsnel'son, A.A.: *Blizhniy poryadok v tvërdykh rastvorakh* (Short-Range Order in Solid Solutions) (Nauka, Moscow 1977)

Kennard, O., Watson, D.G. (eds.): *Molecular Structures and Dimensions. Bibliography 1935–1974* (Oosthoek's Uitgevers Mij, Utrecht 1970–1975) Vols. 1–6

Kitaigorodsky, A.I.: *Organicheskaya kristallokhimiya* (Izd-vo Akad. Nauk SSSR, Moscow 1955) [English transl.: *Organic Chemistry and Crystallography* (Consultants Bureau, New York 1961)]

Kitaigorodsky, A.I.: *Molekulyarnye kristally* (Nauka, Moscow 1971) [English transl.: *Molecular Crystals and Molecules* (Academic, New York 1973); German transl.: *Molekülkristalle* (Akademie, Berlin 1979)]

Klassen-Neklyudova, M.V.: *Mekhanicheskoye dvoinikovaniye kristallov* (Izd-vo Akad. Nauk SSSR, Moscow 1960) [English transl.: *Mechanical Twinning of Crystals* (Consultants Bureau, New York 1964)]

Kosevich, A.M.: *Osnovy mekhaniki kristallicheskoi reshotki* (Fundamentals of Crystal Lattice Mechanics) (Nauka, Moscow 1972)

Kovacs, J., Zsoldos, L.: *Dislocations and Plastic Deformation* (Akademina Kiado, Budapest 1973)

Krebs, H.: *Grundzüge der anorganischen Kristallchemie* (Enke, Stuttgart 1968)

Kripyakevich, P.I.: *Strukturnye tipy intermetallicheskikh soedinenii* (Structural Types of Intermetallic Compounds) (Nauka, Moscow 1977)

Ladd, M.F.C.: *Structure and Bonding in Solid State Chemistry* (Ellis Horwood, Chichester 1979)

Landau, L.D., Lifshits, E.M.: *Kvantovaya mekhanika* (Nauka, Moscow 1972) [English transl.: *Course of Theoretical Physics*, Vol 3, Quantum Mechanics. Non-Relativistic Theory, 3rd ed. (Pergamon, Oxford 1977)]

Landau, L.D., Lifshits, E.M.: *Statisticheskaya fizika* (Nauka, Moscow 1964) [Enlish transl.: *Course of Theoretical Physics*, Vol. 5, Statistical Physics, 2nd ed. (Pergamon, Oxford 1969)]

Lebedev, V.I.: *Ionno-atomnye radiusy i ikh znachenie dlya geokhimii* (Ionic-Atomic Radii and Their Importance to Geochemistry) (Izd-vo Leningrad. Univ., Leningrad 1969)

Leibfried, G., Breuer, N.: *Springer Tracts in Modern Physics*, Vol 81, Points Defects in Metals I (Springer, Berlin, Heidelberg, New York 1978)

Levin, A.A.: *Vvedeniye v kvantovuyu khimiyu tvёrdogo tela* (Moscow, Khimiya 1976) [English transl.: *Quantum Chemistry of Solids: The Chemical Bond and Energy Bands in Tetrahedral Semiconductors* (McGraw, New York 1976)]

Lothe, J.: "Dislocations in Anisotropic Media", in *Computational Solid State Physics*, ed. by F. Herman, N.W. Dalton, T. Koehler (Plenum, New York 1972) pp. 425–440

Madelung, O.: *Introduction to Solid-State Theory*, Springer Series in Solid-State Physics, Vol 2 (Springer, Berlin, Heidelberg, New York 1978)

Makarov, E.S.: *Izomorfizm atomov v kristallakh* (Isomorphism of Atoms in Crystals) (Atomizdat, Moscow 1973)

Milburn, G.H.W.: *X-Ray Crystallography* (Butterworths, London 1973)

Moffat, W.G., Pearsall, G.W., Wulff, J.: *Structure and Properties of Materials*, Vol. I Structures (Wiley, New York 1964)

Mott, H.F., Jones, H.: *The Theory of Properties of Metals and Alloys* (Oxford U.P., London 1936)

Nowick, A.S., Burton, J.J. (eds.): *Diffusions in Solids. Recent Developments* (Academic, New York 1972)

Ormont, B.F.: *Vvedenie v fizicheskuyu khimiyu i kristallokhimiyu poluprovodnikov* (Introduction to the Physical Chemistry and Crystal Chemistry of Semiconductors) (Vysshaya Shkola, Moscow 1968)

Osipjan, Yu.A. (ed.): *Defecty v kristallach i ich modelirovanie na EVM* (Crystal Defects and their Computer-simulation) (Nauka, Moscow 1980)

Ovchinnikov, Yu.A., Ivanov, Yu. T., Shkrob, A.M.: *Membrano-aktivnye kompleksony* (Nauka, Moscow 1974) [English transl.: *Membrane Active Complexones* (Elsevier, Amsterdam 1975)]

Pauling, L.: *The Nature of the Chemical Bond*, 3rd ed. (Cornell, Ithaca 1960)

Pearson, W.B.: *The Crystal Chemistry and Physics of Metals and Alloys* (Wiley, New York 1972)

Pimentel, C., Spratley, D.: *Chemical Bonding Clarified through Quantum Mechanics* (Holden-Day, San Francisco 1970)

Problemy kristallogii (Problems of Crystallography). Collection of Papers Dedicated to Academician N.V. Belov at his 80th Birthday (Izd-vo MGU, Moscow 1971)

Ramachandran, G.N. (ed.): *Conformation of Biopolymers* (Academic, London 1967)

Regel, A.R. (ed.): *Dislokatsii i fizicheskiye svoistva poluprovodnikov* (Dislocations and Physical Properties of Semiconductors) (Nauka, Leningrad 1967)

Robertson, J.M. (ed.): *Chemical Crystallography* (University Park, Baltimore 1972)

Schroeder, K.: "Theory of Diffusion Controlled Reactions of Point Defects in Metals", in *Springer Tracts in Modern Physics*, Vol. 87, Point Defects in Metals II (Springer, Berlin, Heidelberg, New York 1980)

Seeger, A. (ed.): *Vacancies and Interstitia in Metals* (North Holland, Amsterdam 1970)

Sirota, N.N. (ed.): *Khimicheskaya svyaz' v kristallakh poluprovodnikov i polumetallov* (Chemical Bond in Crystals of Semiconductors and Semimetals Collected Papers) (Izd-vo Nauka i Tekhnika, Misk 1965)

Sirotin, Yu.I., Shaskol'skaya, M.P.: *Osnovy kristallofiziki* (Fundamentals of Crystal Physics), 2nd ed. (Nauka, Moscow 1979)

Slater, J.C.: *Quantum Theory of Molecules and Solids*, Vol. 2. Symmetry and Energy Bonds in Crystals (McGraw-Hill, London 1965)

Sobelman, I.I.: *Atomic Spectra and Radiative Transitions*, Springer Series in Chemical Physics, Vol. 1 (Springer, Berlin, Heidelberg, New York 1979)

Structure Reports, Vols. 8–35 (Oosthoek's Uitgevers Mij, Utrecht 1956–1974)

Strukturbericht, Vols. I–VII [Supplement to Z. Kristallogr. (Akademische, Leipzig 1931–1943)]

Tanner, B.K.: *X-Ray Topography* (Pergamon, Oxford 1976)

Thomas, G., Goringe, M.J.: *Transmission Electron Microscopy of Materials* (Wiley, New York 1979)

Urusov, V.S.: *Energeticheskaya kristallokhimiya* (Energetic Crystal Chemistry) (Nauka, Moscow 1975)

Urusov, V.S.: *Teoriya izomorfnoi smesimosti* (Theory of Isomorphous Miscibility) (Nauka, Moscow 1977)

Utevsky, L.M.: *Difraktsionnaya elektronnaya mikroskopiya v metallovedenii* (Diffraction Electron Microscopy in Physical Metallurgy) (Metallurgiy, Moscow 1973)

Vainshtein, B.K.: *Difraktsiya rentgenovskikh luchei na tsepnykh molekulakh* (Izd-vo Akad. Nauk SSSR 1963) [English transl.: *Diffraction of X-Rays by Chain Molecules* (Elevier, Amsterdam 1966)]

Volkenshtein, M.V.: *Molekulyarnaya biofizika* (Nauka, Moscow 1975) [English transl.: *Molecular Biophysics* (Academic, New York 1977)]

Weigel, D.: *Cristallographie et structure des solides* (Masson, Paris 1972) Vol. 1

Wells, A.F.: *Structural Inorganic Chemistry*, 3rd ed. (Clarendon, Oxford 1962)

Wilson, A.H.: *The Theory of Metals* (Cambridge University Press, Cambridge 1953)

Wyckoff, R.W.C.: *Crystal Structures*, Vols. 1–4 (Interscience, New York 1963–1968)

Zemann, J.: *Kristallchemie* (de Gruyter, Berlin 1966)

Ziman, J.M.: *Principles of the Theory of Solids* (Cambridge University Press, Cambridge 1964)

Crystallographic Journals

Acta Crystallographica, Section A: Crystal Physics, Diffraction, Theoretical and General Crystallography (Acta Crystallogr. A), published since 1948, divided into Sections A and B in 1968

Acta Crystallographica, Section B: Structural Crystallography and Crystal Chemistry (Acta Crystallogr. B)

American Mineralogist (Am. Mineral.), published since 1916

Bulletin de la Société Francaise de Minèralogie et de Cristallographie (Bull. Soc. Fr. Mineral. Cristallogr.), published since 1878

Crystal Lattice Defects (Cryst. Lattice Defects), published since 1969

Crystal Structure Communications (Cryst. Struct. Commun.), published since 1972

Doklady Akademii Nauk SSSR (Dokl. Akad. Nauk SSSR), published since 1933 [English transl.: Soviet Physics-Doklady (Sov. Phys. Dokl.)]

Fizika Metallov i Metallovedeniye (Fiz. Met. Metalloved.), published since 1955 [English transl.: Physics of Metals and Metallography USSR (Phys. Met. Metallogr.)]

Fizika Tvërdogo Tela (Fiz. Tverd. Tela), published since 1953 [English transl.: Soviet Physics-Solid State (Sov. Phys. Solid State)]

Izvestiya Akademii Nauk SSSR, Seriya Fizicheskaya (Izv. Akad. Nauk SSSR Ser. Fiz.), published since 1936 [English transl.: Bulletin of Academy of Science of USSR. Physical Series (Bull. Acad. Sci. USSR Phys. Ser.)]

Journal of Applied Crystallography (J. Appl. Crystallogr.), published since 1968

Journal of Crystal Growth (J. Cryst. Growth), published since 1967

Journal of Materials Science (J. Mater. Sci.), published since 1966

Journal of Physics C: Solid State Physics (J. Phys. C), published since 1968

Journal of Physics and Chemistry of Solids (J. Phys. Chem. Solids), published since 1956

Journal of Solid State Chemistry (J. Solid State Chem.), published since 1969

Koordinatsionnaya Khimiya (Koord. Khim.), published since 1975 [English transl.: Soviet Journal of Coordination Chemistry (Sov. J. Coord. Chem.)]

Kristall und Technik (Krist. Tech.), published since 1966

Kristallografiya (Kristallografiya), published since 1956 [English transl.: Soviet Physics-Crystallography (Sov. Phys. Crystallogr.)]

Molecular Crystals and Liquid Crystals (Mol. Cryst. Liq. Cryst.), published since 1966

Physica Status Solidi (Phys. Status Solidi), published since 1961

Physical Review, Section B: Solid State (Phys. Rev. B), published since 1893

Structure Reports (Struct. Rep.), published since 1956, Vol. 8 and following

Strukturbericht, Vols. 1–7, published from 1936 to 1943

Uspekhi Fizicheskikh Nauk (Usp. Fiz. Nauk), published since 1918 [English transl.: Soviet Physics-Uspekhi (Sov. Phys. Usp.)]

Zeitschrift für Kristallographie, Kristallgeometrie, Kristallphysik, Kristall-chemie (Z. Kristallogr. Kristallgeom. Kristallphys. Kristallchem.), published since 1877

Zhurnal Strukturnoi Khimii (Zh. Strukt. Khim.), published since 1959 [English transl.: Journal of Structure Chemistry (USSR) (J. Struct. Chem. USSR)]

References

Chapter 1

1.1 H. Krebs: *Grundzüge der anorganischen Kristallchemie* (Enke, Stuttgart 1968)

1.2 H.F. White: Phys. Rev. *37*, 1416 (1931)

1.3 A.C. Wahl: Sci. Am. *222*, 54–70 (1970)

1.4 J.T. Waber, D.T. Cromer: J. Chem. Phys. *42*, 4116 (1965)

1.5 V.F. Bratsev: *Tablitsy atomnykh volnovykh funktsii* (Tables of Atomic Wave Functions) (Nauka, Moscow 1966)

1.6 B.K. Vainshtein: *Sovremennaya kristallografiya. T.I.* Simmetriya kristallov. Metody strukturnoi kristallografii (Nauka, Moscow 1979) [English transl.: *Modern Crystallography I.* Symmetry of Crystals, Methods of Structural Crystallography, Springer Series in Solid-State Sciences, Vol. 15 (Springer, Berlin, Heidelberg, New York 1981)]

1.7 L.A. Shuvalov, A.A. Urusovskaya, I.S. Zheludev, A.V. Zalesskii, B.N. Grechushnikov, I.G. Chistyakov, S.A. Semiletov: *Sovremennaya kristallografiya. T. 4.* Fizicheskiye svoistva kristallov, ed. by B.K. Vainshtein (Nauka, Moscow 1981) [English transl.: *Modern Crystallography IV,* Physical Properties of Crystals, ed. by B.K. Vainshtein, Springer Series in Solid-State Sciences, Vol. 37 (Springer, Berlin, Heidelberg, New York forthcoming)]

1.8 S.S. Batsanov: Usp. Khim. *37*, 778 (1968)

1.9 V.S. Urusov: *Energeticheskaya kristallokhimiy* (Energetic Crystal Chemistry (Nauka, Moscow 1975)

1.10 R. Brill: Solid State Phys. *20*, 1 (1967)

1.11 H. Witte, E. Wölfel: Z. Phys. Chem. Frankfurt am Main *3*, 296 (1955)

1.12 J. Krug, H. Witte, E. Wölfel: Z. Phys. Chem. Frankfurt am Main *4*, 36 (1955)

1.12a R. Hoppe: Angew. Chem. *20*, 63 (1981)

1.13 A. Schmiedenkamp, D.W.J. Cruickshank, S. Scaarup, P. Pulay, I. Hargittai, J.E. Boggs: J. Am. Chem. Soc. *101*, 2002 (1979)

1.14 K. Hermansson, J.O. Thomas: 5th European Crystallographic Meeting Abstracts, Copenhagen, August 13–17, 1979, p. 351

1.15 J. Almlöf, Å. Kvick, J.O. Thomas: 1st European Crystallographic Meeting Abstracts, Bordeaux, September 5–8, 1973

1.16 I.P. Walter, M.L. Cohen: Phys. Rev. B*4*, 1877 (1971)

1.17 E.V. Zarochentsev, E.Ya. Fain: Fiz. Tverd. Tela Leningrad *17 (7)*, 2058 (1975) [English transl.: Sov. Phys. Solid State *17*, 1344 (1975)]

1.18 Y.W. Yang, P. Coppens: Solid State Commun. *15*, 1555–1559 (1974)

1.19 R. Chen, P. Trucano, R.F. Stewart: Acta Crystallogr. A*33*, 823 (1977)

1.20 P. Coppens, A. Vos: Acta Crystallogr. B*27*, 146 (1971)

1.21 J.F. Griffin, Ph. Copens: J. Am. Chem. Soc. *97*, 3496 (1975)

1.22 Z. Berkovitch-Yellin, L. Leiserowitz: J. Am. Chem. Soc. *97*, 5627 (1975)

1.23 R.F. Stewart: J. Chem. Phys. *51*, 4569 (1969)

1.24 F.L. Hirshfeld: Acta Crystallogr. B*27*, 769 (1971)

1.25 F.L. Hirshfeld: Isr. J. Chem. *16*, 198 (1977)

1.26 R.F. Stewart: Acta Crystallogr. A*32*, 565 (1976)

1.27 V.G. Tsirel'son, M.M. Mestechkin, R.P. Ozerov: Dokl. Akad. Nauk SSSR *233*, 108 (1977) [English transl.: Sov. Phys. Dokl. Biophys. *233*, 108 (1977)]

1.28 N.V. Ageev: Izv. Akad. Nauk SSSR Otd. Khim. Nauk *1*, 176 (1954)

1.29 H. Bensch, H. Witte, E. Wölfel: Z. Phys. Chem. *4*, 65 (1955)

1.30 B.K. Vainshtein: Tr. Inst. Kristallogr. Akad. Nauk SSSR *10*, 115 (1954)

1.31 S.K. Sikka, R. Chambaran: Acta Crystallogr. B*25*, 310 (1969)

1.32 F. Iwasaki, Y. Saito: Acta Crystallogr. B*26*, 251 (1979)

1.33 V.I. Simonov, B.V. Bukvetsky: Acta Crystallogr. B*34*, 355 (1978)

1.34 Yu.I. Sirotin, M.P. Shaskol'skaya: *Osnovy kristallofiziki* (Fundamentals of Crystal Physics) (Nauka, Moscow 1975)

1.35 E.F. Bertaut: Acta Crystallogr. A*24*, 217 (1968)

1.36 A.I. Kitaigorodsky: *Molekularnye kristally* (Nauka, Moscow 1971) [English transl.: *Molecular Crystals and Molecules* (Academic, New York 1973); German transl.: *Molekülkristalle* (Akademie, Berlin 1979)]

1.37 W.L. Bragg: Philos. Mag. *40*, 169 (1920)

1.38 V.M. Goldschmidt: Skr. Nor. Vidensk. Akad. Oslo I. *8*, 69 (1926)

1.39 L. Pauling: J. Am. Chem. Soc. *49*, 765 (1927)

1.40 L. Pauling: *The Nature of the Chemical Bond*, 3rd ed. (Cornell University Press, Ithaca 1960)

1.41 A. Landé: Z. Phys. *1*, 191 (1920)

1.42 N.V. Belov, G.B. Boky: "The Present State of Crystal Chemistry and Its Urgent Tasks", in *Pervoye soveshchanie po kristallokhimii. Referaty, doklady* (Proceedings of the 1st Meeting on Crystallochemisty), ed. by G.B. Boky (Izd-vo Akad. Nauk SSSR, Moscow 1954) p. 7 (in Russian)

1.43 B.S. Gourary, F.I. Adrian: Solid State Phys. *10*, 143 (1960)

1.44 O. Inkinen, M. Järvinen: Phys. Kondens. Mater. *7*, 372 (1968)

1.45 D.F.C. Morris: "Ionic Radii and Enthalpies of Hydration of Ions", in *Structure and Bonding*, Vol. 4 (Springer, Berlin, Heidelberg, New York 1968) p. 63

1.46 R.D. Shannon, C.T. Prewitt: Acta Crystallogr. B*25*, 925 (1969)

1.47 R.D. Shannon: Acta Crystallogr. A*32*, 751 (1976)

1.48 J.C. Slater: J. Chem. Phys. *41*, 3199 (1964)

1.49 J.C. Slater: *Quantum Theory of Molecules and Solids. Symmetry and Bonds in Crystals,* Vol. 2 (McGraw-Hill, London 1965)

1.50 V.I. Lebedev: *Ionno-atomnye radiusy i ikh znachenie dlya geokhimii* (Ionic-Atomic Radii and Their Role in Geochemistry) (Izd-vo Leningrad Univ., Leningrad 1969)

1.51 Yu.V. Zefirov, P.M. Zorky: Vestn. Mosk. Univ. Khim. *19*, 554 (1978)

1.52 W. Nowacki, T. Matsumoto, A. Edenharter: Acta Crystallogr. *22*, 935 (1967)

1.53 A.L. Mackay: Acta Crystallogr. *22*, 329–330 (1967)

1.54 A. Magnus: Z. Anorg. Chem. *124*, 288 (1922)

1.55 N.V. Belov: *Struktura ionnykh kristallov i metallicheskikh faz* (The Structure of Ionic Crystals and Metallic Phases) (Izd-vo Akad. Nauk SSSR, Moscow 1947)

1.56 E. Döngels: Z. Anorg. Chem. *263*, 112 (1950)

1.57 S.V. Soboleva, B.B. Zvyagin: Kristallografiya *13*, 605 (1968) [English transl.: Sov. Phys. Crystallogr. *13*, 516 (1969)]

1.58 V.S. Urusov: *Teoriya izomorfnoi smesimosti* (Theory of Isomorphous Miscibility) (Nauka, Moscow 1977)

1.59 E.S. Makarov: *Isomorfizm atomov v kristallakh* (Isomorphism of Atoms in Crystals) (Atomizdat, Moscow 1973)

1.60 V.B. Aleksandrov, L.S. Garashina: Dokl. Akad. Nauk SSSR *189*, 307 (1969)

1.61 A.G. Khachaturyan: Prog. Mater. Sci *22*, 1 (1978)

1.62 E.G. Knizhnik, B.G. Livshitz, Ya.L. Lipetsky: Fiz. Met. Metalloved. *29 (2)*, 265 (1970)

1.63 A. Magneli: Arh. Kem. *1*, 5 (1950)

1.64 S. Iijima: J. Solid State Chem. *14*, 52 (1975)

1.65 R.J.D. Tilley: Chem. Scr. *14*, 147 (1978–79)

1.66 N.D. Zakharov, I.P. Khodzhi, V.N. Rozhansky: Dokl. Akad. Nauk SSSR *249 (2)*, 359 (1979)

Chapter 2

2.1 J. Donohue: *The Structures of the Elements* (Wiley, New York 1974)

2.2 R. Sinclair, G. Thomas: J. Appl. Crystallogr. *8*, 206 (1975)

2.3   D.W.J. Cruickshank: 2nd European Crystallographic Meeting, Collected Abstracts, Keszthely, Hungary, August 26–29, 1974

2.4   H. Schulz, K. Schwarz: Acta Crystallogr. A*34*, 999 (1978)

2.5   I.D. Brown, R.D. Shannon: Acta Crystallogr. A*29*, 266 (1973)

2.6   I.D. Brown: Chem. Soc. Rev. *7*, 359 (1978)

2.7   S.M. Stishov, N.V. Belov: Dokl. Akad. Nauk SSSR *143* (4), 951 (1962)

2.8   N. Thong, D. Schwarzenbach: Acta Crystallogr. A*35*, 658 (1979)

2.9   N. Thong, D. Schwarzenbach: 5th European Crystallographic Meeting, Abstracts, Copenhagen, August 13–17, 1979, p. 348

2.10   L. Bragg, G.F. Claringbull: *Crystal Structures of Minerals* (Bell, London 1965)

2.11   N.V. Belov: *Kristallokhimiya silikatov s krupnymi kationami* (Izd-vo Akad. Nauk, Moscow 1961) [English transl.: *Crystal Chemistry of Large Cation Silicates* (Consultants Bureau, New York 1965)]

2.12   N.V. Belov: *Ocherki po strukturnoi mineralogii* (Essays on Structural Mineralogy) (Nedra, Moscow 1976)

2.13   V.V. Ilyukhin, V.L. Kuznetsov, A.N. Lobachëv, V.S. Bakshutov: *Gidrosilikaty kal'tsiya* (Potassium Hydrosilicates. Synthesis of Single Crystals and Crystal Chemistry) (Nauka, Moscow 1979)

2.14   N.I. Golovastikov, R.G. Matveeva, N.V. Belov: Kristallografiya *20* (4), 721 (1975) [English transl.: Sov. Phys. Crystallogr. *20*, 441 (1976)]

2.15   F. Libau: Acta Crystallogr. *12*, 180 (1959)

2.16   R.K. Rastsvetaeva, V.I. Simonov, N.V. Belov: Dokl. Akad. Nauk SSSR *117* (4), 832 (1967)

2.17   B.B. Zvyagin: *Elektronografiya i strukturnaya kristallografiya glinistykh mineralov* (Nauka, Moscow 1964) [English transl.: *Electron Diffraction Analysis of Clay Mineral Structures* (Plenum, New York 1967)]

2.18   V.A. Drits, B.A. Sakharov: Trudy Geol. Inst. Akad. Nauk SSSR *295*, 3 (1976)

2.19   Guan-Ya-sian, V.I. Simonov, N.V. Belov: Dokl. Akad Nauk SSSR, *149* (6), 1416 (1963)

2.20   W.L. Roth, F. Reidinger, S. La Place: *Superionic Conductors* (Plenum, New York 1976) p. 223

2.20a W.F. Kuhs, E. Perenthaler, H. Schulz, U. Zucker: 6th European Cystallographic Meeting, Abstracts, Barcelona, Spain, July 28.–August 1, 1980, p. 229

2.21   H. Krebs: *Grundzüge der anorganischen Kristallchemie* (Enke, Stuttgart 1968)

2.22   N.N. Sirota, E.M. Gololobov, A.U. Sheleg, N.M. Olekhnovich: Izv. Akad. Nauk SSSR Neorg. Mater. *2* (10), 1673 (1965)

2.23   L. Pauling, M.L. Huggins: Z. Kristallogr. *87*, 205 (1934)

2.24   R.M. Imamov, S.A. Semiletov, Z.G. Pinsker: Kristallografiya

15 (2), 287 (1970) [English transl.: Sov. Phys. Crystallogr. *15*, 239 (1970)]

2.25   S.A. Semiletov: Kristallografiya *21* (4), 752 (1976) [English transl.: Sov. Phys. Crystallogr. *21*, 426 (1976)]

2.26   E. Mooser, W.B. Pearson: Phys. Rev. *10*, 492 (1956)

2.27   L. Pauling: "The Nature of Bonds of Transient Metals with Bioorganic and Other Compounds", in *Itogi i perspektivy razvitiya bioorganicheskoi khimii i molekulyarnoi biologii* (The Results and Future of the Development of Bioorganic Chemistry and Molecular Biology), ed. by Yu. Ovchinnikov, M.N. Kolosov (Nauka, Moscow 1978) p. 3 (in Russian)

2.28   V.G. Kuznetsov, P.A. Kozmin: Zh. Strukt. Khim. *4*, 55 (1963)

2.28a   A. Simon: "Structure and Bonding with Alkali Metal Suboxides", in *Structure and Bonding*, Vol. 36, Inorganic Chemistry and Spectroscopy (Springer, Berlin, Heidelberg, New York 1979) p. 81

2.29   J.D. Dunitz, L.E. Orgel, A. Rich: Acta Crystallogr. *9*, 373 (1956)

2.30   A.J. James, W.G. Hamilton: Science *139*, 106–107 (1963)

2.31   I.L. Karle, J. Karle: Acta Crystallogr. *16*, 969 (1963)

2.32   F.L. Hirschfeld, S. Sandler, G.M.J. Schmidt: J. Chem. Soc. 4, 2108 (1963)

2.33   R.L. Avoyan, A.I. Kitaigorodsky, Yu.T. Struchkov: Zh. Strukt. Khim. *5* (3), 420 (1964)

2.34   J. Preuss, A. Gieren: Acta Crystallogr. B*31* (5), 1276–1282 (1975)

2.35   I.L. Dubchak, V.E. Shklover, M.M. Levitsky, A.A. Zhdanov, Ju.T. Struchkov: Zh. Strukt. Khim. *21* (6), 103 (1980)

2.36   S. Abrahamsson, D.C. Hodgkin, E.N. Maslen: Biochem. J. *86*, 514 (1963)

2.37   A.I. Kitaigorodsky: *Organicheskaya kristallokhimiya* (Organic Crystal Chemistry) (Izd-vo Akad. Nauk SSSR, Moscow 1955)

2.38   P.M. Bel'sky, P.M. Zorky: Kristallografiya *15* (4), 704 (1970) [English transl.: Sov. Phys. Crystallogr. *15*, 607 (1971)]

2.39   V.K. Be'lsky, P. M. Zorky: Acta Crystallogr. A*33*, 1004 (1977)

2.40   P.M. Zorky, V.K. Be'lsky, S.G. Lazareva, M.A. Porai-Koshits: Zh. Strukt. Khim *8*, 312 (1967)

2.41   P.M. Zorky, B.K. Bel'sky: "The Structure of the Crystals of Tolane and Its Structural Analogs", in *Kristallografiya i mineralogiya* (Crystallography and Mineralogy) Trudy Fëdorovskoi yubileinoi sessii, Leningrad, 1969 (Publ. Leningradskogo Gornogo In-ta, Leningrad 1972) (in Russian)

2.42   A.I. Kitaigorodsky: *Moleculyarnye kristally* (Molecular Crystals) (Nauka, Moscow 1971)

2.43   R.P. Shibaeva, V.F. Kaminsky: Kristallografiya *23*, 1183 (1978) [English transl.: Sov. Phys. Crystallogr. *23*, 669 (1978)]

2.44   A.A. Shevyrëv, L.A. Muradzhan, V.I. Simonov. JETP Lett. *30*, 107 (1979)

2.45   V.I. Smirnova, L.N. Zeibot, N.E. Zhukhlistova, G.N. Tishchenko,

V.I. Andrianov: Kristallografiya *21* (3) 525 (1976) [English transl.: Sov. Phys. Crystallogr. *21*, 291 (1976)]

2.46 B.K. Vainshtein, G.M. Lobanova, G.V. Gurskaya: Kristallografiya *19* (3), 531 (1974) [English transl.: Sov. Phys. Crystallogr. *19*. 329 (1975)]

2.47 R.P. Daubeny, C.W. Dunn, L. Brown: Proc. R. Soc.. London A*226*, 531 (1954)

2.48 C.W. Dunn: Proc. R. Soc. London *180*, 40–66 (1942)

2.49 C. Lelliott, E.D.T. Atkins, J.W. F. Juritz, A.M. Stephen: Polymer *19*, 363 (1978)

2.50 A. Keller: "The Morphology of Crystalline Polymers", in Chimica dellemacromolecole, corso Estivo tenute a Verrena, Villa Monstero, September 18–30, 1961

2.51 V.A. Marikhin, L.P. Myasnikova: *Nadmolekulyarnaya struktura polimerov* (The Supramolecular Structure of Polymers) (Khimiya, Leningrad 1977)

2.52 A. Keller: J. Polym. Sci. Polym. Symp. *51*, 7 (1975)

2.52a G. Fridel: Ann. Phys. *19*, 273 (1922)

2.53 B.K. Vainshtein, I.G. Chistyakov: Pramana Suppl. No. *1*, 79 (1975) [15th Int. Conf. on Liquid Crystal, Raman Research Inst., Bangalore]

2.54 W.R. Krigbaum, Y. Chatani, P. Barber: Acta Crystallogr. B*26*, 97 (1970)

2.55 S. Chandrasekhar, B.K. Sadashiva, K.A. Suresh, N.V. Madhusudana, S. Kumar, R. Shashidhar, G. Venkatesh: J. Phys. Paris Colloq. C*3 40*, Suppl. 4, 121 (1979)

2.55a S. Chandrasekhar: Mol. Cryst. Liq. Cryst. *63*, 171 (1981)

2.55b H. Gasparoux: Mol. Cryst. Liq. Cryst. *63*, 231 (1981)

2.56 B.K. Vainshtein, E.A. Kosterin, I.G. Chistyakov: Dokl. Akad. Nauk SSSR *199* (2), 323 (1971)

2.57 B.K. Vainshtein, I.G. Chistyakov, E.A. Kosterin, V.M. Chaikovsky: Dokl. Akad. Nauk SSSR *174* (2), 341 (1967)

2.58 C. Brink-Shoemaker, D.W.J. Cruickshank, D.C. Hodgkin, M.J. Kamper, D.P. Ming: Proc. R. Soc. London A*278*, 1 (1964)

2.59 G.V. Gurskaya: *Struktury aminokislot* (Nauka, Moscow 1966) [English transl.: *The Molecular Structure of Amino Acids* (Consultants Bureau, New York 1968)]

2.60 G.N. Tishchenko: "The Structure of Linear and Cyclic Oligopeptides in Crystals" in *Itogi Nauki i Tekhniki*. Ser. Kristallokhimiya (Advances in Science and Technology. Crystal Chemistry), Vol. 13 (VINITI, Moscow 1979) p. 189 (in Russian)

2.61 B.K. Vainshtein, L.I. Tatarinova: Kristallografiya *11* (4), 562 (1966) [English Transl.: Sov. Phys. Crystallogr. *11*, 494 (1966)]

2.62 L. Pauling, R.B. Corey: Proc. R. Soc. London B*141*, 10 (1953)

2.63 G.N. Tishchenko, Z. Karimov, B.K. Vainshtein, A. V. Evstratov, V.T. Ivanov, Yu.A. Ovchinnikov: FEBS Lett. *65* (3), 315 (1976)

2.64 F.A. Momany, L.M. Carruthers, H.A. Sherage: J. Phys. Chem. *78*, 1621 (1974)

2.65 G.N. Ramachandran, V. Sasiserharan: Adv. Protein Chem. *23*, 283 (1968)

2.66 L. Pauling, R.B. Corey, H.R. Branson: Proc. Natl. Acad. Sci. U.S.A. *37*, 205 (1951)

2.67 D.W. Fawcett: *The Cell, Its Organelles and Inclusions; An Atlas of Fine Structure* (Saunders, Philadelphia 1966)

2.68 V.V. Barynin, B.K. Vainshtein, O.N. Zograf, S.Ya. Karpukhina: Mol. Biol. Moscow *13* (5), 1189 (1979) [English transl.: Mol. Biol. USSR *13*, 922 (1979)]

2.69 J.C. Kendrew: Science *139*, 1259 (1963)

2.70 J.C. Kendrew, R.E. Dickerson, B.E. Strandberg, R.G. Hart, D.R. Davies, D.C. Phillips, V.C. Shore: Nature London *185*, 422 (1960)

2.71 H. Frauenfelder, G.A. Petsko, D. Tsernoglou: Nature London *280*, 558 (1979)

2.72 M.F. Perutz: Proc. R. Soc London B*173*, 113 (1969)

2.73 M.F. Perutz: Sci. Am. *239*, 92 (1978)

2.74 M.F. Perutz: J. Mol. Biol. *13*, 646 (1965)

2.75 B.K. Vainshtein, E.H. Harutyunyan, I.P. Kuranova, V.V. Borisov, N.I. Sosfenov, A.G. Pavlovsky, A.I. Grebenko, N.V. Konarëva: Nature London *254*, 163 (1975)

2.76 E.H. Haratyunyan, I.P. Kuranova, B.K. Vainshtein, W. Steigemann: Kristallografiya *25* (1), 80 (1980) [English transl.: Sov. Phys. Crystallogr. *25*, 43 (1980)]

2.77 R.E. Dickerson: Sci. Am. *226*, 58 (1972)

2.78 M. Levine, H. Muirhead, D.K. Stammers, D.I. Stuart: Nature London *271*, 626 (1978)

2.79 M. Levitt, C. Chothia: Nature London *261*, 552 (1976)

2.80 J.S. Richardson: Nature London *268*, 495 (1977)

2.81 I.I. Weber, L.N. Johnson, K.S. Wilson, D.G.R. Yeates, D.L. Wild, J.A. Jenkins: Nature London *274*, 433 (1978)

2.81a D.W. Banner, A.C. Bloomer, G.A. Petsko, D.C. Phillips, C.I. Pogson, I.A. Wilson, P.H. Corran, A.J. Furth, J.D. Milman, R.E. Offord, J.D. Priddle, S.G. Waley: Nature London *255*, 609 (1975)

2.82 J.S. Richardson, K.A. Thomas, B.H. Rubin, D.C. Richardson: Proc. Natl. Acad. Sci. U.S.A. *72*, 1349 (1975)

2.83 B.K. Vainshtein, V.P. Melik-Adamyan, V.V. Barynin, A.A. Vagin, A.I. Grebenko: Krystallografiya *26* (5), 1003 (1981)

2.83a B.K. Vainshtein, V.R. Melik-Adamyan, V.V, Barynin, V.V. Vagin, A.I. Grebenko: Nature London *293*, 411 (1981)

2.84 C. Chotia, M. Levitt, D. Richardson: Proc. Natl. Acad. Sci. U.S.A. *74*, 4130 (1977)

2.85 B.W. Mattheus, P.B. Sigler, R. Henderson, D.M. Blow: Nature London *214*, 652 (1967)

2.86 M.J. Adams, M. Buchner, K. Chandrasekhar, G.C. Ford, M.L.

Hackert, A. Liljas, P.Jr. Lentz, S.T. Rao, M.G. Rossmann, I.E. Smiley, J.L. White: In *Protein – Protein Interactions,* Colloquium der Gesellschaft für Biologische Chemie in Mosbach, Baden, Vol. 23 (Springer, Berlin, Heidelberg, New York 1972)

2.87   V.V. Borisov, S.N. Borisova, N.I. Sosfenov, A.A. Vagin, Yu.V. Nekrasov, B.K. Vainshtein, V.M. Kochkin, A.E. Braunshtein: Dokl. Akad. Nauk SSSR *250* (4), 988 (1980) [English transl.: Sov. Phys. Dokl. Biochem. *250*, 45 (1980)]

2.88   A. Liljas, M.G. Rossmann: Annu. Rev. Biochem. *43*, 475 (1974)

2.89   W.N. Lipscomb: Acc. Chem. Res. *3*, 81–89 (1970)

2.90   N.S. Andreeva, A.A. Fëdorov, A.E. Gushchina, R.R. Riskulov, N.E. Schutzkever, M.G. Safro: Mol. Biol. Moscow *12*, 922 (1978) [English transl.: Mol. Biol. USSR *12*, 704 (1978)]

2.91   O.B. Ptitsyn, A.V. Finkelshtein: Proc. Fed. Eur. Biochem. Soc. Congress 1978

2.92   D.C. Phillips: Sci. Am.: *217*, 78 (1966)

2.93   B.W. Matthews, S.Y. Remingtone: Proc. Natl. Acad. Sci. U.S.A. *71*, 4178 (1974)

2.94   V.V. Borisov, S.N. Borisova, G.S. Kachalova, N.I. Sosfenov, B.K. Vainshtein, Yu.M. Torchinsky, A.E. Braunshtein: J. Mol. Biol. *125*, 275 (1978)

2.95   V.L. Tsuprun, N.A. Kiselëv, B.K. Vainshtein: Kristallografiya *23* (4), 743 (1978) [English transl.: Sov. Phys. Crystallogr. *23*, 417 (1978)]

2.96   N.A. Kiselëv, E.V. Orlova, V.Ya. Stel'mashchuk: Dokl. Akad. Nauk SSSR *246* (6), 1508 (1979) [English transl.: Sov. Phys. Dokl. Biophys. *246*, 118 (1979)]

2.97   D.J. De Rosier, R.M. Oliver: Cold Spring Harbor Symp. Quant. Biol. *36*, 199–203 (1972)

2.98   R. Henderson, P.N. Unwin: Nature London *257*, 28 (1975)

2.99   R.E. Franklin, R.G. Gosling: Acta Crystallogr. *8*, 151 (1955)

2.100 M.H.F. Wilkins, A.R. Stokes, H.R. Wilson: Nature London *171*, 739 (1953)

2.101 F.H.C. Crick, J.D. Watson: Proc. R. Soc. London A*223*, 80 (1954)

2.102 R. Langridge, W.E. Seeds, H.R. Wilson, C.W. Hooper, M.H.F. Wilkins, L.D. Hamilton: J. Biophys. Biochem. Cytol. *3*, 767 (1957)

2.103 E.J. O'Brien: Acta Crystallogr. *23*, 92 (1967)

2.104 M.A. Mokul'sky, K.A. Kapitonova, T.D. Mokul'skaya: Mol. Biol. *6*, 883 (1972) [English transl.: Mol. Biol. USSR *6*, 716 (1972)]

2.105 A.H.-J. Wang, F.J. Kolpak, G.J. Quigley, J.L. Crawford, J.H. van Boom, C. van der Mazel, A. Rich: Nature London *282*, 680 (1979)

2.106 S.H. Kim, F.L. Suddath, G.J. Quigley, A. McPherson, J.L. Sussmann, A.H.J. Wang, N.C. Seeman, A. Rich: Science *185*, 435 (1974)

2.107 J.D. Robertus, Y.E. Ladner, Y.T. Finch, D. Rhodes, R.S. Brown, B.F.C. Clark, A. Klug: Nature London *250*, 546 (1974)

2.108 N.A. Kiselëv, V.Ya. Stel'mashchuk, K. Nissler: Dokl. Akad. Nauk SSSR *247*(1), 237 (1979)
2.109 N.A. Kiselëv, V.Ya. Stel'mashchuk, M.J. Lerman, O.Yu. Abakumova: J. Mol. Biol. *86*, 577 (1974)
2.110 P.K. Wellauer, I.B. Dawid: J. Mol. Biol. *89*, 397 (1974)
2.111 J.T. Finch, L.C. Lutter, D. Rhodes, R.S. Brown, B. Rushton, M. Levitt, A. Klug: Nature London *269*, 29 (1977)
2.112 R. Franklin, K. Holmes: Acta Crystallogr. *11*, 213 (1958)
2.113 D.L.D. Caspar, A. Klug: Cold Spring Harbor Symp. *27*, 1–24 (1962)
2.114 A.C. Bloomer, J.N. Champness, G. Bricogne, R. Staden, A. Klug: Nature London *276*, 362 (1978)
2.115 J.N. Champness, A.C. Bloomer, G. Bricogne, P.J.G. Butler, A. Klug: Nature London *259*, 20 (1976)
2.116 J.T. Finch, A. Klug: Nature London *183*, 1709 (1959)
2.117 A. Klug, J.T. Finch: J. Mol. Biol. *11*, 403 (1965)
2.118 A. Klug, J.T. Finch: J. Mol. Biol. *31*, 1 (1968)
2.119 J.T. Finch, A. Klug: J. Mol. Biol. *15*, 344–364 (1966)
2.120 A. Klug, W. Longely, R. Leberman: J. Mol. Biol. *15*, 315 (1966)
2.121 R.A. Crowther, L.A. Amos, J.T. Finch, D.J. De Rosier, A. Klug: Nature London *226*, 421 (1970)
2.122 S.C. Harrison, A.J. Olson, C.E. Schutt, F.K. Winkler, G. Bricogne: Nature London *276*, 362 (1978)
2.123 B.K. Vainshtein: Usp. Fiz. Nauk *109* (3), 455 (1973)
2.124 B.K. Vainshtein: "Electron Microscopical Analysis of the Three-Dimensional Structure of Biological Macromolecules", in *Advances in Optical and Electron Microscopy*, Vol. 7 (Academic, New York 1978)
2.125 B.A. Crowther, A. Klug: Annu. Rev. Biochem. *44*, 161 (1975)
2.126 L.A. Amos, A. Klug: J. Mol. Biol. *99*, 51 (1975)
2.127 A.M. Mikhailov, J.A. Andriashvili, G.V. Petrovsky, A.S. Kaftanova: Dokl. Akad. Nauk SSSR *239*, 725 (1978)
2.128 A.M. Mikhailov, N.N. Belyaeva: Dokl. Akad. Nauk SSSR *250*, 222 (1980) [English transl.: Sov. Phys. Dokl. Biophys. *250*, 5 (1980)]
2.129 B.K. Vainshtein, A.M. Mikhailov, A.S. Kaftanova: Kristallografiya *22* (2), 287 (1977) [English transl.: Sov. Phys. Crystallogr. *22*, 163 (1977)]

Chapter 4

4.1   L.D. Landau: Phys. Z. Sowjetunion *8*, 113 (1935) (In Russian)
4.2   V.L. Ginzburg: Dokl. Akad. Nauk SSSR *105*, 240 (1955)
4.3   T.R. Volk, E.I. Gerzanich, V.M. Fridkin: Izv. Akad. Nauk SSSR, Ser. Fiz. *33*, 348 (1969)

4.4.   C.W. Garland, B.B. Weiner: Phys. Rev. *3*, 1634 (1971)
4.5   I.B. Bersuker, B.G. Vekhter: Fiz. Tvёrd. Tela Leningrad *9*, 2652 (1967) [English transl.: Sov. Phys. Solid State *9*, 2084 (1968)]
4.6   V.M. Fridkin: Pis'ma Zh. Eksp. Teor. Fiz. *3* (6) 252 (1966)
4.7   V.M. Fridkin: *Segnetoelektriki-poluprovodniki* (Nauka, Moscow 1976) [English transl.: *Ferroelectric Semiconductors* (Plenum, New York (1980)]
4.8   L.D. Landau, E.M. Lifshits: *Statisticheskaya fizika* (Nauka, Moscow 1964) [English transl.: *Course of Theoretical Physics,* Vol. 5, Statistical Physics, 2nd ed. (Pergamon, Oxford 1969)]
4.9   V.M. Indenbom: Kristallografiya *5*, 115 (1960) [English transl.: Sov. Phys. Crystallogr. *5*, 106 (1960)]
4.10   Y.L. Indenbom: Izv. Akad. Nauk SSSR Ser. Fiz. *24*, 1180 (1960)
4.11   A.P. Levanyuk, D.G. Sannikov: Usp. Fiz. Nauk *112*, 561 (1974)
4.12   E.M. Lifshits: JETP, *11*, 269 (1941)

Chapter 5

5.1   A.A. Chernov, E.I. Givargizov, K.S. Bagdasarov, V.A. Kuznetsov, L.N. Demyanets, A.N. Lobachev: *Sovremennaya kristallografiya.* T.3. Obrazovaniye kristallov, ed. by B.K. Vainshtein (Nauka, Moscow 1980) [English transl.: *Modern Crystallography III,* Formation of Crystals, ed. By B.K. Vainshtein, Springer Series in Solid-State Sciences, Vol. 36 (Springer, Berlin, Heidelberg, New York forthcoming)
5.2   V.A. Phillips: Acta Metall. 20, 1147 (1972)
5.3   S.L. Chang: Appl. Phys. A 26, 221 (1981)

Subject Index

Additivity of crystallochemical radii 70
Amino acids 215
Aspartate-transaminase 249
Atomic orbitals (AO) 3
   linear combinations 38
Azoxyanisole 213

Bacteriophages 271
Bafertisite 156
Berg – Barret method 403
Beryl 149
Bloch theorem 275
Bloch wave 400
Bond
   chemical 2, 15, 83
   conjugation 32
   covalent 16, 26, 39, 169
   donor – acceptor 29
   heteropolar 16
   homopolar 16
   hydrogen 16, 55, 187
   intermediate-order 33
   ionic 2, 16, 19, 143
   metal-molecular 170
   metallic 16, 49
   multiple 31
   van der Waals 16, 53
Born – Haber cycle 61
Born – von Karman conditions 276, 296
Boron 131
Borrmann effect 401
Brillouin zones 39, 141, 276, 282
Burgers vector 350, 383, 399

Carbon 133
Catalase 230
Cellulose 226
Chymotripsin 244
Close packing 103, 129, 185, 265
   cubic 105
   different numbers of layers 105
   hexagonal 104
   holes (interstices) 107
      octahedral 108

      tetrahedral 108
   principle 93
   space groups 104
Collagen 225
Compounds
   clathrate (inclusion) 189
   complex 165
   electron 140
   heterodesmic 91
   homodesmic 63, 91
   intermetallic 142
Configurational entropy 340
Conformation plot for proteins 222
Coordination 90, 99
   number (c.n.) 99
   polyhedron 100, 106
   sum 289
Corundum 111
Crowdion 345
Curie principle 94
Curie temperature 322
Curie – Weiss law 328
Cylindrical function of interatomic
   distances 213
Cytochromes 237

Debye equation of state 324
Debye temperature 297, 299
Decoration of dislocations 411
Defects of lattice 339
   bivacancies 345
   Frenkel 341
   linear (one-dimensional) 339
   point (zero-dimensional) 339, 342, 347,
      397
   Schottky 341, 346
   stacking fault 339, 369, 399
   surface (two-dimensional) 339
   volume (three-dimensional) 339
Degree of ionization 23
Deltahedra 268
Diamond 44, 70, 133, 308
Dislocations
   core 358, 373

Dislocations (cont.)
  curved   365
  density   378
  edge   350, 373
  loop   351
    prismatic   345
  partial   368, 372
  perfect (ordinary)   367, 394
  polygonal   360
  reactions   358
  screw   350, 371
  twinning   392
Dispersion interaction   53
Dispersion surface   400
Displacement function   200
Displacement peak   349
Domains
  antiphase   372, 389, 395
  in proteins   237, 242
Dulong – Petit law   299

Effective charge   144
Electron configuration   12
Electron density
  in covalent bond   41, 44, 48
  difference   146
    deformation   32, 45, 48
    valence   56, 176
  of atom   3, 13, 50
    radial   6, 13, 24
Electron energy bands   39, 274, 280
Electron energy spectrum   16, 274, 280
Electron shell   9
Electronegativity (EN)   21
Electrostatic-valence (Pauling) rule   146
Elements, crystal structure   127
Enantiomers   179
Enantiomorphous forms   215
Energy
  affinity   19
  atomization   61, 66
  band   39, 274
    forbidden   293
  cohesion   61
  complete dissociation   61
  configurational   313
  crystal   66
  free   61, 340
  interchange   118
  ionic   64
  ionization   19, 61
  lattice   61, 65
  potential   17
  sublimation   61
Energy constant (EC)   65

Energy gap   322
Energy level   9
Exchange integral   28, 59
Exiton   339
Extinction
  bands   398
  contrast   401
  length   398

Feldspar   153
Fermi surface   50, 291
Ferrocene   171
Ferroelectric   320
  improper   334
Fibrillar crystals   198
Figurative point   303
Fluctuon   339

Garnet   145
Germanium   40, 133, 293
Gibbs – Helmholtz equation   316
Graphite   47, 133, 308
Grüneisen constant   324
Grüneisen ratio   294, 324

Hartree – Fock method   8
Heat capacity   299
Heat of sublimation   61
Helicoidal ordering   60
Helium   129
Hemoglobin   231, 234
High polymers
  amorphous   203
  crystals   196
  disordering   200
  lyotropic   214
  microfibrillar structure   198
  ordering   190, 202
  structure   191, 195
Hybridization   29
Hydrogen
  atom   3
  molecule   34
  structure   129
Hydrophobic interaction   216

Inert element (noble gases)   129
  compound   172
Interatomic distances   69, 161
  cylindrical function   213
  equilibrium   73
Interbranch scattering   402
Internal rotation barrier   174
Interstitial atom   339, 345, 397
Ionic fraction of bond   21, 23

Iron   129
Isoenergetic surface   290
Isomorphism   114, 118
    heterovalent   119
    isovalent   119
Isostructural crystal   114

Jahn – Teller effect   169, 317
Jones zones   288

Keratin   224
Klausius – Clapeyron equation   304

Lang method   403
Lattice sums   63
Laves phases   142
Leghemoglobin   231, 236
Ligand   166
    field   168
Linear combinations of atomic orbitals
    (LCAO) method   38
Lipids   215
Liquid crystals   204
    cholesteric   207
    discotic   211
    domains   205
    nematic   205
    ordering   204
    smectic   207, 216
Local symmetry   184
Lonsdeilite   135
Lysozyme   246

Madelung's constant   64
Magnetic structure   60
Manganese   132
Maximum-filling principle   92
Membranes   251
Mesomorphous phases (liquid crystals)   204
Mica   113
Moiré method   397
    x-ray   406
Molecular orbitals (MO)   33
Mooser – Pearson rule   162
Multielectron atom   8
Muscle   251
Myoglobin   231

Naphthalene   181
Nickel arsenide   97, 109
Noble gases, solidified   129
Nodal surface   5
Noncrystallographic ordering   190
Noncrystallographic symmetry   184
Nucleic acids   253

Nucleosomes   260

Orbital
    antibonding   34
    atomic   3
    bonding   34
    hybridized   30
    molecular   33
    spin   8
    symmetry   4
Ordering
    helicoidal   60
    magnetic   58
    noncrystallographic   190
    paracrystalline type   201
Organic molecules   173
Orientation factor   355, 359
Overlapping integral   28, 34

Packing
    coefficient   93
    of molecules   179, 185
Paracrystalline ordering   201
Paraffin   112
Pauli principle   281
Peptides   219
Perovskite   144, 306
Phase diagram   115
    p-T   130
Phase transitions   302
    displacement type   304
    ferroelastic   334
    ferroelectric   320, 334
    ferromagnetic   334
    first-order   304
    homeomorphous   308
    order – disorder type   304, 312
    photostimulated   321
    polymorphous   129, 308
    second-order   304, 325
Phonon   297
Plank function   298
Polaron   339
Polyethylene   197, 199
Polygonization   384
Polymorphous modifications   129, 302
Polypeptide chain   218, 222
Polysaccharides   194
Potential
    ionization   19
    repulsive   20
Prosthetic group   231
Proteins   215, 227, 243
    domains   237, 242
    enzymatic function   246

Proteins (cont.)
  fibrous   224
  globular   226
  homology   242
  structure   216, 223, 237
    primary   218
    quarternary   224
    secondary   219, 222
    tertiary   224
Pseudopotential method   51

Quantum-mechanical uncertainty
  relation   280
Quantum numbers   3
Quartz   147, 388

Radial distribution of electrons in atom   6
Radiation
  doping   349
  growth   349
  swelling   349
Radius
  atomic   70
  Bohr   4
  covalent   162
    octahedral   162
    tetrahedral   161
  crystallochemical   70
    additivity   70
  intermolecular   86
  ionic   74
    atomic – ionic   84
    classical   82
    physical   82
  orbital   12, 14, 23
  root mean square (rms)   13
  strong-bond   88
  weak-bond   88
Reconstruction, three-dimensional   268
Ribosomes   258
Rotation of molecules   200

SCF-LCAO-MO-method   38
Schrödinger equation   3, 275, 382
Seidozerite   149
Self-consistent-field (SCF) method   8
Self-organization principle   233
"Shish-kebab" structure of polymers   198
Short-range decomposition   123
Short-range order   123
Silica   147
Silicates   147
  layer   154
  skeleton   153
Silicium   45, 70, 133, 293

Snoek effect   347
Sodium chloride   22, 74, 86, 109
Solid electrolytes   157, 438
Solid solution
  disordered   138
  interstitial   120
  ordering   138
  substitutional   115, 138
  subtractional   120, 140
Spin density   58
Spin orbital   8
Spinel-type oxides   145
Stepanov's effect   347
Structural class   183
Structure
  chain   111
  coordination   101
  covalent   135, 158
  helicoidal   60, 337
  heterodesmic   91
  homodesmic   91
  incommensurate (modulated)   124, 337
  insular   135
  intermetallic   137
  ionic   143
  isodesmic   101
  layer   113
  magnetic   60
  metal-covalent   131
  molecular   111
  of compounds, classification   127
  of elements   127
  polygonal   60
  resonance   33
  skeleton   102
Superionic conductors   156
Symmetry
  crystal   93
  interaction (Curie principle)   94
  local (supersymmetry)   184
  molecules   177
  noncrystallographic   184
Syntheses of electron density
  difference deformation   31, 47
  difference valence   47

Takagi equations   404
Tensor
  dislocation-density   375, 381
  elastic-distortion   352, 375
  elastic-moduli   355
  inverse-effective-mass   277, 280
Tetrahedron
  stacking-fault   369
  Thompson   359, 368

Thermal conductivity   301
Thermal peak   349
Tolan   183
Trans-influence effect   167
Tungsten   131
Turn of molecules   200
Twinning operation   387, 390
Twins   335, 339, 367, 395
  irrational   386
  polysynthetic   387

Vacancies   339, 397
Valence-bond (VB) method   28
Valence electron distribution   45
Vegard's law   116
Vibration branches
  acoustic   296

optical   296
Vibration, "soft" mode   320
Viruses
  spherical (icosahedral)   265
  structure   261
  tobacco mosaic (TMV)   261
  turnip yellow mosaic (TYMV)   266

Wave function   3, 27
  full   8
  multielectron   8
  one-electron   8

Zeolites   155
Zones
  Brillouin   39, 276, 282
  Jones   288

B. K. Agarwal

X-Ray Spectroscopy

An Introduction

1979. 188 figures, 31 tables. XIII, 418 pages
(Springer Series in Optical Sciences, Volume 15)
ISBN 3-540-09268-4

"...Even with its high density of information, I found
the book easily readable. It begins with a classical
description of X-ray production and then introduces
only simple quantum mechanical treatments as they
are necessary (although references direct one to more
detailed treatments). This general procedure is used
throughout, providing a rather physical interpretation
to the more complicated issues. Much of the jargon
is presented in a clear, concise manner so that the book
also serves as a fine reference text. This book would
serve well as a graduate-level textbook."

J. Am. Chem. Soc.

Z. G. Pinsker

Dynamical Scattering of X-Rays in Crystals

1978. 124 figures, 12 tables. XII, 511 pages
(Springer Series in Solid-State Sciences, Volume 3)
ISBN 3-540-08564-5

"The author, in his preface, indicates that this is an up-
dated version of his earlier book published in 1974.
But it offers a great deal more. Not only is there a considerable amount of extra material in this new edition,
with a new bibliography... but also many of the original
figures and tables have been redrawn and recalculated.
For example, tables which in the Russian edition contained some approximate entries now appear with calculated values, and in many places different values of the
optical parameters... have been used in this new
edition... Clearly, this book contains the most up-to-date
and comprehensive review of the subject at present...
it is to be recommended as the most up-to-date and
comprehensive review at present available."

Optica Acta

Synchrotron Radiation

Techniques and Applications

Editor: C. Kunz
With contributions by numerous experts

1979. 162 figures, 28 tables. XVI, 442 pages
(Topics in Current Physics, Volume 10)
ISBN 3-540-09149-1

Contents:
C. Kunz: Introduction – Properties of Synchrotron
Radiation. – *E. M. Rowe:* The Synchrotron Radiation
Source. – *W. Gudat, C. Kunz:* Instrumentation for Spec-
troscopy and other Applications. – *A. Kotani,
Y. Toyozawa:* Theoretical Aspects of Inner-Level
Spectroscopy. – *K. Codling:* Atomic Spectroscopy. –
E. E. Koch, B. F. Sonntag: Molecular Spectroscopy. –
D. W. Lynch: Solid-State Spectroscopy.

X-Ray Optics

Applications to Solids

Editor: H.-J. Queisser
With contributions by numerous experts

1977. 133 figures, 14 tables. XI, 227 pages
(Topics in Applied Physics, Volume 22)
ISBN 3-540-08462-2

Contents:
Introduction: Structure and Structuring of Solids. –
High Brilliance X-Ray Sources. – X-Ray Lithography. –
X-Ray and Neutron Interferometry. – Section Topo-
graphy. – Live Topography.

Springer-Verlag Berlin Heidelberg New York

Computer Processing of Electron Microscope Images

Editor: P. W. Hawkes
With contributions by numerous experts

1980. 116 figures, 2 tables. XIV, 296 pages
(Topics in Current Physics, Volume 13)
ISBN 3-540-09622-1

Contents:
P. W. Hawkes: Image Processing Based on the Linear
Theory of Image Formation. – *W. O. Saxton:* Recovery
of Specimen Information for Strongly Scattering
Objects. – *J. E. Mellema:* Computer Reconstruction of
Regular Biological Objects. – *W. Hoppe, R. Hegerl:*
Three-Dimensional Structure Determination by Elec-
tron Microscopy (Nonperiodic Specimens). – *J. Frank:*
The Role of Correlation Techniques in Computer Image
Processing. – *R. H. Wade:* Holographic Methods in
Electron Microscopy. – *M. Isaacson, M. Utlaut, D. Kopf:*
Analog Computer Processing of Scanning Transmission
Electron Microscope Images.

Optical Data Processing

Applications

Editor: D. Casasent
With contributions by numerous experts

1978. 170 figures, 2 tables. XIII, 286 pages
(Topics in Applied Physics, Volume 23)
ISBN 3-540-08453-3

Contents:
D. Casasent, H. J. Caulfield: Basic Concepts. –
B. J. Thompson: Optical Transforms and Coherent Pro-
cessing Systems – With Insights From Cristallo-
graphy. – *P. S. Considine, R. A. Gonsalves:* Optical Image
Enhancement and Image Restoration. – *E. N. Leith:*
Synthetic Aperture Radar. – *N. Balasubramanian:*
Optical Processing in Photogrammetry. – *N. Abramson:*
Nondestructive Testing and Metrology. – *H. J. Caulfield:*
Biomedical Applications of Coherent Optics. –
D. Casasent: Optical Signal Processing.

H. Niemann

Pattern Analysis

1981. 112 figures. XIII, 302 pages
(Springer Series in Information Sciences, Volume 4)
ISBN 3-540-10792-4

Contents:
Introduction. – Preprocessing. – Simple Constituents. –
Classification. – Data. – Control. – Knowledge Repre-
sentation, Utilization, and Acquisition. – Systems for
Pattern Analysis. – Things to Come. – References. –
Subject Index.

T. Pavlidis

Structural Pattern Recognition

1977. 173 figures, 13 tables. XII, 302 pages
(Springer Series in Electrophysics, Volume 1)
ISBN 3-540-08463-0

Contents:
Mathematical Techniques for Curve Fitting. –
Graphs and Grids. – Fundamentals of Picture Segmen-
tation. – Advanced Segmentation Techniques. – Scene
Analysis. – Analytical Description of Region Bound-
aries. – Syntactic Analysis of Region Boundaries and
Other Curves. – Shape Description by Region Ana-
lysis. – Classification, Description and Syntactic
Analysis.

Springer-Verlag Berlin Heidelberg New York